# PARTICLES IN GASES AND LIQUIDS 2

## Detection, Characterization, and Control

# PARTICLES IN GASES AND LIQUIDS 2

## Detection, Characterization, and Control

Edited by

K. L. Mittal
IBM US Technical Education
Thornwood, New York

SPRINGER SCIENCE+BUSINESS MEDIA, LLC

Library of Congress Cataloging-in-Publication Data

Symposium on Particles in Gases and Liquids: Detection, Characterization, and Control (2nd : 1989 : Boston, Mass.)
Particles in gases and liquids 2 : detection, characterization, and control / edited by K.L. Mittal.
p. cm.
"Proceedings of the Second Symposium on Particles in Gases and Liquids: Detection, Characterization, and Control, held at the Twentieth Annual Meeting of the Fine Particle Society, August 21-25, 1989, in Boston, Massachusetts"--T.p. verso.
Includes bibliographical references and index.
1. Colloids--Congresses. 2. Particles--Congresses. I. Mittal, K. L., 1945- . II. Fine Particle Society. Meeting (20th : 1989 : Boston, Mass.) III. Title.
QD549.S986 1989
541.3'45--dc20 91-2287
CIP

Proceedings of the Second Symposium on Particles in Gases and Liquids: Detection, Characterization, and Control, held at the Twentieth Annual Meeting of the Fine Particle Society, August 21–25, 1989, in Boston, Massachusetts

**ISBN 978-1-4899-3546-5** **ISBN 978-1-4899-3544-1 (eBook)**
**DOI 10.1007/978-1-4899-3544-1**


**Originally published by Plenum Press, New York in 1990**
**Softcover reprint of the hardcover 1st edition 1990**

PREFACE

This book chronicles the proceedings of the Second Symposium on Particles in Gases and Liquids: Detection, Characterization and Control held as a part of the 20th Annual Fine Particle Society meeting in Boston, August 21-25, 1989. As this second symposium was as successful as the prior one, so we have decided to hold symposia on this topic on a regular (biennial) basis and the third symposium in this series is scheduled to be held at the 22nd Annual Meeting of the Fine Particle Society in San Jose, California, July 29-August 2, 1991.

As pointed out in the Preface to the prior volume in this series[1] that recently there has been tremendous concern about yield losses due to unwanted particles, and these unwelcome particles can originate from a legion of sources, including process gases and liquids. Also all signals indicate that in the future manufacture of sophisticated and sensitive microelectronic components (with shrinking dimensions) and other precision parts, the need for detection, characterization, analysis and control of smaller and smaller particles will be more intensified.

This symposium was organized with the same basic objectives in mind as the prior one, but to provide an update on the R&D activity taking place in the arena of particle detection, characterization and control in various gases and liquids. The printed program comprised a total of 43 papers dealing with many and varied aspects of particles in gases and liquids. There were many positive comments from the attendees and there was consensus that the idea of initiating this series of symposia was an excellent one.

Before commenting on the present volume, it should be recorded that all manuscripts were rigorously peer reviewed and most of them required substantial revision before inclusion in the volume. In other words, this book is not merely a collection of "as-received" manuscripts rather it is a set of properly reviewed and edited papers.

The present volume contains 31 papers divided into four parts as follows: Part I. General Overview Papers; Part II. Particle Generation and Deposition; Part III. Particles in Gases: Detection, Characterization and Control; and Part IV. Particles in Liquids: Detection, Characterization and Control. The topics covered include: modeling of particle deposition in cleanroom environments; sampling of airborne particles; concepts of fractal geometry; operator-generated particles; optical particle counters; wiping materials; robots as a potential source of particulate contamination; particle emissions from air ionization equipment; in-situ particle monitoring; removal of particles; and packaging high purity chemicals.

I certainly hope this and the earlier volume[1] will provide the reader bountiful current information on the detection, characterization and control of particles in gases and liquids.

Acknowledgments: First, thanks are due to the Fine Particle Society for sponsoring this event. I am thankful to the appropriate management of IBM Corporation for permitting me to organize this symposium and to edit this volume. My special thanks are due to reviewers for their time and efforts in providing critical reviews, as comments from peers are sine qua non to maintain the high standard of publications. Also I would like to express my appreciation to my wife, Usha, for helping in many ways as well as in rendering home environment conducive to editing this book. Without the support, interest and contribution of the authors, there would have been no book, and I extend my special thanks to them.

K.L. Mittal
IBM U.S. Technical Education
500 Columbus Ave.
Thornwood, NY 10594

REFERENCES

1. K.L. Mittal, editor, Particles in Gases and Liquids 1: Detection, Characterization and Control, Plenum Press, New York, 1989.

CONTENTS

## PART I. GENERAL OVERVIEW PAPERS



## PART II. PARTICLE GENERATION AND DEPOSITION

## PART III. PARTICLES IN GASES: DETECTION, CHARACTERIZATION AND CONTROL

## PART IV. PARTICLES IN LIQUIDS: DETECTION, CHARACTERIZATION AND CONTROL

# PART I. GENERAL OVERVIEW PAPERS

# MODELING OF PARTICLE DEPOSITION IN CLEANROOM ENVIRONMENTS: CURRENT STATUS

Jay R. Turner

Chemical Engineering Department
Washington University
St. Louis, MO 63130

Recently many models have been presented to describe particle deposition in cleanroom environments. The models can be classified according to the scale of flow – cleanroom scale, equipment scale, or product scale. For the latter case, a detailed flow field is numerically generated or an idealized flow field is assumed. Particle deposition rates, in the presence of various external forces, are subsequently calculated by solving either the equation of convective diffusion or the particle equation of motion. A wide range of deposition mechanisms can be included: to account for actual operating conditions; and to assess the effectiveness of anti-contamination schemes.

In this work, we review the approaches taken by several authors to characterize particle deposition on the product scale. Emphasis is placed on models for particle deposition from laminar stagnation flow impinging onto a flat surface. Model development is stressed, without delving into quantitative details, to familiarize the reader with the fundamental differences between modeling approaches. A more quantitative look is taken at the retardation of particle deposition by thermophoresis.

## INTRODUCTION

Particulate contamination of critical components during electronic device manufacture reduces product yield. Sources of particles include process liquids and gases, equipment wear, shedding and reentrainment by personnel, penetration across filters at air intakes, and infiltration through leaks when rooms are inadequately positive-pressurized. Particles have the potential to become a contaminant only after depositing on the sensitive substrate. Scenarios for particle deposition are: removal from a particle-laden gas or liquid; and transfer via direct contact with a particle-laden solid. Both particle source strength and deposition efficiency dictate the relative importance of a given source/transport combination. The relative importance of various combina-

*Particles in Gases and Liquids 2,* Edited by K.L. Mittal
Plenum Press, New York, 1990

tions depends on the size range designated as potential contaminants. Thus, the dominant sources and mechanisms for 1 $\mu m$ particles may be different from the dominant sources and mechanisms for 0.1 $\mu m$ particles. Effective contamination control schemes should include periodic reevaluation of various source/transport scenarios. One class of scenarios is the deposition of particles from the bulk environment of the cleanroom or workstation. In response to this problem, cleanrooms and workstations have been designed with flow geometries such as laminar horizontal flow, laminar vertical flow or mixed flow.

Various models have been developed to estimate contamination in cleanroom environments. Two missions for modeling can be defined. In the first case, modeling is used as a design tool prior to cleanroom construction. Geometries are sought which: allow rapid venting or dilution of air downstream of a particle source; and minimize the exposure of product surfaces to particle-laden air parcels. In a second case, modeling is used as a diagnostic tool for optimization and troubleshooting of cleanroom operation. These efforts are used to suggest: modifications in cleanroom/process operating conditions; modifications in processing protocol (e.g. device handling and storage between processing steps and movement of personnel in the room); and modifications in the future design of cleanrooms and cleanroom components. The end product of models can be quantitative data or a set of heuristics.

A key question concerns the model scale. Two scales can be defined: a global scale which treats the entire cleanroom as a whole (imposing certain practical limitations with regard to detail); and a local scale which considers only the region in the neighborhood of the contamination - sensitive surface. Both scales require a description of the fluid (and perhaps temperature) field and also a description of particle transport. Global models permit a view of the integrated operation including the consequences of various particle sources and sinks. Assumptions are often made that the particles follow the fluid streamlines; this is valid for transport throughout the bulk of the room (exceptions may arise when modeling transport in process equipment featuring high-speed flows). These models clearly demonstrate how particles are transported to the vicinity of a surface. However, they can become extremely complicated (in terms of computational effort) when particle transport mechanisms other than convection are included. Local scale models are useful for studying particle deposition due to various combinations of transport mechanisms. This approach decouples the area near the surface from the remainder of the room; various phenomena throughout the room are lumped into a description of the flow- and concentration fields at a suitable distance upstream of the surface. These upstream parameters can be obtained from global modeling or experiments. How far "upstream" is not *a priori* clear – far enough away from the surface to avoid biasing by the surface (i.e. in the freestream) but not biased by other objects in the room. Local models often permit the use of simplified descriptions for the fluid flow- and temperature fields.

In this study, we review the literature for local scale modeling of particle deposition in cleanroom environments. Despite this rather narrow focus, the results often can be applied or extended to describe transport in many processing steps which feature drastically different operating conditions (e.g. contamination during CVD processing). Emphasis is placed on particle deposition from laminar stagnation flow. Little attention is given to incorporating the vast literature from related fields of aerosol transport and deposition (e.g. filtration by cylindrical elements, cascade impactor design); rather, we focus on works directly targeted at describing cleanroom deposition. New calculations are presented in the discussion of thermophoretic transport.

## MODELING GOALS AND APPLICATIONS

Modeling results can be used to compare the relative importance of various transport mechanisms and quantitatively calculate particle deposition rates. Deposition rate calculations have been discussed by Cooper[1] and examples are given by Liu et al.[2] and Welker[3]. Particles of diameter $d_p$ deposit onto a surface with flux $J$ $(\#/cm^2/sec)$,

$$J(d_p) = k_p(d_p)\,[n_b(d_p) - n_s(d_p)] \tag{1}$$

where $k_p$ is the particle size-dependent mass transfer coefficient $(cm/sec)$ and $n_b, n_s$ are the size-dependent particle number concentrations in the fluid bulk (freestream) and at the collector surface, respectively. Surfaces are usually assumed to be perfect sinks for particles ($n_s = 0$). Total deposition fluxes, $J_T$, are calculated by integrating Equation 1 over the range of killer-sized particles. Often this range encompasses all particles larger than a critical size, $d_{p,c}$,

$$J_T = \int_{d_{p,c}}^{\infty} k_p(d_p)\frac{dn_b(d_p)}{ddp}dd_p \tag{2}$$

This approach requires a particle size distribution, $n_b(d_p)$, for freestream conditions and also the mass transfer coefficient for deposition. The former are experimentally determined; the latter are calculated by modeling the deposition process. Cleanroom literature has borrowed the term "deposition velocity" from allied fields such as atmospheric deposition science. The deposition velocity, $v_d$, is identical to the mass transfer coefficient defined by Equation 1.

Two formats are commonly used to present particle deposition velocities. In one case, deposition velocities are explicitly given as a function of particle size, $v_d(d_p)$. This format allows direct substitution into Equation 2 to calculate total deposition rates. A second approach casts the results in dimensionless form. This is more fundamental and most models take advantage of the computational efficiency gained by nondimensionalization. In this case, particle deposition velocities are expressed in terms of the Sherwood number,

$$Sh \propto \frac{k_p R}{D_p} \tag{3}$$

where $R$ is the characteristic length of the collector and $D_p$ is the particle diffusivity $(cm^2/sec)$. Sherwood numbers are presented as a function of the Peclet number, $Sh(Pe)$, where

$$Pe \propto \frac{U_\infty R}{D_p} \tag{4}$$

and $U_\infty$ is the fluid freestream velocity $(cm/sec)$. The particle size appears in the Peclet number implicitly through the diffusivity. Care must be taken to correctly define the characteristic length, $R$. Confusion can arise when the idealized collector is an infinite surface because a characteristic collector length does not exist[4]. In this case, an alternative length scale must be defined and properly used to nondimensionalize the governing equations.

A dimensionless formulation is attractive because the independent variables are contracted into a few key groups. For example, a given Peclet number represents a surface in $U_\infty - R - d_p$ space. The price for information contraction is the need to calculate $v_d(d_p)$ from $Sh(Pe)$ if using Equation 2 to estimate total deposition rates. Also note that $Pe$ is not linearly dependent on $d_p$ for all particle sizes; the true dependence is $Pe \propto d_p/C_c$ where $C_c$ is the slip correction factor ($C_c \neq 1$ for particles smaller than the mean free path, where $C_c$ increases with decreasing particle size).

A branching point is reached early in any effort to model cleanroom deposition processes. One branch leads to simple models, for example superposition of deposition velocities for various transport mechanisms. This approach is useful for rapid estimation of deposition rates. A second branch leads to more detailed models (added refinements and/or enhanced generality). These models are needed to determine the range of validity of the simple models (thus the approaches are complementary) and serve as a framework for extending the results to other geometries and operating conditions. Examples for both classes will be discussed.

## LAMINAR STAGNATION FLOW MODELING

Most local scale cleanroom models describe particle deposition onto a flat surface from laminar stagnation flow. This idealization is valid for many geometries and operating conditions encountered in cleanroom environments. Global-scale models and flow visualization verify the existence of stagnation regions attached to many contamination-sensitive surfaces. Both axisymmetric and planar stagnation flows are possible. Axisymmetric patterns are characteristic of flow over a free-standing wafer and flow impinging on a square table. Planar patterns describe flow impinging onto a workbench with symmetric exhaust to front and rear, and flow over a rectangular table with large length-to-width ratio (e.g. tables used for lithography lay-ups). Deposition trends are qualitatively similar for the two geometries; quantitative results differ only slightly to reflect the velocity profile differences of the planar and axisymmetric flows.

Several formulations can be used to describe the fluid- and temperature fields and also particle transport. Fluid- and temperature fields are calculated: using a boundary layer approximation[5]; and by full solution to the Navier-Stokes equations and energy conservation equation[6]. The boundary layer approximation simplifies the flat-surface stagnation flow analysis by assuming the fields are radially uniform at the outer edges of the respective boundary layers. Through proper scaling the governing partial differential equations are reduced to ordinary differential equations in a single spatial dimension. This approach assumes that deposition at the stagnation point is characteristic of the entire surface; the neglect of collector edge effects is a shortcoming of the 1-D boundary-layer approximation.

Particle transport is described by: solution of the equation of motion for a single particle; or solution of the convective diffusion equation. The former yields particle trajectories and the latter yields the particle concentration field. The 1-D boundary layer approximation is often used for the particle concentration field because it reduces the problem to one dimension. Trade-offs between the various approaches will be demonstrated throughout the remainder of this review.

If the convective diffusion equation is used, governing equations for the fluid-, temperature- and concentration fields can be solved simultaneously or sometimes sequentially. Sequential solution is possible if the concentration field can be decoupled from the flow- and temperature fields, which is certainly the case for cleanroom applications where the aerosol is extremely dilute. If the particle trajectory equation is used, the fluid- and temperature fields are solved first, followed by the particle trajectory equation.

The following sections summarize various models for cleanroom particle deposition. Studies are categorized: primarily by the type of particle transport equation; and subsequently by the mechanisms investigated. This organization emphasizes general model development over specific test cases. Most models can be easily extended to include the various transport mechanisms. It is beyond the scope of this paper to review the relevant forces which act on the particles (see e.g. papers appearing in Mittal[7,8]). Readers unfamiliar with the nature and quantification of such forces should consult the literature (see e.g. Cooper[1]; Fissan and Turner[9]).

## MODELING EFFORTS

### Mass Transfer Correlation

Liu and Ahn[10] estimated the deposition velocity of uncharged particles to a free-standing wafer (axisymmetric flow) by modifying the mass transfer correlation of Sparrow and Geiger[11]. The correlation was developed for naphthalene sublimation from a disk and expressed as a $Sh - Re$ relation. Liu and Ahn[10] assumed $Sh \propto Sc^{1/3}$ to include the dependence of mass transfer on diffusivity. The correlation describes deposition of point particles by convective diffusion only. In *cgs* units, the correlation for the axisymmetric stagnation point deposition velocity is

$$v_{d,0} = 0.834(\nu)^{-1/6}\left(\frac{U_\infty}{D_W}\right)^{1/2}(D_P)^{2/3} \tag{5}$$

where $D_W$ is the wafer diameter ($cm$) and $\nu$ is the fluid kinematic viscosity ($cm^2/sec$). Based on naphthalene sublimation analysis, Equation 5 describes deposition over most of the surface but underestimates deposition at the wafer edge. For the deposition velocity averaged over the entire wafer, use the coefficient 1.08 rather than 0.834 in Equation 5. Liu and Ahn[10] further assumed linear superposition of the settling- and convective diffusion velocities to obtain the characteristic U-shaped overall deposition velocity curve.

### Convective Diffusion Equation

Given the fluid velocity field, the convective diffusion equation can be solved to find the particle concentration field[12],

$$\vec{v}_p \cdot \nabla n = D_p \nabla^2 n - \nabla \cdot \vec{c} n \tag{6}$$

where $\vec{v}_p$ is the particle velocity ($cm/sec$) and $\vec{c}$ the migration velocity in an external force field ($cm/sec$). Using this approach it is straightforward to include Brownian diffusion but difficult to rigorously account for inertial effects. Usually one assumes $\vec{v}_p = \vec{u}$ (where $\vec{u}$ is the fluid velocity) and thus neglects inertial effects; this assumption is valid only if the Stokes number is well below a critical value which depends on geometry and operating conditions (see e.g. Friedlander[12]; Peters et al.[4]). Solution of Equation 6 requires that the concentration field be specified: far from the substrate (freestream conditions); and also at the substrate surface (usually assume the surface is a perfect sink for particles).

Full Solution to Navier-Stokes Equations. Liu et al.[2] numerically solved the Navier-Stokes equations using the fluid dynamics code "SIMPLE"[6]. They obtained an exact solution for the convective diffusion of (inertialess) point particles to a free-standing wafer in the presence of gravity. These results validated the linear superposition assumption for the case of settling- and convective diffusion velocities. They also extended the analysis to include spatially-uniform Coulombic forces.

The methodology for modifying a fluid dynamics code (such as SIMPLE) to handle particle transport, according to Equation 6, has been discussed by Stratmann and Whitby[13]. They present the changes needed to treat generalized external forces which act on the particles. Applications of their efforts will be discussed in later sections.

Boundary Layer Approximation. The convective diffusion equation also has been solved with the 1-D boundary layer approximation for the velocity- and concentration fields (Cooper et al.[14]; Peters et al.[4]; Turner et al.[15]). This approximation reduces the governing equations to one dimension oriented normal to the surface; thus, the system of partial differential equations becomes ordinary differential equations. Cooper et al.[14] and Peters et al.[4] treated axisymmetric stagnation flow while Turner et al.[15] studied planar stagnation flow. Results were presented for an infinite plane sheet and semi-infinite plate, respectively. Several subcases have been discussed.

An approximate analytical solution exists for the convective diffusion of point particles when the concentration boundary layer is much smaller than the hydrodynamic boundary layer (large Schmidt numbers). The so-called Levich-Smoluchowski solution has the general form

$$v_{d,0} = \alpha(\nu)^{-1/6}\left(\frac{U_\infty}{R}\right)^{1/2}(D_P)^{2/3} \tag{7}$$

where $\alpha$ is a constant for a given geometry. For deposition onto a semi-infinite plate with half-width $R$ the constant $\alpha$ is 0.661 (Turner et al.[15]); for deposition onto a finite disk with diameter $R$ the constant $\alpha$ is 0.876 (Cooper et al.[14]). The latter solution agrees with the mass transfer correlation of Liu and Ahn[10], Equation 5 .

Convective diffusion in the presence of spatially uniform external fields (e.g. gravity and sometimes Coulombic forces) is described by an integral solution to the deposition velocity (Peters et al.[4]; Turner et al.[15]). For engineering calculations, linear superposition of the convective diffusion velocity (Equation 7) and particle migration velocity, $c_{ext}$, is valid if the net external force reinforces diffusion. Examples are presented by Cooper et al.[14]. However, linear superposition cannot always be used if the net external force counters diffusion. Figure 1 shows $v_d$ vs. $d_p$ for various external force strengths which reinforce diffusion (net force directed towards the surface); in all cases there is close agreement between the integral solution and linear superposition. At sufficiently large external force strengths, the deposition velocity approaches the particle migration velocity; diffusion is negligible in this regime (this implies $Sh \neq Sh(Pe)$ as shown by Cooper et al.[14]). Figure 2 shows the case for net external forces directed away from the surface; the linear superposition assumption can result in large error. Also note that the integral solution becomes invalid when the diffusion boundary layer becomes thicker than the hydrodynamic boundary layer (small Schmidt numbers or large external forces directed away from the surface).

Spatially varying external forces (e.g. image forces and thermophoresis) have also been studied. Approximate analytical solutions are available for attractive particle-surface forces subject to rather severe limitations (Peters et al.[4]). The general solution method requires numerical solution to the governing ordinary differential equations. Documented approaches include simulation by the: shooting method (Peters et al.[4]); finite differences (Turner et al.[15]); and collocation (Turner et al.[16]). The important case of thermophoresis will be treated later in more detail.

Particle Equation of Motion

Given the fluid velocity field, the particle equation of motion can be solved to track the behavior of individual particles. The general form is

$$\frac{d\vec{v}_p}{dt} + \beta(\vec{v}_p - \vec{u}) = \frac{\vec{F}}{m_p} + \vec{A}(t) \tag{8}$$

where $\vec{v}_p, \vec{u}$ are the particle and fluid velocities ($cm/sec$), respectively, $\beta$ the friction coefficient per unit particle mass ($sec^{-1}$), $m_p$ the particle mass ($g$), $\vec{F}$ the external force acting on the particle ($g \cdot cm/sec^2$) and $\vec{A}(t)$ the random (Brownian) force per unit mass ($cm/sec^2$).

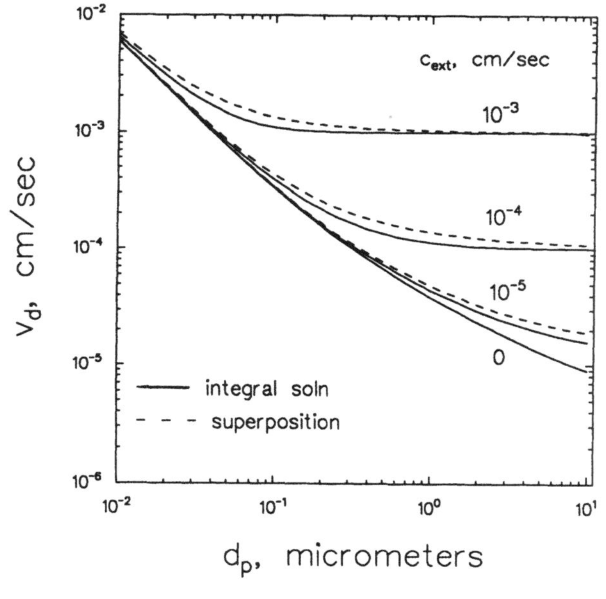


Figure 1. Deposition velocity due to convection, diffusion and an external force directed towards the surface. Superposition includes convective diffusion velocity and migration velocity in external force field. Simulation for planar stagnation flow with $a = 1\ sec^{-1}$ and $\nu = 0.15\ cm^2/sec$.

The second term accounts for inertial effects (inertialess particles for $\vec{v}_p = \vec{u}$). The last term, $\vec{A}(t)$, accounts for Brownian motion and usually requires solution by a stochastic method because $\vec{A}(t)$ follows a Gaussian process (Gupta and Peters[17]); this complicates the computational effort. When solving Equation 8, we must specify the particle's initial position above the substrate and also define crtieria for particle collection by the surface.

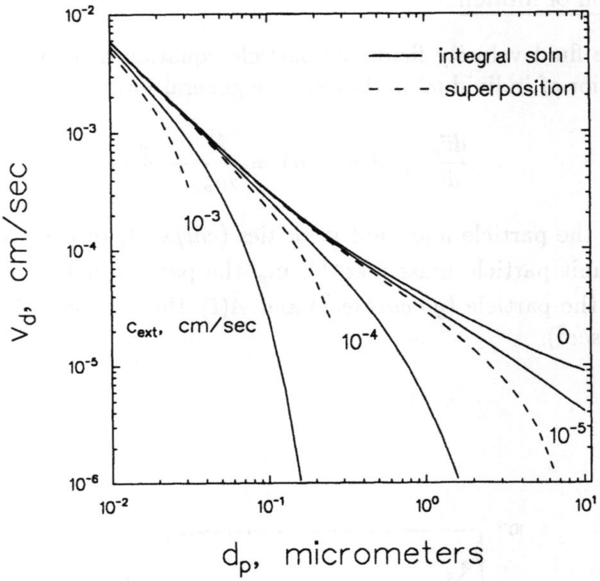


Figure 2. Deposition velocity due to convection,diffusion and an external force directed away from the surface. Superposition includes convective diffusion velocity and migration velocity in external force field. Simulation for planar stagnation flow with $a = 1\ sec^{-1}$ and $\nu = 0.15\ cm^2/sec$.

Boundary Layer Approximation. Peters et al.[4] provided a detailed treatment of the Brownian dynamics approach to modeling deposition from stagnation flows by integration of Equation 8. They presented a full theoretical development of the governing equations and results for Monte Carlo simulations with simultaneous diffusion and inertia. Results were compared: quantitatively to solutions to the convective diffusion equation for Brownian transport; and qualitatively to the analysis of Fernandez de la Mora and Rosner[18] for the diffusion-inertial regime. While computationally intensive, this approach is very powerful. It can be easily extended to include other external forces which act on the particles. Perhaps its greatest value will arise in modeling cleanroom processing equipment which feature high speed flows. The analysis has shown that inertially-enhanced deposition can become significant well below the critical Stokes number used to characterize the onset of inertial effects in a diffusion-free system.

Full Solution to Navier-Stokes Equations. Abuzeid and Busnaina[19] and Busnaina and Abuzeid[20] solved the particle equation of motion in conjunction with full solutions to the Navier-Stokes equations for various flow geometries. Like the analysis of Peters et al.[4], they modeled diffusion as a white noise (Gaussian) process; thus, for each set of operating conditions several simulations are required to converge to a "most probable" solution. Preliminary studies addressed the accuracy of finite difference schemes for modeling 2-D flow fields relevant to cleanroom environments (Busnaina and Abuzeid[20]). Subsequently, they have focused on differences between

Eulerian (advection-diffusion equation) and Lagrangian (particle equation of motion) simulations in laminar stagnation flow (Abuzeid and Busnaina[19]) and turbulent channel flow (Abuzeid et al.[21]).

Other Approaches

Peterson et al.[22] used a "hybrid" 2-D model to include inertial effects in the convective diffusion equation. They locally solved for the particle velocity in the absence of diffusion using the equation of motion,

$$\frac{d\vec{v}_p}{dt} + \beta(\vec{v}_p - \vec{u}) = \frac{\vec{F}}{m_P} \tag{9}$$

A perturbation formulation in terms of the Stokes number was used to account for inertia, and boundary layer approximations were used for the fluid velocity- and temperature fields. With the generated estimates for the local particle velocity, $\vec{v}_p$, they used a finite differences scheme to solve the convective diffusion equation in the absence of external forces,

$$\vec{v}_p \cdot \nabla n = D_p \nabla^2 n \tag{10}$$

To date, this 2-D analysis has been used: in a model intercomparison study against the truly one-dimensional boundary layer model and also full two-dimensional solutions to the Navier-Stokes equations using the modified SIMPLE code (Peterson et al.[22]); and to study a particle beam located above a wafer (Stratmann et al.[23]).

Comments

Up to this point, we have described various approaches taken to model particle deposition onto a flat surface from stagnation flow. Let us briefly comment on several of the conclusions reached by these efforts for typical cleanroom/workbench operating conditions. Details are given in many of the papers cited above.

Uncharged particles deposit primarily by convective diffusion and sedimentation. A minimum in the $v_d - d_p$ curve exists in the particle size range $0.1 - 1\ \mu m$ . Convective diffusion dominates the deposition of smaller particles; sedimentation dominates the deposition of larger particles. As the fluid freestream velocity increases, $v_d$ in the diffusion-controlled regime increases and the minimum shifts to larger particle sizes.

Electrostatic forces can greatly enhance particle deposition rates. Deposition due to attractive Coulombic forces can dominate over diffusion for submicron particles even if the electric field strength is relatively weak (in terms of common cleanroom static charge build-up levels). Deposition rates for typical cleanroom operating conditions may be several orders of magnitude larger than the no-electrostatics case. Even if the substrate surface potential is neutralized there still exist image forces; these forces may enhance deposition by up to a factor of five in the submicron size range.

The net effect of electrostatic forces is difficult to quantify. A rigorous analysis requires flux calculations analogous to Equation 2 using $v_d(d_p, n_e)$ and $n_b(d_p, n_e)$ where $v_d$ and $n_b$ are now surfaces in size-charge space. Using the above modeling approaches we can calculate $v_d(d_p, n_e)$. However, the particle concentration distributed in size-charge space is unknown for cleanroom environments. To date, flux calculations have assumed: the aerosol is in Boltzmann charge equilibrium; or all particles carry the same charge (usually expressed as a multiple of the root-mean-square charge for an aerosol in Boltzmann equilibrium). These cases are useful for sensitivity studies but do not reveal the quantitative impact of electrostatic forces on particle deposition.

Relatively little experimental data is available to validate the models. The data of Hayakawa et al.[24] have been discussed by Liu et al.[2]; many unknowns in the system operating conditions render a comparison inconclusive. Various investigators are currently conducting experiments which will be used to validate the models.

Other flow geometries also warrant attention. Complicated geometries usually require a full solution to the Navier-Stokes equations to compute the velocity profiles. However, there are other simplified geometries which are amenable to analysis as performed for the flat-surface stagnation flow. There exists vast literature concerning deposition to cylinders and spheres. These results can be placed, as needed, in a cleanroom modeling context. Peterson et al.[22] considered flow over a wedge at arbitrary angles of attack; this is a generalization of the stagnation flow case where the angle is ninety degrees. Similar is jet flow, impinging at arbitrary angles of attack, which may be important for many processing systems. Another subcase of the wedge flow is flow parallel to a flat plate (in this case, wedge flow at zero attack angle). Liu and Ahn[10] discuss deposition to horizontally and vertically oriented wafers in parallel flow.

Finally, the role of turbulence needs to be addressed. Turbulence may arise as: an additional component superimposed on the basically laminar flow; and as the only component for convective transfer. The former case has been incorporated into global models (see e.g. Kuehn [25]). In local scale modeling, however, we require a detailed description of turbulence in the vicinity of the surface; this deserves further attention (Fujii et al.[26]). Mixed-flow cleanrooms may in some ways resemble large vessels where the gas is turbulently mixed (either by movement or intake/exhaust). A few efforts have considered particle deposition in mixed vessels both experimentally (Donovan et al.[27]; Shimada et al.[28,29]) and theoretically (Shimada et al.[28,29]; Turner and Fissan[30]). However, the extension to cleanroom operating environments has yet to be fully exploited.

Now let us look in more detail at the role of thermophoresis on particle transport and deposition.

## THERMOPHORESIS

Surface heating induces a thermophoretic force which drives particles away from the surface. This has been proposed[31] as a method for reducing particle contamination. Two key issues are at hand: the effect of heating on the fluid flow field; and the motion of particles in the thermophoretic field. Large temperature gradients and/or low velocities result in a heat/momentum coupling via the buoyancy term in the momentum equation. However, if the temperature gradient is small and/or convection rate is high, the energy and momentum equations can be decoupled. In this case, the flow field can be determined independently, and the temperature profile subsequently obtained by solving the energy equation (the so-called advection problem). The temperature gradient appears in the particle transport equation as an external force acting on the particle. For a dilute aerosol, the thermophoretic migration velocity, $V_T$, is

$$V_T = -K\frac{\nu}{T}\nabla T \tag{11}$$

where $K$ is the thermophoretic coefficient. The term $\nabla T/T$ is obtained as a function of position from the computed temperature profile. Unfortunately, the correct form for $K$ is still subject to debate owing to the complexity of the particle/thermal field interactions. Several expressions have been proposed, with the most-accepted being (Batchelor and Shen[32]),

$$K = \frac{1.5\left(\frac{k_g}{k_p} + 2.18\mathrm{Kn}\right)\left(1 + \mathrm{Kn}\left[1.2 + 0.41\exp(-0.88/\mathrm{Kn})\right]\right)}{(1 + 3.42\mathrm{Kn})\left(1 + 2\frac{k_g}{k_p} + 4.36\mathrm{Kn}\right)} \tag{12}$$

where $Kn$ is the particle Knudsen number, and $k_g$, $k_p$ the gas and particle thermal conductivities, respectively.

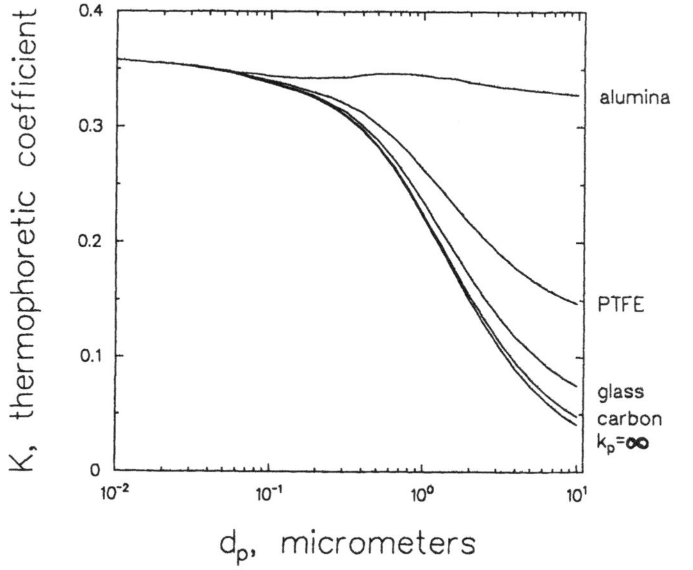


Figure 3. Thermophoretic coefficient for particles of various materials in air at 20 °$C$. The ratio of thermal conductivities (gas-to-particle) are: alumina, 0.38; PTFE, 0.097; glass, 0.028; and carbon, 0.0056.

Figure 3 shows the dependence of $K$ on particle size in air at room temperature. Knudsen aerosols ($Kn >> 1$) exhibit $K$ approximately 0.36 independent of particle size and composition; continuum aerosols ($Kn << 1$) exhibit a dependence on composition due to the thermal conductivity ratio. The curve for alumina in Figure 3 is representative of many oxides and other types of insulators (e.g. asbestos and fiberglass). The thermal conductivity of polytetrafluoroethylene (PTFE) is indicative of many polymer materials (such as polyethylene, polyamide). Many conducting materials lie in the range $0 < k_g/k_p < 10^{-2}$ (e.g. carbon, iron, copper); in this regime the curves are close to the limiting case of infinite particle conductivity, $k_g/k_p = 0$, which often can be used as a reasonable approximation. For particles with larger conductivity the proper values for the thermophoretic coefficient must be calculated. In theory, for continuum aerosols K can also increase drastically with increasing particle size; however, this is rarely realized for applications of practical interest and is probably relevant to cleanroom operations only when processing in a pure hydrogen atmosphere. For almost all applications of interest, the thermal migration velocity will decrease with increasing particle size.

This development assumes that the particles are in thermal equilibrium with the surrounding gas. However, a shearing flow will cause the particles to rotate; if the characteristic time for heat transfer, $\tau_{pT}$, is greater than the characteristic time for particle rotation, $\Omega_f^{-1}$, a non-equilibrium situation exists. The particle maintains a uniform surface temperature which is not in equilibrium with the local gas temperature and the thermophoretic force is reduced. Chomiak and Gupta[33] present a criterion for the onset of reduction in the thermophoretic migration velocity due to particle rotation in boundary layer flow over a flat plate. The analogous relation for boundary layer plane stagnation flow is, at the surface,

$$\Omega_{f,w}^{-1} = \frac{1}{2}\left(\frac{\partial u_y}{\partial x}\right)_w = \frac{1}{2}U_\infty\left(\frac{a}{\nu}\right)^{1/2}\left.\frac{d^2\phi}{d\eta^2}\right|_{\eta=0} = \frac{\phi''(0)}{2}U_\infty\left(\frac{a}{\nu}\right)^{1/2} \tag{13}$$

where $u_y$ is the fluid velocity normal to the surface ($cm/sec$), $x$ and $y$ are the coordinates parallel and normal to the surface, respectively ($cm$), $a$ is the fluid velocity gradient ($sec^{-1}$), $\phi$ the dimensionless velocity normal to the surface ($-u_y/\sqrt{a\nu}$), and $\eta$ the dimensionless coordinate normal to the surface ($-y\sqrt{a/\nu}$). For plane stagnation flow $\phi''(0) = 1.2326$ . Given a cleanroom workbench geometry with $U_\infty = 50\ cm/sec$ and a bench half-depth of $50 cm$ , then $a = U_\infty/R = 1\ sec^{-1}$ , and the reciprocal characteristic time for rotation is $80\, sec^{-1}$ . Following Chomiak and Gupta[33], the regime for significant rotational effects on thermophoretic transport is

$$\tau_{pT}\Omega_{f,w} > 1 \tag{14}$$

Figure 4 shows the characteristic time for heat transfer for unit density spheres with equal gas- and particle specific heats (data from Chomiak and Gupta[33]). For the above example, rotational effects are important only for particles larger than approximately $100\mu m$ in diameter. However, this effect may become important for much smaller particles in certain high-speed-flow processing applications.

As with other transport mechanisms, deposition models which account for thermophoresis in cleanroom applications have focused on the stagnation flow geometry. Stratmann et al.[34] considered particle transport to a slightly heated surface in plane stagnation flow. Using boundary layer approximations for the velocity- and temperature fields they solved the advection problem to find the temperature gradient at the surface and correlated the result as a function of Prandtl number,

$$\theta'_w = \left(\frac{d\theta}{d\eta}\right)_w = -0.5692\ Pr^{0.377} \tag{15}$$

where $\theta$ is the dimensionless temperature ($[T_w - T]/[T_w - T_\infty]$) and $Pr$ the Prandtl number ($C_p\mu/k_g$). Subsequently, they solved the force balance on a single particle (particle equation of motion) to estimate the thickness of the dust-free space in the boundary layer above the surface, $\delta_{df}$. The asymptotic formula for point particles subjected to convection and thermophoresis is

$$\delta_{df} = 0.9610\left(\frac{\nu}{a}\right)^{1/2} K^{1/2}\left(\frac{T_W - T_\infty}{T_\infty}\right)^{1/2} Pr^{0.189} \tag{16}$$

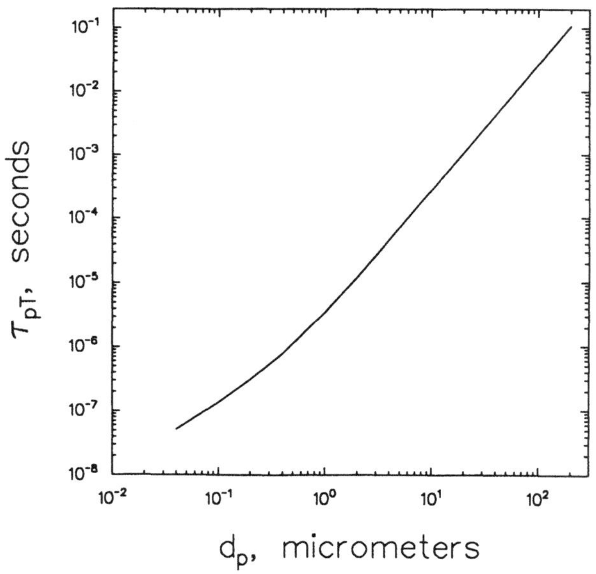


Figure 4. Characteristic time for heat transfer, $\tau_{pT}$, of unit density spheres in air at $20\,°C$. Data from Chomiak and Gupta[33].

They validated this approximate solution with detailed numerical simulations. This analysis demonstrated the structure of the particle field in the case of thermophoresis-induced transport. However, it is necessary to extend the analysis to include additional forces which act upon the particles.

The first extension is given by Friedlander et al.[35] who noted that thermophoresis protects surfaces only over a finite range of particle sizes. At the upper end, gravity dominates over the weakening thermal barrier (see Figure 3) and at the lower end, particles diffuse across the dust-free space. Thus, they proposed to determine the conditions for the onset of diffusive penetration. In contrast to Stratmann et al.[34], this analysis uses the particle concentration field rather than a force balance on a single particle. They obtained an approximate solution (valid for mild surface heating and Schmidt numbers much larger than unity) for the thermophoretic blowing effect,

$$\frac{v_d}{v_{d,\nabla T=0}} = \frac{\Gamma(1/3)}{\sum_{n=0}^{\infty} \frac{(3\Delta^2)^{n/3}}{n!} \Gamma\left[\frac{n+1}{3}\right]} \tag{17}$$

where $\Delta = [\phi''(0)\ Sc\ \delta^3]/2$ and $\delta = \delta_{df}\sqrt{a/\nu}$. This solution was validated against full numerical solutions to the boundary layer equations. Let us define the criteria for the critical particle size below which significant leakage occurs as $\frac{v_d}{v_{d,\nabla T=0}} = 10^{-3}$ ; that is, a three order-of-magnitude reduction in the deposition velocity relative to the isothermal case. This occurs at $\Delta = 10.5$ and the resulting criterion for the onset of deposition is

$$\delta^3\, Sc\, -\, 17 = 0 \tag{18}$$

where both the boundary layer thickness and Schmidt number are functions of particle size. The critical size is given by the lower curve in Figure 5 as a function of surface temperature.

In the diffusion-free regime, the influence of sedimentation and a constant Coulomb force can be included by extending the analysis of Stratmann et al.[34]. In this case, we find

$$\delta_{df} = 1.2738 \left(\frac{\nu}{a}\right)^{1/2} \left[0.5692\, K \left(\frac{T_W - T_\infty}{T_\infty}\right) Pr^{0.377} + \frac{c_{ext}}{\sqrt{a\nu}}\right]^{1/2} \tag{19}$$

where $c_{ext}$ is positive in the direction away from the surface. When $c_{ext} = 0$, Equation 19

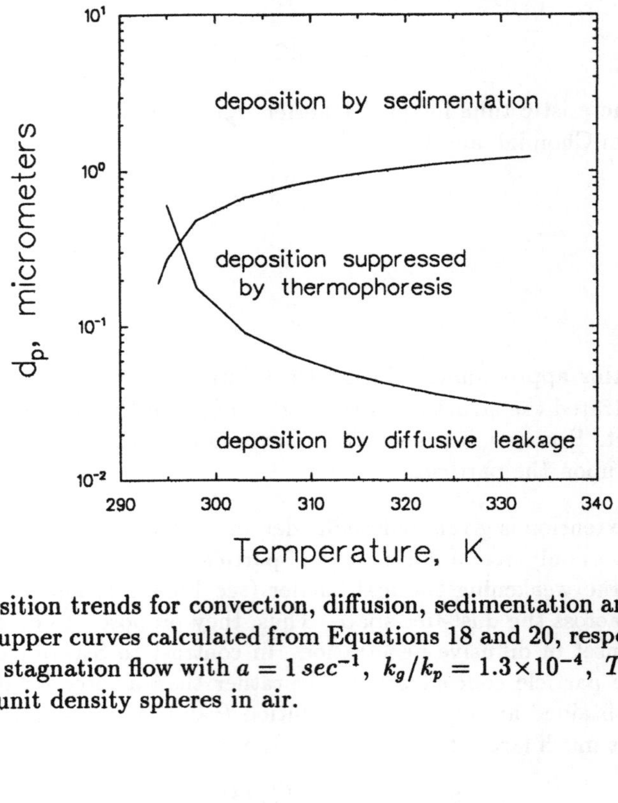


Figure 5. Deposition trends for convection, diffusion, sedimentation and thermophoresis. Lower and upper curves calculated from Equations 18 and 20, respectively. Simulations for planar stagnation flow with $a = 1\, sec^{-1}$, $k_g/k_p = 1.3\times10^{-4}$, $T_\infty = 293\, K$, $\nu = 0.15\, cm^2/sec$, unit density spheres in air.

reduces to Equation 16. For sedimentation (and no Coulomb force), a critical size exists above which deposition is not suppressed; this size is calculated by setting $\delta_{df} = 0$ (note point particles were assumed in the initial derivation and should be retained in this analysis). This means the term in brackets of Equation 19 is merely set equal to zero and solved for the particle size which appears in $K$ and $c_g$,

$$\frac{\mid c_g \mid cos\theta}{\sqrt{a\nu}} - K \left(\frac{T_w - T_\infty}{T_\infty}\right) Pr^{0.377} = 0 \tag{20}$$

Figure 5 shows the critical particle size above which thermophoresis cannot suppress particle deposition by sedimentation (plate normal to gravitational field). Thus, the center region of Figure 5 is a window in particle-size space for which deposition can be successfully suppressed by the applied surface heating.

Turner et al.[16] and Stratmann et al.[23] presented $v_d - d_p$ plots for various transport mechanisms in addition to thermophoresis. Turner et al.[16] extended the analysis of the 1-D boundary layer equations to show the effect of thermophoresis on particle deposition velocity. A typical curve is shown in Figure 6 (there are minor differences with respect to the simulations of Turner et al.[16] due to the use of a different definition for $K$). Figure 7 shows the simulations of Figure 6 scaled to the deposition velocity in the absence of thermophoresis. Note the excellent agreement between the full model results and the approximate solutions for critical-sized particles, discussed above. The deposition-free particle-size window broadens with increased surface heating. Turner et al.[16] demonstrated the augmentation of the thermophoretic effect; as the Coulomb interaction strength increases, the deposition-free particle-size window narrows and eventually disappears. It is disconcerting that the studied Coulombic force strengths (for heating up to forty degrees above ambient) are well within the range suspected to arise in cleanrooms. Stratmann et al.[23] considered the case of a particle beam above a surface and reached similar conclusions.

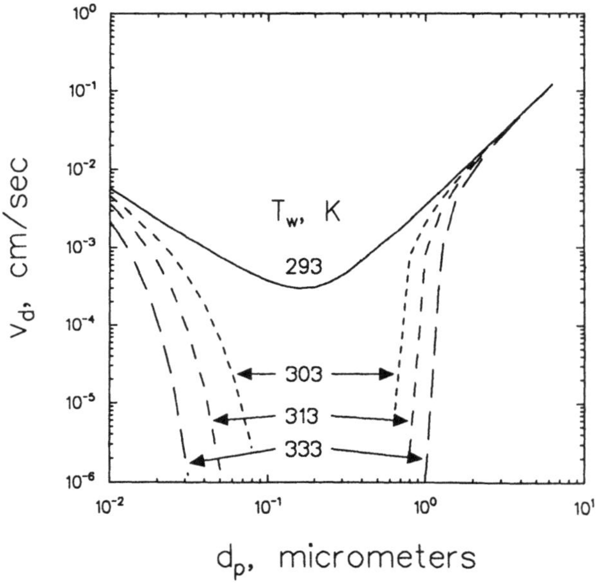


Figure 6. Deposition velocity for convection, diffusion, sedimentation and thermophoresis. Simulations for operating conditions of Figure 5 using the full 1-D boundary layer analysis of Turner et al.[16].

## CONCLUSIONS

A robust battery of models are available to describe particle deposition to a flat surface from laminar stagnation point flow. We have classified the models according to the form of the governing equations used in model development. Sophisticated models exist for detailed analysis; simple approximations can be used for engineering calculations. Areas for future work include: extension to other relevant flow geometries; experimental verification of the existing models; analysis of particle charge distributions in cleanrooms; and incorporation of turbulence effects.

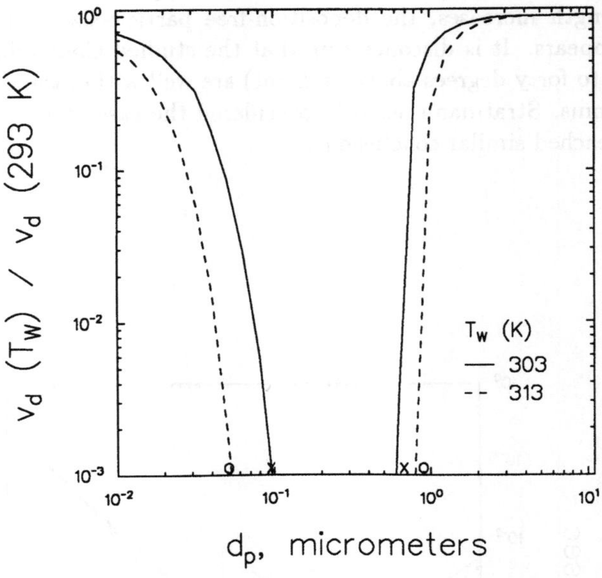


Figure 7. Deposition velocity for convection, diffusion, sedimentation and thermophoresis normalized to the case of no thermophoresis. Simulations for operating conditions of Figure 5. Curves are numerical solutions to the full 1-D boundary layer equations; ×'s and o's are solutions to Equations 18 and 20 for temperatures 303 K and 313 K, respectively.

We have elaborated on the role of thermophoresis in preventing particle deposition. Simple equations have been presented to estimate the range of particle sizes for which substrate heating can effectively prevent particle deposition (in the absence of electrostatic effects). Substrate heating appears to be a viable method for reducing contamination only if the net external force acting on the particles is small. This restriction may require some form of surface electric potential reduction in combination with substrate heating.

| | |
|---|---|
| $a$ | fluid velocity gradient, $sec^{-1}$ |
| $\vec{A}(t)$ | random (Brownian) force per unit mass, $cm/sec^2$ |
| $\vec{c}$ | local particle migration velocity in an external force field, $cm/sec$ |
| $C_c$ | Cunningham slip correction factor |
| $C_p$ | fluid heat capacity, $g \cdot cm/sec^2/K$ |
| $d_p$ | particle diameter, $cm$ |
| $d_{p,c}$ | smallest killer-particle size, $cm$ |
| $D_p$ | particle diffusion coefficient, $cm^2/sec$ |
| $D_w$ | wafer diameter, $cm$ |
| $\vec{F}$ | external force acting on particle, $g \cdot cm/sec^2$ |
| $J$ | particle flux to surface, $\#/cm^2/sec$ |
| $k_j$ | thermal conductivity of medium $j$, $g \cdot cm/sec^3$ |
| $k_p$ | particle mass-transfer coefficient, $cm/sec$; also particle thermal conductivity |
| $K$ | thermophoretic coefficient |
| $Kn$ | Knudsen number, $2\lambda/d_p$ |
| $m_p$ | particle mass, $g$ |
| $n$ | particle number concentration, $\#/cm^3$ |
| $n_e$ | number of elementary charge units on a particle |
| $Pe$ | Peclet number, Equation 4 |
| $Pr$ | Prandtl number, $C_p\mu/k_g$ |
| $R$ | characteristic length of collector, $cm$ |
| $Sc$ | Schmidt number, $\nu/D_p$ |
| $Sh$ | Sherwood number, Equation 3 |
| $t$ | time, $sec$ |
| $T$ | fluid temperature, $K$ |
| $\vec{u}$ | local fluid velocity, $cm/sec$ |
| $u_y$ | fluid velocity component normal to surface, $cm/sec$ |
| $U_\infty$ | fluid freestream velocity, $cm/sec$ |
| $\vec{v}$ | local particle velocity, $cm/sec$ |
| $v_d$ | particle deposition velocity, $cm/sec$ |
| $V_T$ | particle migration velocity due to thermophoretic force, $cm/sec$ |
| $x$ | coordinate parallel to surface, $cm$ |
| $y$ | coordinate normal to surface, $cm$ |
| greek | |
| $\beta$ | friction coefficient per unit mass, $1/sec$ |
| $\Gamma(z)$ | gamma function of argument $z$ |
| $\delta$ | dimensionless dust-free layer thickness, $\delta_{df}\sqrt{a/\nu}$ |
| $\delta_{df}$ | dust-free layer thickness, $cm$ |
| $\eta$ | dimensionless coordinate normal to surface, $-y\sqrt{a/\nu}$ |
| $\theta$ | dimensionless temperature; also angle between surface and gravitational field |
| $\lambda$ | fluid mean free path, $cm$ |
| $\mu$ | fluid dynamic viscosity, $g/cm/sec$ |
| $\nu$ | fluid kinematic viscosity, $cm^2/sec$ |
| $\tau_{pT}$ | characteristic time for particle-medium heat transfer, $sec$ |
| $\phi$ | dimensionless fluid velocity normal to surface, $-u_y/\sqrt{a\nu}$ |
| $\Omega_f$ | particle angular velocity, $1/sec$ |

subscripts

| | |
|---|---|
| $b$ | fluid bulk (freestream) |
| $ext$ | external force field |
| $g$ | gas; gravity |
| $p$ | particle |
| $s$ | surface |
| $T$ | total |
| $w$ | wall (surface) |
| 0 | stagnation point |
| $\infty$ | fluid freestream |
| $\nabla T = 0$ | absence of thermophoresis |

superscripts

| | |
|---|---|
| ′ | derivative with respect to $\eta$ |

ACKNOWLEDGEMENTS

The author gratefully acknowledges previous collaboration and subsequent correspondence with Heinz J. Fissan, University of Duisburg. Much of the material presented in this paper was compiled during the author's tenure at the University of Duisburg as a fellow of the Deutscher Akademischer Austauschdienst (DAAD).

Jay Turner is currently a Monsanto Fellow in the Chemical Engineering Department at Washington University.

REFERENCES

1. D.W. Cooper, Aerosol Sci. Technol., 5, 287 (1986).
2. B.Y.H. Liu, B. Fardi and K.H. Ahn, in "1987 Proceedings – Institute of Environmental Sciences", 461 (1987).
3. R. Welker, in "1988 Proceedings – Institute of Environmental Sciences", 449 (1988).
4. M.H. Peters, D.W. Cooper and R.J. Miller, J. Aerosol Sci., 20, 123 (1989).
5. H. Schlichting, "Boundary-Layer Theory", McGraw-Hill, NY, 1979.
6. S.V. Patankar, "Numerical Heat Transfer and Fluid Flow", McGraw-Hill, NY, 1980.
7. K.L. Mittal, editor, "Particles on Surfaces 1: Detection, Adhesion and Removal", Plenum Press, NY, 1988.
8. K.L. Mittal, editor, "Particles on Surfaces 2: Detection, Adhesion and Removal", Plenum Press, NY, 1989.
9. H.J. Fissan and J.R. Turner, in "1988 Proceedings – Institute of Environmental Sciences", 400 (1988).
10. B.Y.H. Liu and K.H. Ahn, Aerosol Sci. Technol., 6, 215 (1987).
11. E.M. Sparrow and G.T. Geiger, J. Heat Trans., 107, 321 (1985).
12. S.K. Friedlander, "Smoke, Dust and Haze", McGraw-Hill, NY, 1977.
13. F. Stratmann and E.R. Whitby, J. Aerosol Sci., 20, 437 (1989).
14. D.W. Cooper, M.H. Peters and R.J. Miller, Aerosol Sci. Technol., 11, 133 (1989).
15. J.R. Turner, D.K. Liguras and H.J. Fissan, J. Aerosol Sci., 20, 403 (1989).
16. J.R. Turner, H.J. Fissan and D.K. Liguras, J. Aerosol Sci., 19, 797 (1988).
17. D. Gupta and M.H. Peters, J. Colloid Interface Sci., 104, 375 (1985).
18. J. Fernandez de la Mora and D.E. Rosner, J. Fluid Mech., 125, 379 (1982).
19. S. Abuzeid and A.A. Busnaina, Modeling of particle transport in clean rooms, presented at the 20th meeting of the Fine Particle Society, Boston, MA, 1989.
20. A.A. Busnaina and S. Abuzeid, in "1989 Proceedings – Institute of Environmental Sciences", 245 (1989).

21. S. Abuzeid, A.A. Busnaina and G. Ahmadi, Eulerian and Lagrangian simulations of particle deposition in a turbulent channel flow, submitted to Aerosol Science and Technology.
22. T.W. Peterson, F. Stratmann and H.J. Fissan, J. Aerosol Sci., 20, 683 (1989).
23. F. Stratmann, H.J. Fissan and T.W. Peterson, J. Environ. Sci., 39 (Nov/Dec 1988).
24. I. Hayakawa, S. Fujii and K.Y. Kim, in "1986 Proceedings – Institute of Environmental Sciences", 483 (1986).
25. T.H. Kuehn, J. Aerosol Sci., 19, 1405 (1988).
26. S. Fujii, K.Y. Kim, G. Xie and I. Hayakawa, in "1988 Proceedings – Institute of Environmental Sciences", 443 (1988).
27. R.P. Donovan, A.C. Clayton and D.S. Ensor, in "1987 Proceedings – Institute of Environmental Sciences", 473 (1987).
28. M. Shimada, K. Okuyama and Y. Kousaka, J. Aerosol Sci., 20, 419 (1989).
29. M. Shimada, K. Okuyama, Y. Kousaka, Y. Okuyama and J.H. Seinfeld, J. Colloid Interface Sci., 128, 157 (1989).
30. J.R. Turner and H.J. Fissan, Chem. Eng. Sci., 44, 1255 (1989).
31. R.P. Donovan, A.C. Clayton and D.S. Ensor, Dependence of particle deposition velocity on silicon wafer surface temperature, presented at the meeting of the Electrochemical Society, Honolulu, HI, 1987.
32. G.K. Batchelor and C. Shen, J. Colloid Interface Sci., 107, 21 (1985).
33. J. Chomiak and A.K. Gupta, J. Aerosol Sci., 20, 1 (1989).
34. F. Stratmann, H.J. Fissan, A. Papperger and S.K. Friedlander, Aerosol Sci. Technol., 9, 115 (1987).
35. S.K. Friedlander, J. Fernandez de la Mora and S. Gokoglu, J. Colloid Interface Sci., 126, 351 (1988).

# SAMPLING OF AIRBORNE PARTICLES FOR CONTAMINATION ASSESSMENT

**Madhav B. Ranade**

Particle Technology, Inc.
Bldg.335, Paint Branch Drive, College Park, MD20742

**Y. C. Chang and J. W. Gentry**

Department of Chemical Engineering
University of Maryland
College Park, MD 20742

Air or gas-borne particles are encountered in numerous industrial operations and their measurement is required in situations including process monitoring, microcontamination control, environmental emissions, and occupational safety monitoring. In microelectronic manufacturing, for example, particles adversely affect the product yield. Ultraclean rooms and environments are often essential and monitoring of small submicrometer particles at very low concentration levels is required to ensure proper operation of such facilities. In other examples, a sample may be collected on a substrate for mass loading measurement or transported for particle size measurement using real-time instruments such as light scattering counters. Preservation of mass and size characteristics in the sample transfer is essential for truly representative sampling.

## IMPORTANT FACTORS IN SAMPLING OPERATION

In any sampling operation three components are usually involved which are aerosol environment, sampling interface and analyzer. As indicated in Figure 1, the aerosol environment may be nearly static or slowly moving as in indoor sampling for occupational health and contamination monitoring, or flowing in open atmosphere, or in process duct and emission stacks. Even in the otherwise static systems considerable convective motion may be present.

The sampling interface transports the particulates to the sensor or analyzer. The accuracy of the aerosol size concentration measurement is determined by the detection efficiency as well as overall efficiency of the sampling interface. The aspiration efficiency into the sampler inlet and the transport efficiency through the interface may both be size dependent and the particle size and concentration as

*Particles in Gases and Liquids 2,* Edited by K.L. Mittal
Plenum Press, New York, 1990

seen by the analyzer will not be representative of the aerosol.

Factors which determine the aspiration and transmission efficiencies are also indicated in Figure 1.

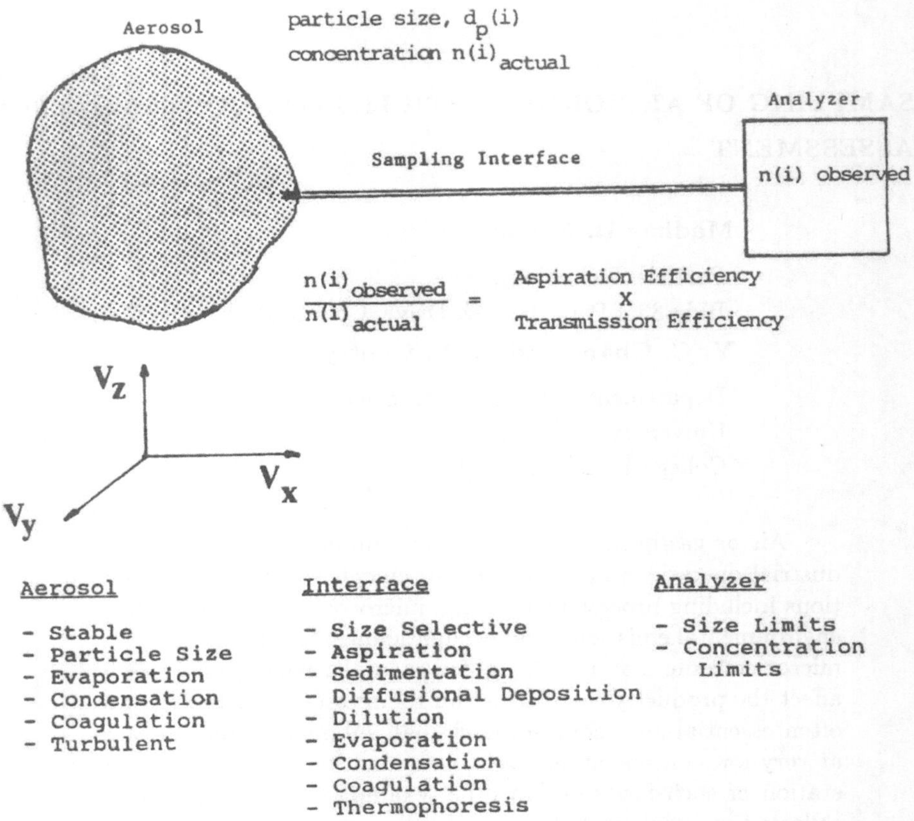


**Figure 1. Aerosol sampling system components and important factors for obtaining representative sample.**

## Aspiration Efficiency

The aspiration efficiency of a sampler inlet is determined by the aerosol particle size, the aerosol flow field, geometry of the sampler nozzle, and the tendency of the particles to bounce off the nozzle walls into or away from the sample flow. Stevens[1] has reviewed the underlying theories of aspiration and experimental data and concluded that the efficiency can be calculated to within 10% accuracy for thin-walled nozzles.

The aspiration efficiency of a sampler inlet is dependent upon whether it is "thin "or "thick "walled. Most laboratory open atmosphere samplers are thick-walled and particle aspiration is controlled by the Stokes number:

$$Stk = \frac{\rho_p d_p^2 U}{18\mu L}$$

where $\rho_p$ is the particle density, $d_p$ is the particle diameter, $U$ is the gas velocity, $\mu$ is gas viscosity, and $L$ is the characteristics dimension of the sampler. For thin-walled sampler inlets the inside diameter is the characteristic diameter, while for thick-walled samplers the actual sampler dimension may determine $L$.

Recently several new experimental and theoretical results have been reported on the thin-walled sampler inlets. Okazaki et al.[2] and Okazaki and Willeke[3] presented results which indicate that at Stokes number less than 0.3 no errors are introduced if the sample is withdrawn along the flow direction. For larger Stokes numbers, combined inertial and gravitation losses become significant. Both the intensity and the scale of turbulence may affect the aspiration efficiency (Weiner et al.[4] ).

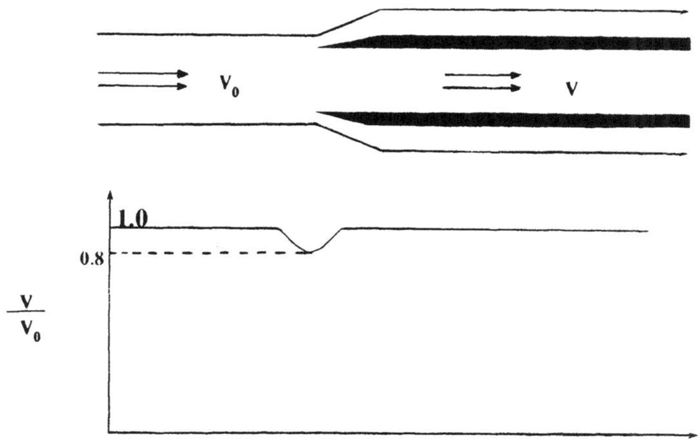


**Figure 2. Effect of nozzle shape on sample velocity**

In ducted flows, isokinetic sampling by thin-walled probes is always preferable. The term isokinetic implies that the airflow direction as well as air speed is preserved at the sample inlet. Sampling by a sharp edge orifice is shown schematically in Figure 2. Even in very sharp orifices a flow discontinuity at the edge results in non-uniform conditions and the ratio between the probe inlet velocity $V$ and the main stream velocity $V_0$ may be of the order of 0.8 at the entry. Leschonski and Rothele[5] show that in practical systems the thickness of the wall can prevent true isokinetic conditions and present curves for achieving isokinetic conditions for different thickness values using differential pressures between the inside and outside walls other than zero (as used in the so-called "null or zero pressure probes ").

Nozzle orientation in the flow or the yaw can also influence the sampling effectiveness since the definition of velocity includes direction as well as speed. An example of the magnitude is shown in Figure 3. For small yaw angle, the observed concentration is given by the equation[6]

$$\frac{C_a}{C_0} = 1 - 4 sin\theta . Stk/\pi$$

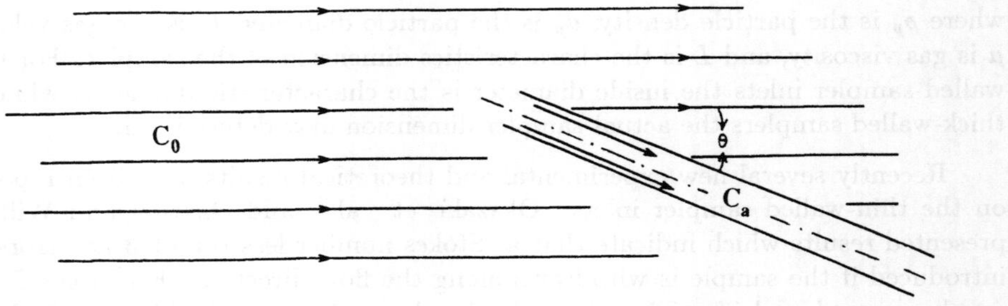


**Figure 3. Effect of nozzle orientation on isokinetic condition**

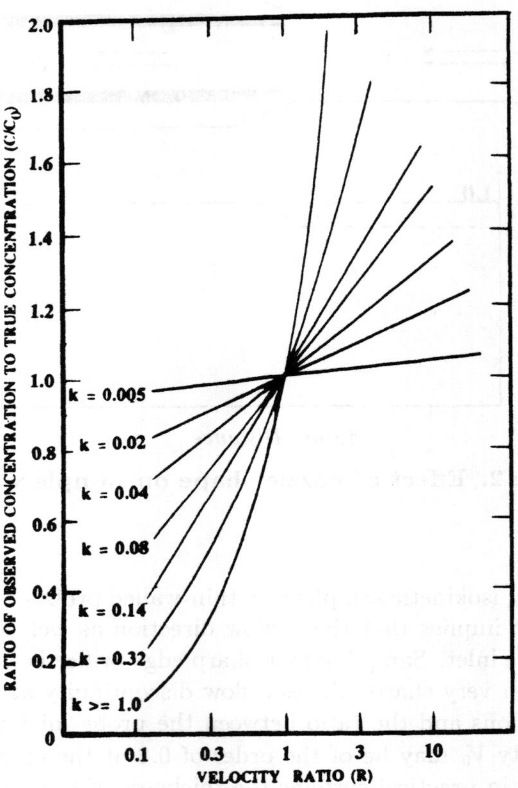


**Figure 4. Errors due to anisokinetic sampling**

in which $C_a$ is aerosol concentration in the sampling tube, $C_0$ is the concentration in the stream and $\theta$ is the yaw angle.

For a yaw of 10 degrees, the ratio of observed to true concentration is seen to be 86 percent for 10 $\mu m$ particle at a wind speed of 100 feet per second.

The error introduced by anisokinetic conditions is shown in Figure 4[7] and is dependent on parameter $k$ which is indicative of the inertia–or particle size and is given by

$$k = \frac{C_p \rho_p V_0 d_p^2}{18 \mu D}$$

where $D$ is probe diameter, $\mu$ is gas viscosity and $C_p$ is Cunningham correction factor. The horizontal axis is the velocity ratio defined by

$$R = \frac{V_0}{V}$$

Let us consider an example, with $V_0$ equal to 100 feet per second and sampling using a 0.5 *inch* diameter probe. The value of $k$ is calculated to be 0.007, 0.175, 0.7 for 1.0, 5.0 and 10.0 $\mu m$ particles respectively. As seen from Figure 4 , a 10 percent decrease in velocity ratio indicates decrease of 1 percent, 6 percent and 12 percent, respectively, for the three particle sizes from the true concentration. Generally speaking particles below 5 $\mu m$ are not subject to significant errors due to anisokinetic conditions. Larger than 5 $\mu m$ particles are subject to errors and the magnitude of errors increases with particle size increase.

Open area samplers used for ambient particulate measurements in support of the new U.S. PM-10 standard (PM-10 particles smaller than 10-$\mu m$) require a size selective cut at the inlet to remove large particles (Kashdan et al.[8], Ranade et al.[9]). This is achieved as a combined effect of aspiration and transport efficiencies in thick-walled inlets. Samplers for respirable particulates for occupational health monitoring also are thick-walled type and either a cyclone or an impactor is used for inertial pre-separation of larger particles.

## Transmission Efficiency

The aerosol sample after entry in the inlet may be subject to several size-dependent losses, including inertial collection in bends, sedimentation, diffusion to the walls, and deposition of electrically charged particles. In sampling hot gasses using cold-wall probes, thermophoretic losses may also be significant. Okazaki et al.[2] and Okazaki and Willeke [3] present results for inertial and gravitational losses. In general, long sample probes can cause significant losses of particles above a few micrometers. If a continuous size analyzer is to be used these losses may eventually block the probes.

Ranade et al.[9] describe a porous-walled probe with an inwardly transpiring sheath airflow as shown in Figure 5. Up to 100% particle loss of 50 $\mu m$ particles was reported in 0.5 inch diameter, 5 feet long probe without transpiration. The losses were eliminated when the transpiration velocity exceeded the particle settling velocity.

Such transpiration systems can be used to dilute the aerosol sample when particle concentration is high. Analyzers based on light scattering are usually limited to low concentration because of potential coincidence errors. Dilution factors of 1 : 1,000 were achieved using transpiring air by Ranade and Lawless [10]. Part of the aerosol sample is filtered and recycled for dilution of the sample. In this arrangement the gaseous composition is preserved and particle size and concentration

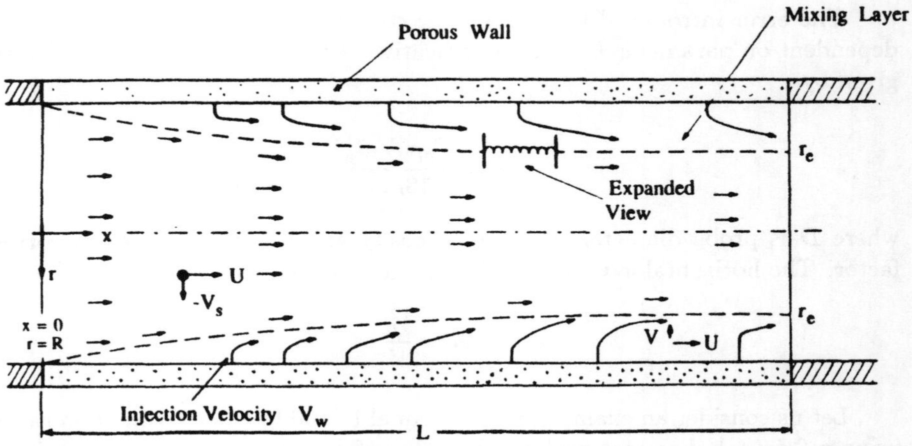


**Figure 5. Transpiration flow scheme**

changes due to condensation and evaporation are minimized.

## COLLECTION ON SURFACES

When the particles have to be deposited on a surface for microscopy, weighing, or chemical analysis several techniques may be useful.

Various types of filter media such as fibrous filters, membrane filters, or sieve-type filters such as the Nuclepore® filter are commercially available and can be used to collect particles quantitatively for subsequent examination and analysis. Characteristics of several filter media are shown in Table I[11].

Inertial impactors are also used to collect particles on surfaces. As seen in Figure 6, particles separate from air stream because of inertia and may deposit on the solid surface. A typical size-dependent collection efficiency is shown in Figure 7[12]. The collection efficiency of 50 percent is called the "cut point "characteristic of an impactor and is related to the Stokes number as shown by several workers. The actual impactor efficiency curve differs from the ideal because of non-uniform velocity profile in the nozzle and other aerodynamic effects.

Particle bounce from surfaces is potentially a serious source of error and must be checked generally for particles larger than 2-3 $\mu m$. Use of greased surfaces can prevent particle bounce; however, they can get overloaded over time depending on aerosol concentration.

Cascading of several impaction stages with progressively decreasing cut diameter is commonly used to obtain particle-size distribution.

Other collection schemes such as the electrostatic precipitator can be used over a wide-size range for depositing samples on microscope slides and electron microscope slides. The TSI electrostatic sampler shown in Figure 8 uses an instantaneous square wave voltage pattern. During this "off "cycle the aerosol fills the chamber and is deposited during the "on "cycle. This operation avoids size segregation of particles in the electric field along the length of the collection chamber because

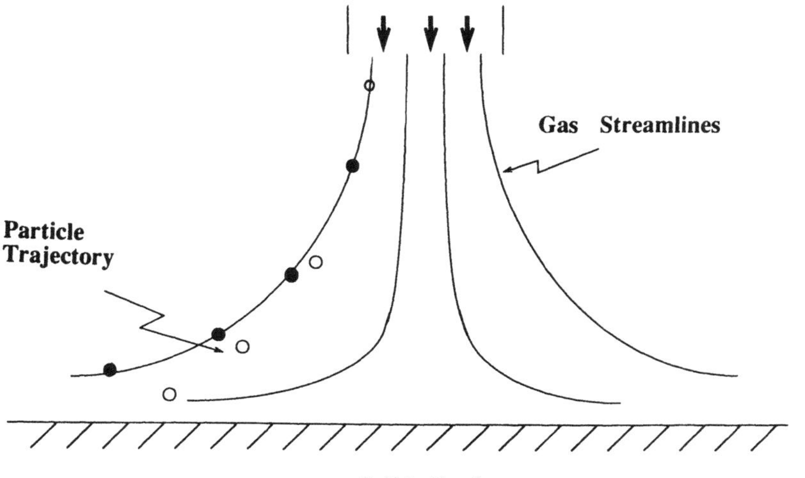


**Figure 6. Schematic of an inertial impactor**

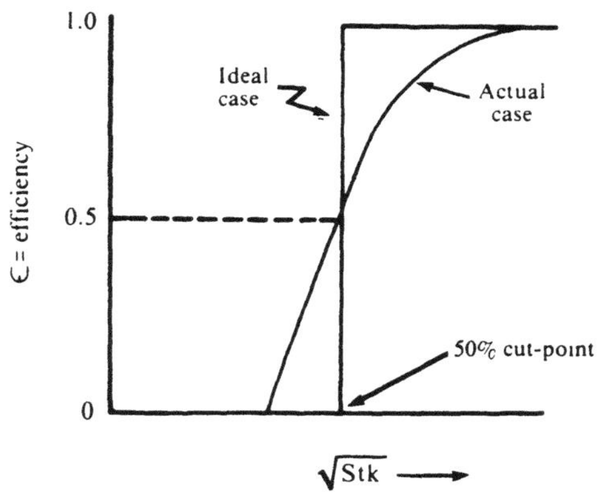


**Figure 7. Typical impactor stage efficiency curve**

almost entire volume in the collection chamber is precipitated out instantaneously during the "on "cycle.

## SUMMARY

Representative samples of open and ducted aerosols can be conveyed to an analyzer or a sensor or a collection device such as filters if proper sampling nozzles and sampling conditions are used. Isokinetic conditions are generally necessary for particles larger than a few micrometers. For large particles, deposition in the sampling lines may also be significant. Use of porous-walled transpiration probe

**Table I. Characteristics of some common aerosol sampling filter materials**

| Filter | Type | Material | Thickness (mm) | Fiber Diameter or Pore Diameter (micrometer) | Pressure Drop [a] (cm $H_2O$) | Relative Efficiency [b] |
|---|---|---|---|---|---|---|
| Whatman 41 | Fiber | Cellullose | 0.19 | 3 - 20 | 25 | 72 |
| Nuclepore | CPM [c] | Polycarbonate | 0.01 | 0.8 [d] | 60 | 90 |
| Microsorban | Fiber | Polystyrene | 1.5 | 0.7 | 30 | 99.5 |
| MSA 1106B | Fiber | Glass | 0.23 | 0.1 - 4 | 20 | 99.93 |
| Millipore AA | Membrane | Cellulose ester | 0.15 | 0.8 | 97 | 99.98 |

[a] At $U_0 = 27$ cm/s.
[b] For $d_p = 0.3$ micrometer.
[c] CPM = capillary pore membrane.
[d] 0.4 micron is recommended.

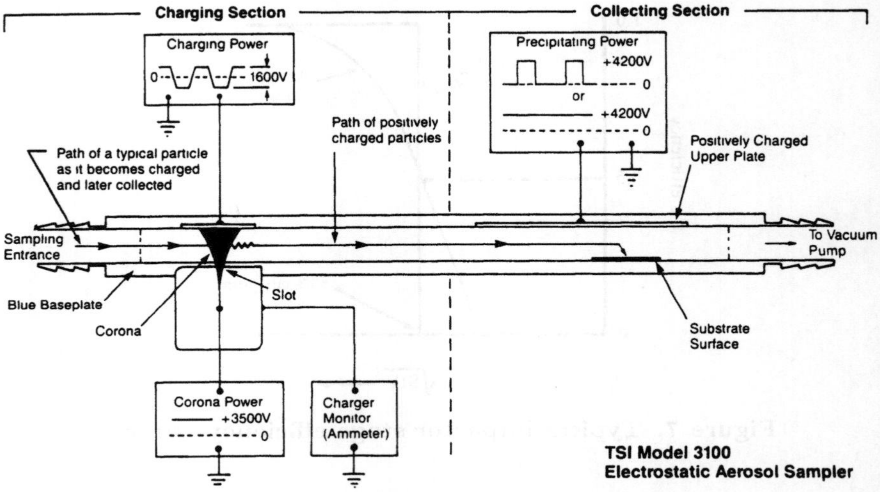


Figure 8. TSI electrostatic sampler

can minimize particle losses and preserve the size distribution of the aspired sample. On-line dilution is also achieved allowing the use of common particle size analyzers such as the optical particle counters for real-time size analysis of high concentration aerosols.

Particle collection devices are used to deposit samples on suitable substrates for microscopy or for other analytical techniques and diagnostic information about their source can be obtained.

## REFERENCES


1. D.C. Stevens 'Review of aspiration coefficients of thin-walled sampling nozzles, ' J. Aerosol Sci., 17(4), 729-743 (1986).
2. K. Okazaki, R.W. Wiener and K. Willeke, 'Isoaxial aerosol sampling: non-dimensional representation of overall sampling efficiency, ' Env. Sci. Technol., 21(2), 178-182 (1987).
3. K. Okazaki and K. Willeke, 'Transmission and deposition behavior of aerosols in sampling inlets, ' Aerosol Sci. Technol., 7, 275-283 (1987).
4. R.W. Weiner, K. Okazaki and K. Willeke, 'Influence of turbulence on aerosol sampling efficiency, ' Atmospheric Environment, 21(5), 1181-1185 (1987).
5. K. Leschonski and S. Rothele, 'A new probe for highly-concentrated gas-solid-flow systems, ' unpublished results (1988).
6. N.A. Fuchs, 'Review papers – Sampling of aerosols, ' Atmospheric Environment, 9, 697-707 (1975).
7. C. Orr, Jr. and E.Y.H. Keng, 'Sampling and particle-size measurement, ' in "Handbook on Aerosols, " R. Dennis editor, Technical Information Center Office of Public Affairs, U.S. Energy Research and Development Administration (1976).
8. E.R. Kashdan, M. B. Ranade, L. J. Purdue, and K. H. Rehme, 'Interlaboratory evaluation of two inlets for sampling particles less than 10 $\mu m$, ' Env. Sci. Technol., 20(9) , 911-916 (1986).
9. M.B. Ranade, D.K. Werle and D.T. Wasan, 'Aerosol transport through a porous sampling probe with transpiration air flow, ' J. Colloid Interface Sci., 56(1), 42-52 (1976).
10. M.B. Ranade and P.A. Lawless, 'A porous-tube dilution interface for continuous monitoring of particulates, ' Presented at the 74th AIChE Meeting, New Orleans, LA (1981).
11. W.C. Hinds, "Aerosol Technology, " John Wiley & Sons (1982).
12. P.C. Reist, "Introduction to Aerosol Science, " Macmillan Publishing Co., New York (1984).

# THE ROLE OF BACTERIAL BIOFILMS IN CONTAMINATION OF PROCESS FLUIDS BY BIOLOGICAL PARTICULATES

Marc W. Mittelman and David C. White

Institute for Applied Microbiology
University of Tennessee
10515 Research Dr. Ste. 300
Knoxville, TN 37932-2756

Bacterial contamination of various industrial process fluids creates a significant threat to a wide spectrum of critical products. Semiconductor devices, parenteral fluids, medical devices, and potable waters are subject to bioparticulate contamination. Bacteria utilize a number of specialized adaptive mechanisms to grow and replicate in purified water and other process fluids. Adhesion to surfaces is, for critical manufacturing operations, the most significant survival mechanism associated with bacterial growth and replication. In many industrial systems, the majority of bacteria are associated with surfaces. The bacterial biofilm is the major source of bulk-phase bioparticulates. The organic (e.g., endotoxins) and ionic (e.g., chloride) constituents of bacterial cells and their associated biomass are yet another source of contamination. The tendency of bulk-phase bacteria to colonize surfaces, form adhesive biofilms, and then re-colonize the bulk phase, presents a number of challenges to bioparticulate detection and treatment. This review presents some of the prevailing theories for bacterial adhesion to surfaces. In addition, some novel strategies for bacterial detection and treatment are considered.

## INTRODUCTION

Bacteria possess a number of adaptive mechanisms for responding to those physicochemical factors which define their environment. These factors include nutrient availability, pH, Eh, temperature, organic and ionic content, and the presence of antagonistic agents. Depending upon the types and numbers present, bacteria alter their physiology or physical state in response to the environment. Organic and inorganic acid production, heavy metal binding, transformation of xenobiotics, and extracellular polysaccharide production are important adaptive tools for bacteria in this regard. In most ecosystems, these activities are dependent upon the ability of bacteria to attach to surfaces.

*Particles in Gases and Liquids 2,* Edited by K.L. Mittal
Plenum Press, New York, 1990

That bacteria are a form of self-replicating organic and ionic particulate material distinguishes them from other, abiological, contaminants. In natural aquatic systems, the majority of bacteria are attached to surfaces. Indeed, surface area is a major limiting factor for microbial growth in nearly every freshwater and marine environment. The ratio of planktonic (free-floating) to sessile (attached) bacteria is a function of several interrelated factors. These include surface energetics, materials of construction, topology, hydraulic factors, and biofilm chemistry. It is the biofilm which gives rise to biological particulates and by-products of metabolism which are responsible for biological fouling activities.

The impact of these microenvironmental alterations on various surfaces and fluid handling systems of industrial importance can be significant. Biological fouling can be defined in terms of its effects on various products and processes. Mechanical blockage of flowing systems, corrosion activities, product contamination, and impedance of heat transfer processes result from bacterial adhesion processes. The economic effects of these activities can be staggering: it is estimated, for example, that biological fouling of condenser tubes in power generating operations results in annual losses exceeding 1 billion dollars in North America alone.

Despite their omnipresence in most fluid process systems, relatively little is known about those factors which contribute to bacterial growth and replication in these dynamic environments. This review addresses the role that attachment processes play in the survival, growth, and replication of bacteria in environmental and industrial systems.

## BACTERIAL ATTACHMENT TO SURFACES

### Adaptive Advantages

There are several adaptive advantages which have been ascribed to a sessile existence. Zobell[1] proposed that solid surfaces not only act to concentrate nutrients by adsorption, but also retard the diffusion of exoenzymes away from the cell--thus promoting the uptake of substrates which must be hydrolyzed extracellularly. Several workers have demonstrated that attachment processes are, in part, a response to nutrient availability. Decreasing bulk phase carbon-source concentrations in an aqueous system promote the attachment of marine[2,3] and freshwater bacteria[4]. Geesey et al.[5] showed that the predominant bacterial population in pristine mountain streams was associated with surfaces. Bacteria in industrial purified water systems, which have many similarities to natural oligotrophic ecosystems, also show a preference for surfaces[6]. Oligotrophic environments have been defined by some workers as containing <1 mg $l^{-1}$ total organic carbon. However, each bacterium and each consortium possess a unique range of requirements for nutrients. Concentrations falling outside this range will result in growth inhibition. Attachment appears to be an important adaptive mechanism in what Morita[3] has called the "starvation-survival" of bacteria in oligotrophic environments. The ability of sessile bacteria to elaborate a polyanionic, extracellular polymeric matrix may be an important factor in the concentration of trace nutrients from the bulk phase environment.

The ability of bacteria to form consortia in aqueous and terrestrial environments is essential for the survival of many species.

Microorganisms isolated from natural and industrial environments rarely exist as pure cultures. Rather, they live in diverse societies composed of individuals with various physiological and morphological characteristics. Animate and inanimate surfaces provide a matrix for the formation of these societies[7]. Many sulfate-reducing, methanogenic, methyltrophic, and sulfur-oxidizing bacteria are only found on surfaces in close association with other microorganisms of diverse physiological activities which provide the narrow range of nutrients these microbes can use. For example, the sulfate-reducing bacterium (SRB), Desulfobacter spp., although one of the most commonly detected SRB, is difficult to recover as a monoculture. It is, however, easily grown in a consortium[8]. Substrate transfer is facilitated by the close association of organisms within these diverse populations[9]. The role of consortial members in the formation and activity of biofilms is an important piece of information missing from adhesion studies. For example, classification of microorganisms involved at various stages of surface succession would be of value. Recent studies from our laboratory indicate that a differential sensitivity to oxidizing biocides exists within a consortium of biofilm bacteria on mild steel[10].

Sessile bacteria are afforded a measure of protection from antagonistic agents present in bulk phase aqueous environments. Protection from lytic bacteria such as Bdellovibrio sp.[11], the toxic effects of heavy metals[12], and bactericidal agents[13,14] are important advantages afforded to sessile organisms. This protective feature is a significant factor in many disease processes and biological fouling activities. Bacteria associated with biofilms are more resistant to antibiotic treatments which would otherwise prove effective against free-living populations[7]. Apparent biocide resistance to potable water treatments is a serious problem throughout the world. LeChevallier, et al.[15] showed that attachment of Klebsiella pneumoniae to glass slides increased resistance to free chlorine disinfection by a factor of 150. While the high concentrations and multiple sites of action associated with many biocides may, for example, overwhelm plasmid-mediated resistance, some biofilm organisms may be resistant by virtue of their ability to form polyanionic extracellular polymeric substances (EPS). Unfortunately, most biocide efficacy studies employ planktonic bacterial populations as challenge organisms. Thus, treatments for biofouling problems very often are ineffective and, in some instances, can exacerbate existing fouling activities[16].

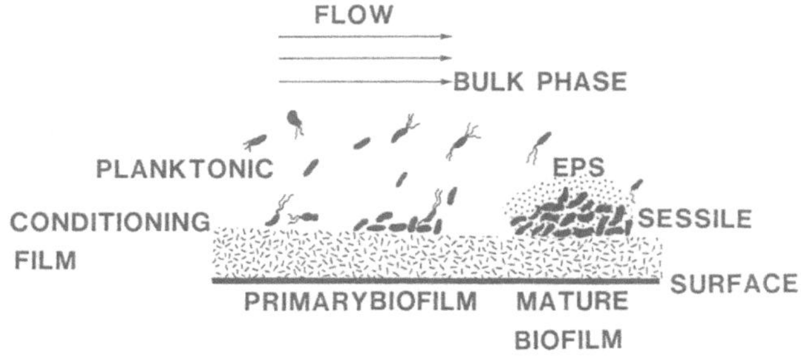


Figure 1.Fouling biofilm development in a flowing system. Development of a fouling biofilm is preceded by a conditioning film which is then colonized by an initial fouling community.

## Biofilm Development

Bacterial attachment and the subsequent formation of a biofilm appears to take place in a three-stage process (figure 1). In the first stage, surfaces are rapidly coated with an organic "conditioning" film. In the blood, this film might consist of proteinaceous compounds such as albumin[17,18]; on teeth, mucopolysaccharides[19]; and in freshwater environments, humic substances[20]. In studies of blood interactions with foreign substances, it was found that within 5 s most surfaces become uniformly coated with strongly adherent proteinaceous films[21]. Rapid modification of clean surfaces in marine waters by glycoprotein and polysaccharide compounds has also been reported[22,23].

In the second stage of adhesion, single bacterial cells are transported to surfaces (see below) and reversible bonds are formed between the cell wall and surface. Bacterial EPS appear to mediate the attachment of primary colonizers to organic conditioning films associated with animate and inanimate surfaces (figure 2). Corpe[23] described an "acid polysaccharide" which acts as a polymer bridge between bacterial cell walls and surfaces. It is interesting to note, however, that Allison and Sutherland[24] have presented evidence that extracellular polysaccharide, while important in the development of surfaces films, is not directly involved in the initial adhesion process of freshwater bacteria. The high uronic acid content of many types of EPS probably plays an important role in concentrating essential substrates and preventing antagonistic agents from reaching labile cellular processes.

The nature or third-stage biofilm consists of the organic conditioning film, a succession of colonizing bacterial consortia with their associated EPS, and various biofilm-associated detrital particles and ionic species. It is this structure which gives rise to the planktonic bacteria and their by-products (e.g., endotoxins) which act as self-replicating, particulate contaminants.

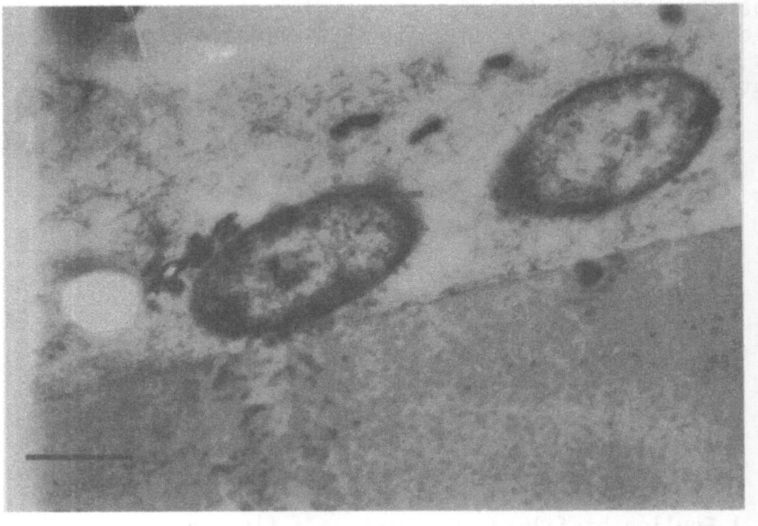

Figure 2.Transmission electron micrograph of bacteria and surrounding extracellular polymeric material associated with a polymeric substratum. Bar=0.5 μm (Courtesy of Dr. Gill Geesey).

The question of whether differing substratum surface properties are communicated to the initial or succeeding organisms through the conditioning film is of great interest. This raises the intriguing possibility that bacteria recognize--and are attracted to--moieties within the conditioning film rather than structures on the underlying surface. Despite recognition of the importance of conditioning films as precursors to biological fouling activities, treatments have not been developed for their control or modification. If, for example, the constituents of a conditioning film elicit a chemotactic response, the chemistry and topology of a given surface might exert a relatively minor influence on biofilm development. The type of conditioning film present may be the major factor involved with the penetration of pathogenic organisms into tissues, tooth surfaces, and implantable medical devices. It may also affect the rate and extent of such surface perturbations as pitting corrosion activities.

## Adhesion Mechanisms

Mechanisms for bacterial surface recognition and initial colonization have yet to be fully elucidated. Some workers have described a chemotactic response by bacteria to surface-associated compounds. Chemotaxis is the movement of an organism in the direction of a chemical gradient. Young and Mitchell[25] have implicated positive chemotaxis in the initial colonization of bacteria to marine surfaces such as ship hulls. The phenomenon of "negative chemotaxis" in response to surface-associated toxic substances has also been described[26,27]. Models of bacterial chemotaxis to surfaces in flowing aqueous systems have been developed[28,29]. Implicit in the chemotaxis mechanistic theory, however, is the ability of bacteria to direct their movement towards the surface (figure 3). Since non-motile bacteria are also associated with fouling biofilms, chemotaxis can only be used to describe one factor involved in the initial attachment mechanism.

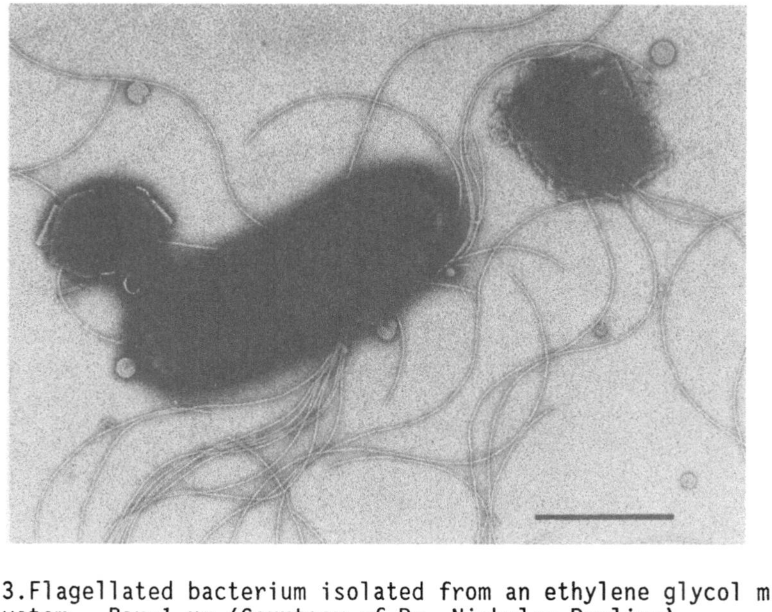

Figure 3.Flagellated bacterium isolated from an ethylene glycol mixed waste system. Bar=1 μm (Courtesy of Dr. Nicholas Dowling).

Impedance of flagellar activity by mechanical contact with surfaces appeared to promote attachment and lateral flagella formation in a marine vibrio[30]. Upon contact with a surface, the polar flagella of Vibrio parahemolyticus ceased to function. Shortly thereafter, lateral flagella formed around the cells, apparently mediating the "irreversible" attachment process. Pilus mediated adhesion to gastrointestinal surfaces is recognized as a major factor in the toxicity of enteropathogenic Escherichia coli in humans and other animals[31]. Pilus negative E. coli are incapable of attaching to these surfaces and producing pathogenic effects. Ridgway and Olson[32] presented scanning electron microscopic evidence which indicated that attachment of bacteria to potable water pipelines was mediated by extracellular fibrillar appendages.

Finally, and perhaps most significantly, physical characteristics of bacteria and substrata such as surface energy[33,34,35,36], hydrophobic effects[37,38], and surface topology[39,40,41] may influence the initial attachment process. The DLVO theory of colloid stabilization, as applied by Marshall et al.[42] to bacterial colloids, holds that the separation between bacteria and adsorbents in an electrolyte solution is dependent upon a balance between attractive and repulsive forces. This relationship can be described by the following equation:

$$V_t=V_r+V_a$$

where, $V_t$=total energy of interaction

$V_r$=energy of repulsion

$V_a$=energy of attraction

The electrical double-layer, which is associated with surfaces in electrolyte or dilute organic solutions, is thought to present a repulsive barrier to bacterial-adsorbent interactions. This repulsion, however, is balanced by the short-range van der Waals forces which attract colloids to surfaces in such environments (figure 4). The

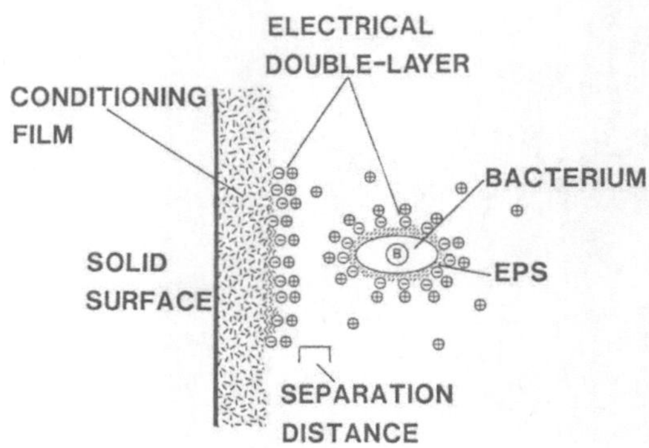


Figure 4. Electrostatic interactions between bacterial cells and surfaces in aqueous environments as inferred from the DLVO Theory.

separation distance is minimized under conditions of low surface potential and bacterial affinity for surfaces is maximized at increased salt concentrations. However, above a salt concentration of 0.1 M, Fletcher[35] found that bacterial attachment began to decline exponentially with increasing cation concentration. Firm ("irreversible") attachment requires some mechanism whereby the finite separation distance imposed by the electrical double-layer is bridged. Corpe[41] and others believe that EPS are involved in this bridging process.

There is also evidence that hydrophobic interactions play a role in controlling adhesion processes. As bacteria approach a surface, this theory holds that water molecules are displaced, creating a lower free energy of interaction. When two surfaces covered by ordered layers of water molecules approach one another, layers of water are released into the bulk phase from between the two surfaces. This results in a decrease in free energy and an increase in entropy. Depending upon the degree of interaction between the cell and a surface, this process can favor adhesion[38]. Surface energetics tends to favor colonization of bacteria to low energy (hydrophobic) surfaces such as Teflon. In general, colonization, but not necessarily irreversible adhesion, is greatest for bacteria possessing hydrophobic cell surfaces interacting with hydrophobic substrata.

The data are often confusing, however, with respect to the mechanisms of bacterial adhesion processes. While surface energy considerations would appear to favor adhesion of bacteria to low energy (hydrophobic) surfaces such as Teflon[43], Absolom et al.[33] pointed out that preferential adhesion to relatively high energy (hydrophilic) surfaces can occur when the surface energy of the bacteria is greater than that of the suspending medium. Clearly, the surface energies of substrata (and their associated condition films), bacterial cell surfaces, and bulk phases are interdependent variables.

## FACTORS INFLUENCING ADHESION PROCESSES

### Environmental Factors

Studies of environmental and substrata variables which influence the initial attachment of bacteria and the subsequent formation of mature biofilms have generally lacked relevance to the actual *in situ* conditions. Previous studies by Fletcher et al.[44,43,35], Dexter et al.[45], Marshall et al.[41], Rosenberg et al.[46,47], and others have focused on physicochemical interactions between bacterial monocultures and glass or polystyrene surfaces. The effects of varying bulk phase conditions, surface characteristics, *in situ* conditioning films, and other environmental factors on attachment processes have not been fully explored. Preliminary work by Corpe[23], Dexter et al.[45], and Fletcher[44] did indicate, for example, that organic conditioning films have an effect on attachment processes. The nature of the conditioning films and their relative importance for initial attachment were not described, however.

Bacterial attachment typically occurs under the influence of some type of hydraulic influence. Characklis[39,48] and McCoy[49] have demonstrated that such fluid hydraulic parameters as turbulence and fluid velocity have an effect on the nature of biofilms. Below a critical shear force, however, the amount of biomass per unit area does not appear to be significantly reduced at increased linear fluid velocities. Indeed, recent studies with 316 stainless steel surfaces in our laboratory (figure 5) have indicated that bacterial attachment and biomass formation are increased at higher shear values under laminar flow (Re <2,000) conditions[50]. Fluid velocities in the range of 5-10 ft $sec^{-1}$ have been recommended by some designers of purified water distribution systems[51].

## Metabolic Effects

A key to the understanding of adhesion processes is the ability to elicit biofilm function from structure and activity. In studies of biomass formation and mineralization, sessile bacteria have usually demonstrated significantly higher activities than have planktonic populations. For example, a natural population of sessile marine bacteria was found to transport ATP 100 times faster than planktonic organisms on a per cell basis[52]. The nature of extracellular polysaccharides also differs between sessile and Valeur et al.[53] demonstrated that a significan difference existed between the ratio of saturated/unsaturated fatty acids and in PHB content between planktonic and sessile bacteria. A subpopulation of the test organism containing a higher ratio of total C18/C16 fatty acids and lower PHB levels selectively adhered to test surfaces. Nichols et al.[54] demonstrated

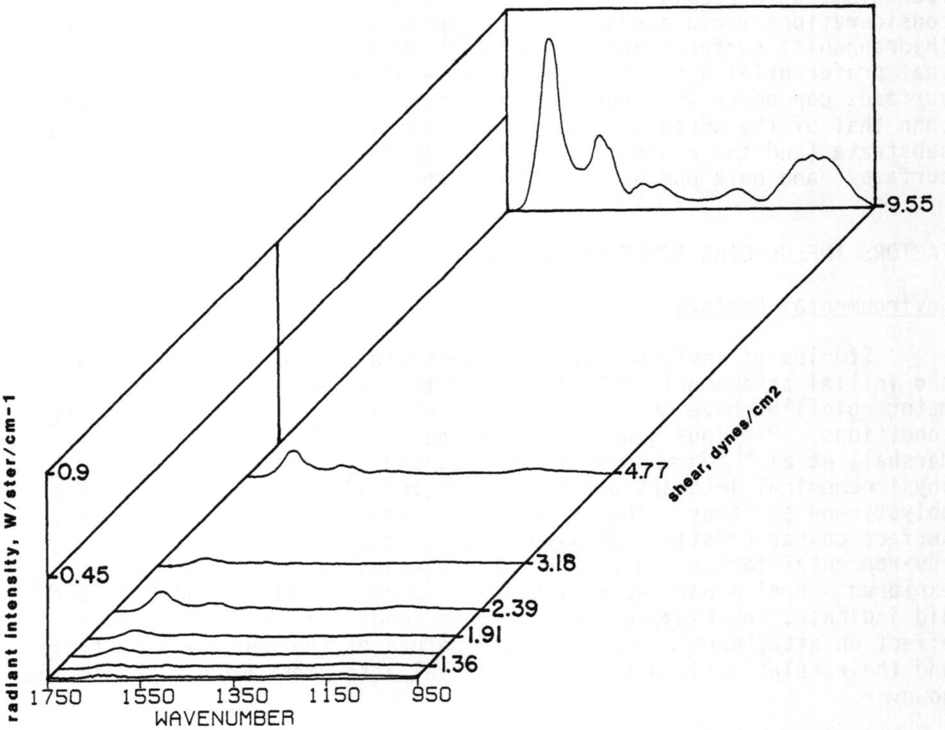


Figure 5.Biomass formation on 316 stainless steel vs. fluid shear as measured by diffuse reflectance FTIR spectroscopy. Amide I (1630 $cm^{-1}$), amide II (1550 $cm^{-1}$) and carbohydrate (1105-1130 $cm^{-1}$) peak absorbances are used as indicators of protein and polysaccharide biomass development.

that significant shifts in the IR spectra of P. atlantica biofilms as measured by FTIR result when the organism is grown under different growth conditions. Environmental factors such as bulk phase C:N ratio and fluid shear also appear to affect the nature and activity of biofilms on surfaces such as stainless steel[50,55].

Recent work by Bartlett et al.[56] has shown that extracellular polysaccharide production by P. atlantica is a variable trait, determined by expression of a now defined "eps" locus. Mucoid colonies ($eps^+$) produce extracellular polysaccharide while the spontaneously reverted crenated colonies ($eps^-$) apparently lack the ability to synthesize large amounts of extracellular polysaccharide. When recombinant plasmids with a locus for extracellular polysaccharide production are transferred to crenated variants, expression of the mucoid ($eps^+$) form results. What effect expression of this eps gene has on attachment characteristics is as yet undetermined. A clearer picture of extracellular polysaccharide function in the development of primary and mature biofilms could lead to a better understanding of those factors which control the transition of bacteria from a planktonic to a sessile existence. An alteration of extracellular polysaccharide composition or quantity in response to some environmental perturbation could be an important factor in this transition process.

## BIOFILM-MEDIATED CONTAMINATION

### Disease Processes

Bacterial biofilms play an important role in disease processes, industrial contamination, and environmental pollution. Antibiotic therapies have, for most part, failed to address the mechanisms associated with initial bacterial colonization and biofilm formation[57]. This gap in the approaches to treating gastrointestinal (e.g., enteropathogenic E. coli infections) and pulmonary diseases (e.g., secondary Pseudomonas infections of cystic fibrosis patients) could be narrowed with further research into these mechanisms. Problems related to bacterial colonization of such invasive medical devices as heart valves, catheters, penile prostheses, and intrauterine devices may eventually be resolved by a mechanistic approach to biofilm control[7]. This would involve studying mechanisms of surface recognition and initial colonization for various materials of construction. Determinations of attachment efficiency under varying conditions of pH, Eh, organic (protein) concentration, etc. would provide important information for medical device designers as well as physicians. For example, knowledge of the effects of these and other variables on attachment efficacy could be used to predict the probability of device colonization in certain patients. This finding would have implications for researchers concerned with microbial adhesion to inanimate and biological matrixes.

### Purified and Potable Waters

High quality purified water is essential to numerous applications in pharmaceutical, dialysis, medical device, microelectronics, aerospace, and chemical processing industries. Bacterial contamination of these systems remains the most significant threat to product and process quality. Attachment of bacteria and subsequent biofilm formation can occur in waters containing levels of ions and organics below the limits of analytical sensitivity. The recognition of surface associated organics by bacteria in extremely oligotrophic environments has not been studied. If chemotaxis does indeed play an important role in initial bacterial colonization, such nutrient-limited environments would be ideal systems for studying recognition and uptake mechanisms. The oligotrophic nature and large surface area:volume ratio associated with purified water systems promotes the attachment of bacteria to surfaces.

Ion exchange resins[58], reverse osmosis and dialysis membranes[59], storage facilities, and distribution systems[60] are subject to biofilm formation. Endotoxin production, product contamination, corrosion activities, and mechanical fouling problems seriously compromise product safety. Novitsky et al.[61] demonstrated that bacterial endotoxins, which are the predominant pyrogenic ("fever-producing") substances in parenteral fluids, preferentially sorb to container surfaces. Removal of these recalcitrant bacterial by-products from surfaces continues to present challenges to pharmaceutical manufacturers. In the microelectronics industry, high bacteria levels in ultrapure deionized water have been associated with yield anomalies in NMOS, CMOS, and bipolar technologies. In addition, biologically mediated failures of back-side metallization adhesion processes have been reported[62]. As particulates, bacteria have also been implicated in "shadowing" of oxide surfaces during photolithographic processes.

The so-called "bacterial regrowth" phenomenon following chlorine treatments of purified and potable water systems is likely a function of biofilm formation rather than resistance in the classical sense of antibiotic resistance[15,59,63,64,65]. A combination of the polyanionic nature of many EPS moieties coupled with the inherent biocide demand associated with extracellular slime matrices acts as an inhibitor of biocide actions. Residual concentrations of oxidizing biocides in estuarine power plant effluents can have serious consequences for marine primary producers. Biocide-mediated oxidation of bromide ions in seawater creates a greater threat to living organisms than do the biocides themselves[66]. The fact that most power plant condenser waters are treated on a continuous or frequent slug dosage basis is an indication of the recalcitrance of fouling biofilms. A treatment strategy which focused on modifications of the primary condition film and/or initial colonizing bacterial population might obviate the need for toxic biocides, thus removing a significant threat to aquatic ecosystems.

## Microbially Influenced Corrosion

The billions of dollars spent in controlling the effects of corrosion and other biologically mediated fouling activities have failed to target the attachment processes which are responsible for the observed effects. Microbially influenced corrosion activities (MIC) are preceded by the development of a biofilm, which can then mediate the formation of chloride concentration cells, development of differential oxygen cells, depolarization of cathodic hydrogen, production of organic acids, and other dissolution processes[13,14,67]. Using electrochemical techniques such as AC impedance spectroscopy, Dowling and White[68] described the relationship between biofilm development on metal surfaces and corrosion activity. Corrosion rates of mild steels were shown to be significantly accelerated in the presence of fouling microorganisms. It has been estimated that leaks in the Alaskan oil pipelines related to MIC activities result in losses of one million dollars per day in oil[69].

Monitoring

The development of real-time, on-line biofilm monitoring devices will lead to a better understanding of those factors which affect the primary stages of bacterial attachment. Ebersole and Ward[70] have described an oscillating piezoelectric quartz crystal microbalance (QCM) for measuring particle adhesion on surfaces. This device, currently under development in our laboratory, may provide a means for detecting initial colonization by bacteria. Perturbations in signal frequency and amplitude measured by the QCM may prove useful for quantitative and qualitative analysis of primary fouling populations. Electrochemical measures of surface activities provide a means for studying interactions between various surfaces and biofilms[68]. Development of real-time image analysis methods for direct observation of surfaces exposed to flowing liquids has been described by Caldwell and Germila[71] and Escher and Characklis[72]. Attenuated total reflectance FTIR spectroscopy has been employed in studies of MIC of copper surfaces in flowing systems. Geesey et al.[73] was able to measure the erosion of molecular layers of copper on a real-time basis. Bacterial exopolymers were implicated in the dissolution process.

A variety of surface scanning analytical techniques have recently been applied to biofilm studies. Chen et al.[74] employed scanning Auger microprobe analysis to study biofouling of seawater condenser tubes at a power plant in Taiwan. They found that O, Na, and S were enriched in areas of stainless steel that were influenced by MIC activities. Stainless steel (#304 plate) surfaces were found to have significantly greater corrosion than the titanium. Ridgway and Olsen[32] utilized X-ray energy microanalysis in studies of pipe surfaces colonized by bacteria. Their work revealed the presence of Si, P, S, Ca, Fe, Zn, Mg, Al, K, and Mn associated with the biofilm.

White[75] has reviewed the use of various signature biomarker analyses for in situ biofilm analysis. Community structure, specific activity, and biomass measures are important indicators of fouling potential. Utilization of fluorescent and $^{32}P$-labeled RNA/DNA probes may provide a means for quantitating members of biofilm consortia. The utility of DNA probes in studies of soil consortia, for example, has been described by Ogram and Sayler[76]. Combined with signature biomarker studies, FTIR analysis, and activity measurements, DNA probe technologies could provide information concerning which groups of biofilm bacteria are active under varying environmental conditions. It is possible that these probes can be used in conjunction with biocide efficacy assays to determine the in situ biocide resistance/sensitivity of specific consortial members. The continued development of ultramicro probes for pH, $O_2$ (redox), and trace ion detection will provide means for monitoring bacterial activity at the surface/biofilm interface. Changes in biofilm community structure, biofilm stability, and biofilm constituents could be rapidly detected with these microprobe tools.

Treatment Considerations

The vast majority of advances in industrial biological contamination control have focused on the treatment of planktonic bacterial populations. More efficient microporous filtration systems[77], ultraviolet sterilization devices[78], and novel ion exchange processes[58], for example, have all addressed the removal of planktonic microorganisms while virtually ignoring the role of biofilms in their generation. The effective treatment of sessile bacteria and their associated biomass usually requires the application of oxidizing biocides and/or mechanical treatments such as circulating sponge-rubber balls. McCoy[79] provides an overview of biofouling treatment processes in industrial water systems.

A patented process developed by J. W. Costerton[80] utilizes a circulating glycol solution to effect ice nucleation within the biofilm. Under controlled thawing conditions, the frozen cells and biomass are stripped from the surface and trapped via side-stream filtration devices. The applicability of this process to a broad industrial market is, however, still to be validated. Future treatment strategies will undoubtedly focus on the prevention of primary bacterial colonization. This will entail alterations in the surface chemistry of both the bulk fluid and the substratum. Although great progress has been made in developing advanced fluid handling materials of construction, mankind still has not developed a surface to which bacteria will not adhere. There exists a real need for process and product compatible biofilm treatments.

## SUMMARY

1. Bacteria are exposed to a number of stressful nutrient and environmental factors in natural and industrial fluid handling systems. These include low nutrient availability, non-optimal pH and Eh, the presence of antagonistic agents, and hydraulic shear.

2. Bacteria have developed a number of adaptive advantages for responding to these stress factors. Attachment to various surfaces appears to be an important mechanism for bacterial survival in fluid systems. The attachment process is characterized by a three step process: formation of a primary conditioning film on the substratum, initial bacterial colonization, and the development of a mature biofilm composed of cellular biomass and entrained particulate matter.

3. Mechanisms for bacterial colonization and adhesion include chemotaxis, hydrophobic interactions, and electrostatic interactions such as short-range van der Waals forces. Surface energetics and bacterial cell surface constituents are important factors in determining the rate and stability of the adhesion process.

4. As the major reservoir for bioparticulates, biofilms play a major role in various disease processes, product contamination problems, and degradation of fluid processing systems. Bioparticulates in semiconductor-grade purified waters have been associated with yield anomalies in NMOS, CMOS and bipolar technologies. Conductance phenomena and shadowing of oxide surfaces during photolithography have also been reported. Microbially mediated corrosion, particulate, and mechanical fouling problems result in economic losses exceed 1 billion dollars per year in the U.S. alone.

5. Emerging monitoring strategies for bacterial biofilms and bioparticulates include an oscillating piezoelectric quartz crystal microbalance, attenuated total reflectance FTIR spectroscopy, scanning Auger microprobe analysis, and signature biomarker analyses using biochemical and genetic markers.

6. Treatments for biological fouling problems must consider the influence of biofilms on bacterial survival. Apparent biocide resistance in purified and other process waters is often the result of a

failure to adequately treat and remove adherent bacteria and biomass from various surfaces. Emerging treatment strategies are addressing structural and molecular characteristics of adhering organisms. Coupled with effective preventative maintenance strategies, process and product compatible biocides will control the influence of bioparticulates and their biofilm reservoirs.

## ACKNOWLEDGMENTS

Part of the work described in this paper was supported by the following grants: U.S. Office of Naval Research, no. N00014-87-K00012; U.S. Office of Naval Research, no. N00014-89-J-3095; and Gas Research Institute, no. 5086-260-1303.

## REFERENCES

1. C.E. Zobell, The effect of solid surfaces on bacterial activity, J. Bacteriol. 46,39-56 (1943).

2. K.C. Marshall, Adhesion and growth of bacteria at surfaces in oligotrophic environments, Can. J. Microbiol. 34,503-506 (1988).

3. R.Y. Morita, Starvation-survival of heterotrophs in the marine environment, Adv. Microb. Ecol. 6,171-198 (1982).

4. C.M. Brown, D.C. Ellwood, and J.R. Hunter, Growth of bacteria at surfaces: Influence of nutrient limitation, FEMS Microbiol. Lett. 1,163-166 (1977).

5. G.G. Geesey, R. Mutch, and J.W. Costerton, Sessile bacteria: An important component of the microbial population in small mountain streams, Limnol. Oceanogr. 23,1214-1223 (1978).

6. M.W. Mittelman, R. Islander, and R.M. Platt, Biofilm formation in a closed-loop purified water system, Med. Device Diag. Ind. 9(10), 50-55,75 (1987).

7. J.W. Costerton, T.J. Marrie, and K.-J Cheng, Phenomena of bacterial adhesion, in "Bacterial Adhesion: Mechanisms and Physiological Significance," D.C. Savage and M. Fletcher, editors, pp. 3-43, Plenum, New York (1985).

8. N.J.E. Dowling and D.C. White, Phospholipid fatty acid and infrared spectroscopic analysis of a sulfate reducing consortium, FEMS Microbiol. Ecol. 53,325-334 (1988).

9. R.A. Mah, D.M. Ward, L. Baresi, and T.L. Glass, Biogenesis of methane, Ann. Rev. Microbiol. 31,309-341 (1977).

10. M.J. Franklin, D.E. Nivens, M.W. Mittelman, A.A. Vass, R.F. Jack, N.J.E. Dowling, R.P. Mackowski, S.L. Duncan, D.B. Ringelberg, and D.C. White, An analogue MIC test system with specific bacterial consortia, to test effectiveness of material selection and countermeasures, Proc. Ann. Meeting of the National Association of Corrosion Engineers, paper no. 518, New Orleans, April, 1989.

11. A.D. Venosa, Lysis of *S. natans* swarm cells by *Bdellovibrio bacteriovorus*, Appl. Microbiol. 29,702-705 (1975).

12. M.W. Mittelman and G.G. Geesey, Copper-binding characteristics of exopolymers from a freshwater-sediment bacterium, Appl. Environ. Microbiol. 49,846-851 (1985).

13. J.W. Costerton and E.S. Lashen, Influence of biofilm on efficacy of biocides on corrosion-causing bacteria, Materials Performance 23(2),13-17 (1984).

14. J.W. Costerton, R.T. Irvin, and K.-J. Chen, The bacterial glyocalyx in nature and disease, Annu. Rev. Microbiol. 35,299-324 (1981).

15. M.W. LeChevallier, C.D. Cawthon, and R.G. Lee, Factors promoting survival of bacteria in chlorinated water supplies, Appl. Environ. Microbiol. 54,649-654 (1988).

16. I. Ruseska, J. Robbins, J.W. Costerton, and E.S. Lashen, Biocide testing against corrosion-causing oil-field bacteria helps control plugging. Oil and Gas J. pp 18-25 (March, 1982).

17. D.R. Absolom, W. Zingg, and A.W. Neumann, Protein adsorption to polymer particles: Role of surface properties, J. Biomed. Mat. Res. 21,161-171 (1987).

18. G. Reid, L.A. Hawthorn, R. Mandetori, R.L. Cook, and H.S. Beg, Adhesion of lactobacilli to polymer surfaces in vivo and in vitro, Microbial Ecol. 16,241-251 (1988).

19. G. Embery, T.G. Heaney, and J.B. Stanbury, Studies on the organic polyanionic constituents of human acquired dental pellicle, Arch. Oral Biol. 31,623-625 (1986).

20. K.C. Marshall, Bacterial adhesion in oligotrophic habitats, Microbiol. Sci. 2,321-326 (1986).

21. R.E. Baier, and R.C. Dutton, Initial events in interactions of blood with a foreign surface, J. Biomed. Mater. Res. 3,191-206 (1969).

22. R.E. Baier, Influence of the initial surface condition of materials on bioadhesion, in "Proc. Third Int. Congr. on Marine Corrosion Fouling," R.F. Acker et al. editors, pp. 633-639, Northwestern University Press, Evanston, IL, 1972.

23. W.A. Corpe, An acid polysaccharide produced by a primary film-forming marine bacterium, Dev. Ind. Microbiol. 11,402-412 (1970).

24. D.G. Allison, and I.W. Sutherland, The role of exopolysaccharides in adhesion of freshwater bacteria, J. Gen. Microbiol, 133,1319-1327 (1987).

25. L.Y. Young and R. Mitchell, The role of chemotactic response to primary microbial film formation, in "Proc. Third Int. Congr. Marine Corrosion Fouling," R.F. Acker et al. editors, pp. 615-624, Northwestern University Press, Evanston, IL, 1972.

26. I. Chet, P. Asketh, and R. Mitchell, Repulsion of bacteria from marine surfaces, Appl. Environ. Microbiol. 15,1043-1045 (1970).

27. L.Y. Young and R. Mitchell, Negative chemotaxis of marine bacteria to toxic chemicals, Appl. Microbiol. 25,972-975 (1973).

28. E.F. Keller and J. Segel, Traveling bands of chemotactic bacteria: A theoretical analysis, J. Theor. Biol. 30,235-248 (1971).

29. I.R. Lepidus, Microbial chemotaxis in flowing water in the vicinity of a source of attractant or repellent, J. Theor. Biol. 85,543-547 (1980).

30. M.R. Belas and R.R. Colwell, Adsorption kinetics of laterally and polarly flagellated Vibrio, J. Bacteriol. 151,1568-1580 (1982).

31. S.L. Gorbach, B.H. Kean, S.G. Evans, and D. Bessudo, Travelers' diarrhea and toxigenic Escherichia coli, New England J. Med. 292,933-936 (1975).

32. H.F. Ridgway and B.H. Olson, Scanning electron microscopic evidence for bacterial colonization of a drinking-water distribution system, Appl. Environ. Microbiol. 41,274-287 (1981).

33. D.R. Absolom, F.Y. Lamberti, Z. Policova, W. Zingg, C. van Oss, and A.W. Neumann, Surface thermodynamics of bacterial adhesion, Appl. Environ. Microbiol. 46,90-97 (1983).

34. R.E. Baier, E.G. Shafrin, and W.A. Zisman, Adhesion: Mechanisms that assist or impede it, Science 162,1360-1363 (1968).

35. M. Fletcher, Attachment of Pseudomonas fluorescens to glass and influence of electrolytes on bacterium-substratum separation distance, J. Bacteriol. 170,2027-2030 (1988).

36. L.J. van Dijk, R. Goldsweer, and H.J. Busscher, Interfacial free energy as a driving force for pellicle formation in the oral cavity, Biofouling 1,19-25 (1988).

37. B. Dahlback, M. Hermannsson, S. Kjelleberg, and B. Norkrans, The hydrophobicity of bacteria--An important factor in the initial adhesion at the air-water interface, Arch. Microbiol. 128,267-270 (1981).

38. M. Rosenberg and S. Kjelleberg, Hydrophobic interactions: Role in bacterial adhesion, in "Advances in Microbial Ecology," K.C. Marshall, editor, Vol. 9, pp. 353-393, Plenum, New York, 1986.

39. W.G. Characklis, Attached microbial growths-II. Frictional resistance due to microbial slimes, Water Res. 7,1249-1258 (1973).

40. W.F. McCoy, J.D. Bryers, J. Robbins, and J.W. Costerton, Observations of fouling biofilm formation, Can. J. Microbiol. 27,910-917 (1981).

41. W.A. Corpe, Microbial surface components involved in adsorption of microorganisms onto surfaces, in "Adsorption of Microorganisms to Surfaces," G. Gitton and K. C. Marshall, editors, pp.105-144, John Wiley, New York, 1980.

42. K.C. Marshall, R. Stout, and R. Mitchell, Mechanisms of the initial events in the sorption of marine bacteria to surfaces, J. Gen. Microbiol. 68,337-348 (1971).

43. M. Fletcher and G.I. Loeb, Influence of substratum characteristics on the attachment of a marine pseudomonad to solid surfaces, Appl. Environ. Microbiol. 37,67-72 (1979).

44. M. Fletcher, The effects of proteins on bacterial attachment to polystyrene. J. Gen. Microbiol. 94,400-404 (1976).

45. S.C. Dexter, J.D. Sullivan, J. Williams, and S.W. Watson, Influence of substrate wettability on the attachment of marine bacteria to various surfaces, Appl. Microbiol. 30,298-308 (1975).

46. M. Rosenberg, D. Gutnick, and E. Rosenberg, Adherence of bacteria to hydrocarbons: A simple method for measuring cell surface hydrophobicity, FEMS Microbiol. Lett. 9,29-33 (1980).

47. M. Rosenberg, Bacterial adherence to polystyrene: A replica method of screening for bacterial hydrophobicity, 42,375-377 (1981).

48. W.G. Characklis, Attached microbial growths I, Attachment and growth, Water Res. 7,1113-1127 (1973).

49. W.F. McCoy and J.W. Costerton, Fouling biofilm development in tubular flow systems, Dev. Ind. Microbiol, 23,551-558 (1982).

50. M.W. Mittelman and D.C. White, The effects of shear on the adhesion and activity of Pseudomonas atlantica on stainless steel, Abstracts of the American Society for Microbiology Annual Meeting, May, 1989, New Orleans, LA.

51. G.T. Nickerson, DI water distribution loop design, Ultrapure Water 3(4),52-55 (1986).

52. R.E. Hodson, A.E. Maccubbin, and L.R. Pomeroy, Dissolved adenosine triphosphate utilization by free-living and attached bacterioplankton, Marine Biol. 64,43-51 (1981).

53. A. Valeur, A. Tunlid, and G. Odham, Differences in lipid composition between free-living and initially adhered cells of a Gram-negative bacterium, Arch. Microbiol. 149,521-526 (1988).

54. P.D. Nichols, J.M. Henson, J.B. Guckert, D.E. Nivens, and D.C. White, Fourier transform-infrared spectroscopic methods for microbial ecology, J. Microbiol. Meth. 4,79-94 (1985).

55. H.W. Fowler and A.J. McKay, The measurement of microbial adhesion in "Microbial Adhesion to Surfaces," R.C.W. Berkeley, J.M. Lynch, J. Melling, P.R. Rutter, and B. Vincent, editors, pp. 143-161, Harwood, Chichester, England, 1980.

56. D.H. Bartlett, M.E. Wright, and M. Silverman, Variable expression of extracellular polysaccharide in the marine bacterium Pseudomonas atlantica is controlled by genome arrangement, Proc. Natl. Acad. Sci. USA, 85,3923-3927 (1988).

57. J.W. Costerton and K.-J Cheng, The role of the bacterial cell envelope in antibiotic resistance, J. Antimicrobial Chemother. 1,363-377 (1975).

58. H.-C. Flemming, Microbial growth on ion exchangers, Water Res. 21,745-756 (1987).

59. H.F. Ridgway, C.A. Justice, C. Whittaker, D.G. Argo, and B.H. Olson, Biofilm fouling of RO membranes--Its nature and effect on treatment of water for reuse, J. Am. Water Works Assoc. 76,94-102 (1984).

60. M.W. Mittelman, Biological fouling of purified water systems, in "Biological Fouling of Industrial Water Systems: A Problem Solving Approach," M.W. Mittelman and G.G. Geesey, editors, pp. 194-233, Water Micro Assoc., San Diego, CA, 1987.

61. T.J. Novitsky, J. Schmidt-Gengenbach, and J.F. Remillard, Factors affecting recovery of endotoxin adsorbed to container surfaces, J. Parent. Sci. Technol. 40,284-286 (1986).

62. R.A. Craven, A.J. Ackermann, and P.L. Tremont, High purity water technology for silicon-wafer cleaning technology, Microcontamination, 4(11),14-21 (1986).

63. M.W. Lechevallier, T.S. Hasenauer, A.K. Camper, and G.A. McFeters, Disinfection of bacteria attached to granular activated carbon, Appl. Environ. Microbiol. 48,918-923 (1984).

64. H.F. Ridgway and B.H. Olsen, Chlorine resistance patterns of bacteria from two drinking water distribution systems, Appl. Environ. Microbiol. 47,61-67 (1982).

65. R.L. Wolfe, N.R. Ward, and B.H. Olsen, Inorganic chloramines as drinking water disinfectants: A review, J. Am. Water Works Assoc. 76(5),74-88 (1988).

66. G.D. Groninger and J.F. Mills, Disinfection of wastewater plant effluents using bromine chloride versus chlorine and its environmental impact, in "Water Chlorination: Environmental Impact and Health Effects," R.L. Jolley, editor, pp. 733-740, vol. 3, Ann Arbor Science, Ann Arbor, MI, 1980.

67. C.J. Thomas, R.G.J. Edyvean, and R. Brook, Biologically enhanced corrosion fatigue, Biofouling 1,65-77 (1988).

68. N.J.E. Dowling and D.C. White, On-line electrochemical monitoring of microbially mediated corrosion, in "Microbial Corrosion," G.J. Cicina, editor, pp. 5-15, proceedings of the 1988 EPRI workshop, report no. R-6345, Electric Power Research Institute, Palo Alto, CA, 1989.

69. R. Lewin, Microbial adhesion is a sticky problem, Science 224,375-377 (1984).

70. R.C. Ebersole and M.D. Ward, Amplified mass immunosorbent assay with a quartz crystal microbalance, J. Am. Chem. Soc. 110,3623-8628 (1988).

71. D.E. Caldwell and J.J. Germila, Evaluation of difference imagery for visualizing and quantitating microbial growth, Can. J. Microbiol. 31,35-34 (1985).

72. A.R. Escher and W.G. Characklis, Microbial colonization of a smooth substratum: A kinetic analysis using image analysis, Proc. Water Wastewater Microbiol. Conf., Feb. 8-11, 1988, Newport Beach, CA, pp. 277-284.

73. G.G. Geesey, M.W. Mittelman, T. Iwaoki, and P.R. Griffiths, Role of bacterial exopolymers in the deterioration of metallic copper surfaces, Materials Performance 25,37-40 (1986).

74. J-R. Chen, S-I Lew, C-J. Huang, C-S Fang, and W-S Tse, Investigations of the biological corrosion of condenser tubes by scanning Auger microprobe techniques, Appl. Surface Sci. 33/34,212-219 (1988).

75. D.C. White, Validation of quantitative analysis for microbial biomass, community structure, and metabolic activity, Arch. Hydrobiol. Beih. 31,1-18 (1988).

76. A. Ogram and G.S. Sayler, The use of gene probes in the rapid analysis of natural microbial communities, J. Indust. Microbiol. 3,281-292 (1988).

77. T.H. Meltzer, "Filtration in the Pharmaceutical Industry," pp. 493-544, Marcel Dekker, New York, 1987.

78. B.F. Severin, M.T. Suidan, and R.S. Engelgbrecht, Kinetic modeling of U.V. disinfection of water, Water Res. 17,1669-1678 (1983).

79. W.F. McCoy, Treatment of biofouling, in "Biological Fouling of Industrial Water Systems: A Problem Solving Approach," M.W. Mittelman and G.G. Gessey, editors, pp. 247-268, Water Micro Assoc., San Diego, CA, 1987.

80. J. W. Costerton, U.S. Patent No. 4,419,248 (1983).

# DESCRIBING FILTERS AND FILTRATION PROCESSES USING THE CONCEPTS OF FRACTAL GEOMETRY

Brian H. Kaye

Department of Physics
Laurentian University
Sudbury, Ontario, Canada

The structure of some aerosols manifests fractal characteristics. The usefulness of the fractal dimension of crimped fibres and complex aerosol fineparticles is outlined. The concept of the Sierpinski fractal of filter structure is presented and the relationship of the fractal dimensions of filters and deposited capture trees of fineparticles is discussed.

## INTRODUCTION

Fractal Geometry, the geometry of rugged and complex structures, was conceived and developed by Benoit Mandelbrot[1,2]. From one perspective it can be viewed as an extension of classical dimensional analysis to describe complex systems in which the classical 1, 2 and 3 dimensional description of systems is augmented by means of a fractional addendum to the topological dimension of a system, this fractional quantity being descriptive of the ruggedness of the system. Thus, one of the simplest applications of fractal geometry of interest to the filtration specialist is the characterization of crimp in fibres used to create filter systems and synthetic felted or woven fabrics.

In Figure 1 fibres with varying amounts of crimp are shown. If these fibres were made of rubber they could all be straightened out so that their topological dimension is one. Topology is the study of relationships between various objects which remain the same when the space in which they are supported is distorted. If the fibres are rigid, one can describe the various structural attributes of the fibres by means of the fractional addendum to their topological dimension. For a discussion of the technique for measuring the fractal dimensions of such rugged lines see reference 3. Varying the structure of fibrous filters using fibres of various crimp structures has not been an active area of filter structure research but could become important as technologists seek to improve respirator filter performance. In such studies fractal dimension description of the crimped fibres used in filter construction could be useful.

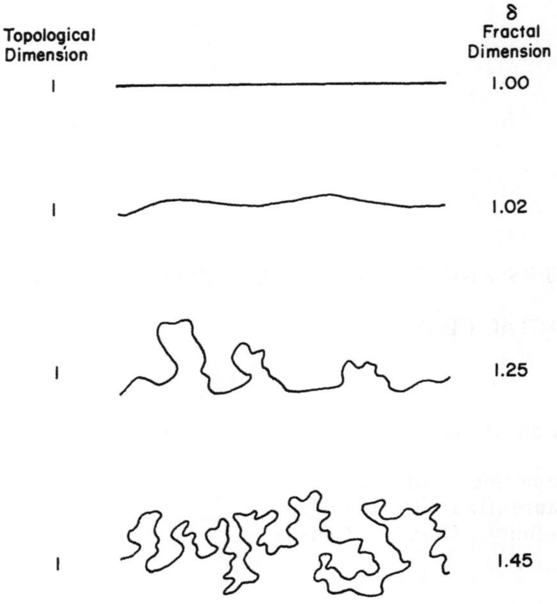


Figure 1. The crimp in a fibre can be characterized by the fractal dimension of the rugged fibre.

FRACTAL STRUCTURE OF FUMED AEROSOLS

Fractal geometry has been widely applied to the study of the structure of aerosols. Thus in Figure 2 the outline of a carbonblack produced from a fume of incomplete hydrocarbon combustion is shown. To estimate the fractal dimension of a rugged profile the perimeter is estimated at a series of resolutions and the changes of perimeter estimate plotted against the magnitude of the resolution parameter. When these data are plotted on log-log scales the resultant data display is known as a Richardson plot in honour of Lewis Fry Richardson, one of the pioneer workers in the study of rugged boundaries. The technique used to estimate the perimeter at various scales of resolution, illustrated in Figure 2, is known as the Equipaced Method for exploring the boundary. In this technique the boundary is digitized into a series of equipaced steps and then polygons are constructed for various magnitude strides around the perimeter. Thus in Figure 2 various polygons for 15 steps, 10 steps, 5 steps and the basic digitized pattern are shown.

The typical Richardson plot for fumed fineparticles shown in Figure 2 manifests a fractal dimension of 1.35 for the coarse resolution inspection and 1.13 for the fine resolution. These are typical of the fractal dimensions displayed by a system formed by a two-stage agglomeration. Thus, the carbonblack profile was probably formed in two stages with the initial formation of relatively densely packed agglomerates with the surface fractal dimension of 1.13. This dimension can be related to the packing of the primary spheres of the fume in the primary agglomeration stage. Subsequent collisions of these agglomerates produce a more open-structured agglomerate with fractal dimensions[3-7] typically in the range of 1.35 to 1.43.

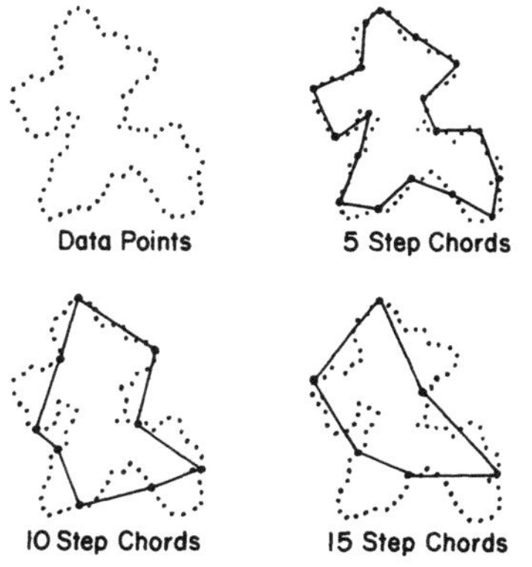


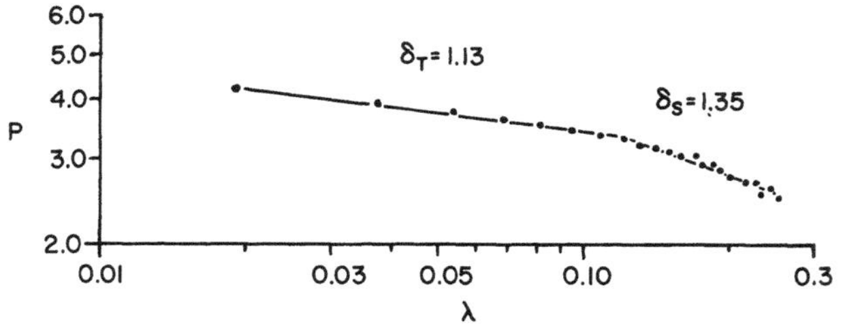


Figure 2. The structure of carbonblack profiles can be characterized by means of two fractal dimensions manifest at different levels of inspection[3].

$\lambda$ : Normalized Expected Chord Length
P : Normalized Perimeter Estimate
$\delta_T$ : Textural Fractal Dimension
$\delta_S$ : Structural Fractal Dimension

By studying the fractal structure of fineparticles captured by filters, the specialist can begin to differentiate between agglomerates formed on the fibres of the filter and those formed in free space by collision dynamics.

## THE SIERPINSKI CARPET AS A MODEL OF FILTER STRUCTURE

Of perhaps greater utility to the filtration specialist is another fractal model which can be used to study the structure of filters. This is the so-called Sierpinski carpet model of a two-dimensional rugged system.

An ideal Sierpinski carpet structure can have various structures depending upon the construction algorithm. The first 3 construction stages of a typical Sierpinski carpet are shown in Figure 3(a). Subareas of the carpet are removed in a systematic manner by dividing the carpet into the appropriate number of squares [9 for the carpet of Figure 3(a)] and removing the central square. This construction algorithm is then repeated on the remaining squares of the preceding order of carpet. Thus by the third stage we are removing 64 of the smaller squares to leave the pattern shown. By carrying out the construction ad infinitum one obtains a system which has zero area

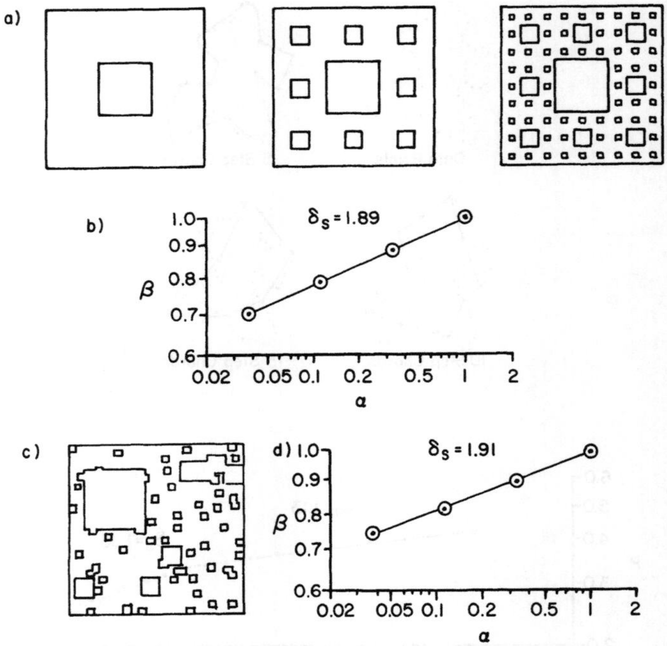


Figure 3. The Sierpinski carpet can be used to create a fractal model of a filter.

(a) Three stages in the construction of an idealized Sierpinski carpet with a fractal dimension of 1.89
(b) Dataline for the rate of disappearance of the ideal Sierpinski carpet shown in (a).
(c) Randomized Sierpinski carpet derived from (a)
(d) Dataline showing that the randomized Sierpinski carpet has a fractal dimension of 1.91.

α : Normalized Resolved Hole Size ( $\sqrt{\text{Area}}$ )
β : Normalized Residual Area of Carpet
$\delta_S$ : Sierpinski Fractal Dimension

and an infinite number of infinitely thin fibres separating the holes in the carpet. To characterize the structure of such a filter one studies the rate at which the area of the carpet disappears as one carries out successive stages of the algorithm. Thus in Figure 3(b) the rate at which the surface disappears as the carpet is constructed is shown. It can be shown that the slope of this line gives the fractal dimension of the carpet by the relationship $\delta = 2\text{-}m$ where m is the slope of the dataline and δ is the fractal dimension of the carpet. For the Sierpinski carpet shown in Figure 3(a) the fractal dimension can be shown to be 1.89.

The appearance of the ideal Sierpinski carpet does not give the impression that one is looking at any real type of filter. The usefulness of the Sierpinski carpet model becomes more apparent when we consider what is known as a statistically self-similar version of the Sierpinski carpet. In this model the idealized Sierpinski carpet is subjected to randomization of the different holes in the carpet to create a system such as that shown in Figure 3(c). It should be noticed that an important difference between the statistically self-similar carpet and the ideal one is that with the overlapping of holes due to their randomization the holes begin to show some interconnectiveness, a feature completely absent from the idealized carpet. It can be shown that the Sierpinski carpet is a two-dimensional analogue of a mathematical construction known as a Menger sponge in three-dimensional space. Again the Menger sponge can be used to model three-dimensional depth filters, but a discussion of this aspect of fractal modelling of filters is beyond the scope of this review. It should, however, be noted that again a Menger sponge as an ideal mathematical curve does not have interconnected pores, whereas the self-similar version has interconnected pore structure[8].

It is interesting to note that Mandelbrot discussed the relevance of the Sierpinski carpet to fractal geometry by reference to the structure of two different kinds of swiss cheese. In his book he also presented a diagram which he said looked like the map of a European city to discuss the application of Sierpinski carpet to a figure made up of crisscrossing randomly positioned lines. Filtration experts seeing this diagram in Mandelbrot's book would anticipate a discussion of the structure of fibrous filters only to discover that Mandelbrot considers his diagram to be an academic novelty of no practical use. However, Kaye and co-workers have shown that Mandelbrot's European city map is a useful model for characterizing the structure of a fibrous filter[3].

Because the self-similar Sierpinski fractal permits the overlapping of holes, the rate of disappearance of the self-similar carpet is smaller than that of the idealized carpet. This is shown by the data used to estimate the fractal dimension of the randomized structure of Figure 3(c) as shown in the graph of Figure 3(d). The fractal dimension of this self-similar carpet is 1.91 as compared to that of the ideal model of 1.89.

In Figure 4 a fibrous filter photographed by Clarenburg and Piekkar[9] is shown. When we measure the size distribution of holes in this filter and transform it into a data plot to estimate the Sierpinski fractal of the system we note the interesting fact that there are two datalines. The upper one represents the coarse hole distribution of the filter and the lower one the small hole distribution. It has been postulated by Kaye and co-workers[8] that the flow through a fibrous filter is dominated by the available coarse holes and that the fractal dimension of the coarse hole distribution will be related to the factor modifying the flow through the filter of given fiber density. This is a modified version of a hypothesis put forward early in the application of fractal geometry to fibrous filters by Ensor[10] which suggested that the fractal dimension of the filter bearing dendritic capture trees could be related to the flow through the filter. At that time the fact that the Sierpinski fractal could be used to describe filter structure and that most real filters appear to exhibit more than one Sierpinski fractal dimension was not known.

Model simulation of the structure of fiber filters proceeding at Laurentian University seems to offer an explanation of this bifractal structure of fibrous filters. Most fibrous filters currently in use are made by a felting process in which the constituent fibres are allowed to fall at random onto a support screen to form the web of fibres which eventually constitutes the felted fabric of the filter. This process can be simulated with a computer by letting short lines fall onto a field of view with the position of arrival being determined by x and y co-ordinates in the field of view being chosen with random number tables. Also the angle of orientation of the fiber is selected at random again using a random number table to specify the orientation of the fibres. Successive stages in the building up of the simulated filter/felt are shown in Figure 5[11]. On the left hand side of these diagrams the appearance of the

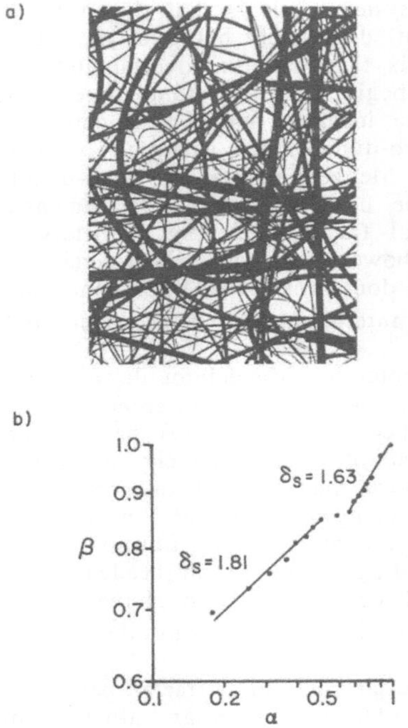


Figure 4. Electron miciograph of a fibrous filter studied by Clarenburg and Piekkar[9].

(a) Filter structure
(b) Sierpinski carpet plot showing bifractal structure of a real filter.

α : Normalized Resolved Hole Size ( $\sqrt{\text{Area}}$ )
β : Normalized Residual Area of Carpet
$\delta_S$ : Sierpinski Fractal Dimension

filter/felt after one has accumulated the stated number of fibres of specified length is shown. The fibres were allowed to have lengths between 25 and 75 pixels with all lengths in this range being equally probable. The Sierpinski fractals for the hole distribution created by this model are summarized in the diagrams. Thus, for a simulated felt containing two hundred fibres the Sierpinski fractal at coarse resolution had a value of 1.78 and a value 1.53 at high resolution. An addition of a further 100 fibres resulted in a Sierpinski fractal at coarse resolution of 1.87. For the smaller holes it was not possible to measure their size distribution with any accuracy at this stage of investigation.

A possible explanation of the persistence of a few large holes as the filter builds up is related to the probability of fibres arriving at the surface intercepting a large hole. Thus, if one looks at a subarea delineated by the

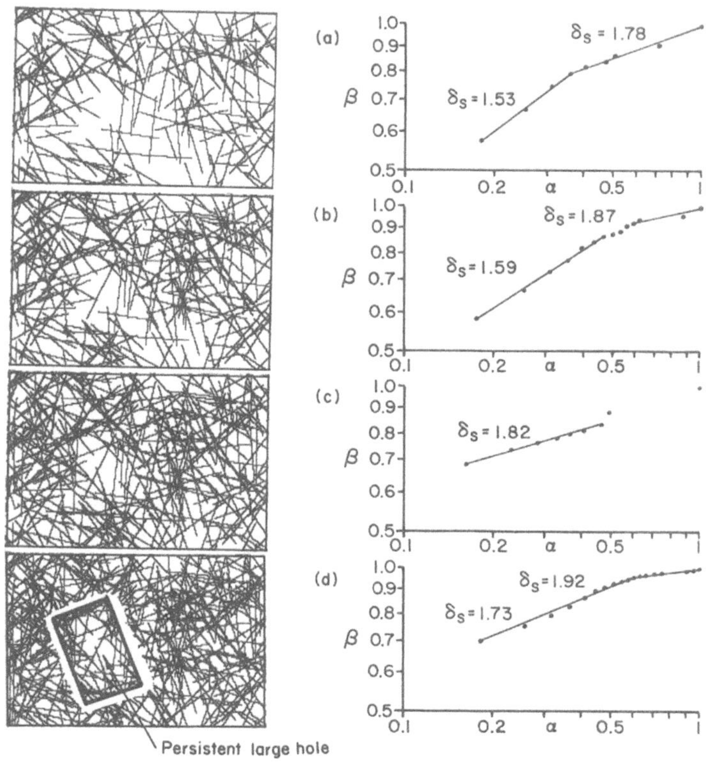


Figure 5. Simulated felted fibres exhibit bifractal structure. Fibre range permitted 25 to 75 pixels long (all equally probable).

(a) felt of 200 fibres
(b) felt of 300 fibres
(c) felt of 400 fibres
(d) felt of 500 fibres

$\alpha$ : Normalized Resolved Hole Size ( $\sqrt{\text{Area}}$ )
$\beta$ : Normalized Residual Area
$\delta_S$ : Sierpinski Fractal Dimension

solid line in Figure 5(d) the area of the large hole is a small fraction of the field of view. The chance of a fibre hitting this major hole with a chance of altering its structure is quite low and proportional to the area of the big hole over the area of the filter. Thus, a random build-up of fibres predicts a well known empirical fact that felted fibres contain a few large holes that dominate their flow behaviour. It was for this reason that early filters used as respirators were often exposed to a coarse smoke before being used in respirators. The concept was that the coarse smoke blocked off the large holes to which they migrated naturally under the air flow conditions of the filter and that subsequently the filters performance was greatly improved. Experiments are under way at Laurentian University to modify the felting process in such a way that arriving fibres have a much greater probability of closing off the few persistent large holes in the filter rather than to be located over more small holes already efficiently blocking air flow. The fact that most of the flow through a filter takes place through a few large holes has recently been demonstrated by R. Trottier and co-workers[12] in their studies of the build-up of deposits in a fibrous filter.

The first application of fractal geometry to the filtration process was reported by Ensor and Mullins[13]. They drew attention to the fact that the build-up of deposits captured by filters by diffusion capture exhibited fractal structure. Ensor and Mullins speculate that the structure of such fractal deposits will modify the flow through the holes of the filter in such a way that the equation controlling the flow will contain a term related to the fractal dimension of the capture trees. Kaye and co-workers have postulated that the fractal structure of the actual deposit will also be related to the capture efficiency of the fibres constituting the filter with respect to the specific dust being filtered[3,12]. For example, in Figure 6 a simulated filter created on a computer by Trottier et al along with the simulated build up of dendritic capture trees in the filter hole when specified capture rules are built into the capture dynamics is shown. By modifying the capture rules one can modify the structure of the simulated dendrites on the filters and by comparing real dendritic capture trees with modeled trees one can interpret the capture dynamics of the filter system.

CONCLUSION

In conclusion, it can be seen that fractal geometry is offering new descriptive tools for characterizing the structure of complex aerosol fineparticles. It can be used to study the structure of both surface and depth filters consisting of random structures and in a study of the capture dynamics on a filter. Fractal geometry is a fast growing branch of applied mathematics

Figure 6. The structure of a fibrous filter and deposits formed under various capture conditions can be simulated on computers using the concepts of fractal geometry. Dendritic capture trees shown above 'grown' by assuming 100% capture of an aerosol fineparticle on contact with a tree growing by diffusion [Brownian Motion] capture[12].

with many applications in various branches of applied science, in particular it is proving to be a useful methodology for describing aerosol systems. In this communication several possible applications of fractal geometry which may prove useful to the aerosol specialist have been reviewed and several possible areas of active future research suggested. It was not possible to include detailed technology for measuring fractals within the scope of this review. The pertinent details for such measurement technology is to be found in the literature referenced in this communication.

REFERENCES


1. The theories of fractal geometry were first presented to the English speaking world by Benoit Mandelbrot in the book he wrote entitled "Fractals: Form, Chance and Dimension", W. Freeman, San Francisco, 1977. Mandelbrot expanded this original essay concerning his ideas into a book which he considers to be the definitive description of his ideas and which is specified in reference 2.
2. B.B. Mandelbrot, "Fractal Geometry of Nature", W.H. Freeman, San Francisco, 1983.
3. B.H. Kaye, "A Randomwalk Through Fractal Dimensions", VCH Publishers, Weinheim, West Germany, 1989.
4. B.H. Kaye, Characterizing the significant structural features of hazardous dusts (in preparation).
5. B.H. Kaye, Characterizing the structure of fumed pigments using the concepts of fractal geometry, in preparation.
6. T. Vicsek, "Fractal Growth Phenomena", published by World Scientific Publishing Company, New Jersey, 1989.
7. R. Jullien, and R. Botet, "Aggregation and Fractal Aggregates", World Scientific Publishing Company, New Jersey, 1987.
8. B.H. Kaye, N.I. Robb, G.G. Clark, Mathematical models of depth filters and their characterization from the perspective of fractal geometry (in preparation).
9. L.A. Clarenburg, and H.W. Piekkar, Aerosol filters - pore size distribution in fibrous filters, Chem. Eng., 22, 1399-1408 (1967).
10. Personal communication with D. Ensor, Research Triangle Institute, Research Triangle Park, North Carolina , 1985.
11. The pictures of the simulated filters are taken from "Fractal Geometric Description of Filters and the Filtration Process"., G.G. Clark, M.Sc thesis, Laurentian University, Canada . (in preparation).
12. R. Trottier, G.G. Clark, and A. Hoffman, Fractal characterization of experimental and simulated dendritic capture tree growths in filter systems, paper presented at the American Association for Aerosol Research, Reno, Nevada, USA, October 9-13, 1989.
13. D.S. Ensor, and M.E. Mullins, The fractal nature of dendrites formed by the collection of particles on fibres, Part.Charact., 2, 77-78 (1985).

# DETECTION AND CONTROL OF PARTICLES IN VACUUM ENVIRONMENTS FOR SEMICONDUCTOR PROCESSING

T. C. Smith

Motorola Semiconductor
Advanced Technology Center
2200 W. Broadway
Mesa, AZ 85202

In VLSI and ULSI wafer processing areas, the detection and control of particles in vacuum processing equipment is crucial for cost-effective manufacturing. This is true because of stringent requirements for low defectivity in order to achieve realistic yields and because more than half the operations are performed in vacuum environments. Sources of particles and the mechanisms of transferring them in vacuum are reviewed, along with the special problems related to each of the various processes. Comparisons are made of the equipment configurations, modes of operation, gas flows, operating pressures, pumping and venting cycles, and important processing parameters. Results of particle contamination studies in some of these vacuum systems are discussed.

Examples of equipment modifications to reduce particulates are presented. Equipment qualification schemes and operational procedures for monitoring the system performance are stressed. The use of in-situ particle detectors for real time monitoring of particle contamination levels is discussed. An overview of the present levels of performance of both processing and measuring equipment is given, along with estimates of the requirements for future applications.

## INTRODUCTION

The impact upon yield of particulate contamination in the various processes used for fabricating VLSI and ULSI circuits has resulted in considerable efforts to reduce particle levels throughout the manufacturing line. These vacuum processes encompass various similarities, yet many differences, in their principles of operation, system configurations, and associated process parameters. Semiconductor processes which utilize vacuum environments include epitaxial film growth, low pressure chemical vapor deposition (LPCVD), plasma and reactive ion etching (RIE), photoresist ashing, ion implantation, evaporation and sputter deposition, and rapid thermal processing (RTP). Equipment manufacturers and users have made modifications to both individual component designs and system operating features to achieve reduced particle counts.

## TRENDS IN VLSI/ULSI WAFER PROCESSING

In wafer fabrication lines throughout the semiconductor industry, the minimum pattern dimension in ICs continuously decreases with each succeeding generation of DRAM product technology.[1] At the same time, the increased level of integration or product functionality in

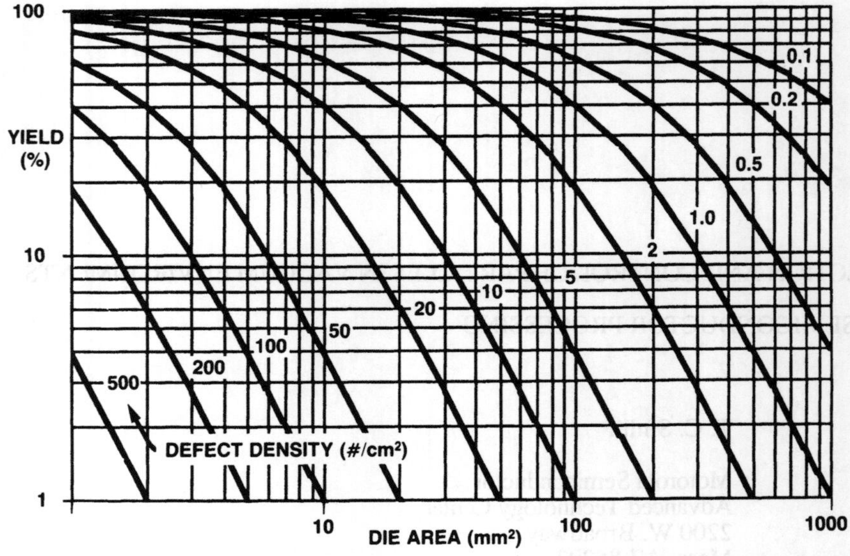


Figure 1. Defect limited IC yield (via Murphy's model) versus die or chip area for various values of the effective defect density. (from Ref. 12).

large microprocessors or gate arrays is driving chip sizes larger, limited only by advances in packaging technology. In addition to these trends, the overall process complexity is also increasing with the emergence of BiCMOS products, which require a combination of bipolar and CMOS process steps. Processing developments such as the use of more than twenty masking levels, up to four levels of interconnect metal, and complex planarizing schemes are becoming prevalent. The defect limited yield as a function of die or chip area and various values of the effective defect density is shown graphically in Figure 1. Particles introduced in any wafer processing step may result in yield loss and reduced reliability. The notion that particles are "killer defects" if their diameter is approximately one-tenth the minimum feature size in the circuit is only a rough standard, because various film thicknesses are smaller than that.[2] Fortunately, not all defects seen in individual process steps are killers, but the total particle budget is indeed small in order to get good yield with large chips. It is now the consensus that the major cause of yield loss is no longer the clean room environment nor the personnel, but contamination caused by particles generated from, and within, the process equipment itself. It is likely that future submicron ULSI fabrication will require ultraclean technologies covering the entire processing environment with radically different process techniques.[3] Trends toward fast-cycle operations with single wafer processing and large wafer sizes will lead to the evolution of "cluster" tools for integrated vacuum processing through numerous steps in one pumpdown cycle.[4]

Several recent reviews treat the overall subject of particle contamination in semiconductor manufacturing and its effect upon yield.[2, 5, 6] A number of papers discuss the generation and monitoring of particles in processing equipment.[7, 8, 9] Some other surveys have specifically reviewed particle control and reduction in vacuum processes.[1, 10, 11] This paper will review the present-day vacuum processes used in silicon wafer fabrication, focussing on the details of the equipment and the process, and outlining operational procedures for monitoring the particle

Table I. Frequency of Process Steps for an Advanced VLSI Product.

| PROCESS STEP | FREQUENCY | # TIMES | VACUUM? |
|---|---|---|---|
| PHOTO LITHOGRAPHY | XXXXXXXXXXXXXXXXXXXXX | 21 | |
| RESIST ASH/STRIP | XXXXXXXXXXXXXXXXXXXXX | 21 | Yes |
| IMPLANT | XXXXXXXXXXXXXX | 14 | Yes |
| DIFF./OXIDE or ANNEAL | XXXXXXXXXXXX | 12 | |
| WET ETCH | XXXXXXX | 7 | |
| DIELECTRIC DEPOSIT | XXXXXXX | 7 | |
| RIE ETCH | XXXXXX | 6 | Yes |
| LPCVD | XXXXX | 5 | Yes |
| METAL | XXXX | 4 | Yes |
| "OTHER" | XXXXXX | 6 | |
| | TOTAL = | 103 | 50 of 103 |

performance of various systems. Table I lists the frequency of process steps for a typical advanced VLSI product with approximately half of them performed in vacuum ambients in this case. The most frequently used vacuum process is photo-resist ashing in an oxygen plasma; this is often followed by a wet chemical cleaning step. It should be obvious that among the vacuum processes, the implant operation deserves considerable emphasis in efforts toward achieving and vigilance in maintaining particle-free processing because of its repeated use.[12]

The particle performance of each process step has come to be regarded as an essential equipment specification and a process parameter which must be characterized, measured, and controlled in a mode which is as close as possible to "real time". This strategy is necessary to achieve high yields at an earlier phase of the product cycle and to make further process improvements by reducing particle counts. Rapid reaction to any out-of-control situations in the fabrication area is imperative to maintain yield goals.

Particles are detected on unpatterned process monitors using surface scanning instruments or counted on patterned product wafers by tedious microscope inspection or with newer laser scanners which are capable of inspecting patterned wafers. An in-situ particle flux monitoring device which uses a solid state laser in a small sensor has been found useful in characterizing the operation of vacuum process equipment.[13] In another configuration, particle flux counts from this type of sensor have been shown to correlate well with both particle counts measured by surface scanners and electrical yield on EPROM devices.[14] An example of the utility of in-line monitoring as it correlates to final test yield has been described for a 1-megabit production line.[15] Figure 2 shows the yield impact of particles in both first and second polysilicon chemical vapor deposition (CVD) operations. The variations in measured defect densities via in-line monitors from microscope foreign matter (MFM) inspection generally track with yield as measured by failure analysis (FA) testing for product lots run through these operations in the same time frame.[15]

There are many similarities but also many differences among the vacuum processes used in wafer fabrication. Depending upon the manufacturer's design and the application, the system might be configured for either single-wafer or batch processing. Similarly, the wafer handling mode may be one in which wafers are exchanged in the clean room ambient at atmospheric pressure or the wafers might enter and leave the process chamber through a vacuum load lock, which keeps the processing chamber isolated from the clean room. Table II lists the operating

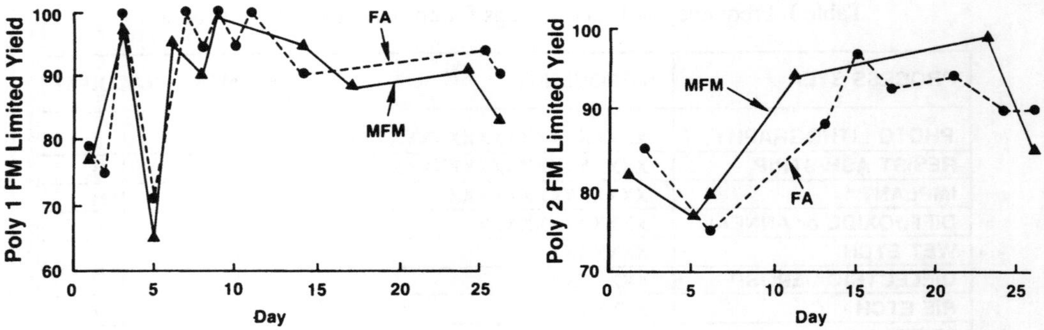


Figure 2. Foreign matter (FM) limited yield at 1st and 2nd polysilicon deposition: failure analysis (FA) and microscopic foreign matter inspection (MFM) yields versus time. (from Ref. 15).

pressure ranges and some distinguishing features of each process. The various thin films deposited and etched in device structures include doped and un-doped oxides, nitrides, polysilicon, metals, silicides and others used as insulating layers or interconnects. It is difficult to say which type of system configuration is most common for certain processes, because there are sometimes more than a dozen vendors who manufacture systems for processing wafers of different sizes in research or production modes. As mentioned above, the trend in processing equipment has been toward single-wafer processing. The necessary features of some processes, such as the presence of a photoresist (PR) masking layer, the chemistry of process gases, operating temperatures, or wafer throughput requirements may place constraints upon the system configuration and the schemes which could be employed to further reduce particle counts.

Table II. Operating Pressure Ranges and Process Features.

| VACUUM PROCESS TECHNOLOGY | RANGE OF OPERATING PRESSURES (Torr) | COMMENTS ON FEATURES UNIQUE TO THIS PROCESS |
|---|---|---|
| Epitaxial film growth | (20 - 200) | Defects caused by *any* size particle! |
| Low Pressure CVD | (0.1 - 2.0) | Temperatures: (400 - 1000) °C |
| Plasma Enhanced CVD | (0.1 - 2.0) | Temperatures: (100 - 600) °C |
| Plasma etching and RIE | (0.1 - 2.0) | w/PR; many chemistries used |
| PR ashing | (0.1 - 0.5) | Oxygen plasma or downstream ash |
| Ion implantation | (1 - 20) x $10^{-6}$ | w/PR; room temperature |
| Sputtering | (1 - 20) x $10^{-3}$ | w/etch, heat, or bias - mostly Al alloys |
| Evaporation | (1 - 100) x $10^{-7}$ | w/heat |
| Rapid thermal processing | (0.1 - 760) | Most applications at atmos. pressure |

For those vacuum processes which require active pumping of process gases, a wide range of operating pressures and types of pumps are used. The American Vacuum Society's recommended practices for pumping hazardous gases, which address concerns related to system design and operation for technical and safety reasons, are also beneficial for maintaining good particle performance.[16] Theoretical and experimental studies of the gravitational fall and the transport of particles in vacuum have been reported without reference to specific processes.[1] In Figure 3, the gravitational settling time is plotted as a function of pressure for various particle diameters. Results of particle measurements made on wafers using a laser surface scanning tool have led to recommendations for controlling the pumpdown and venting sequences to avoid turbulence. Observations on particles added as a function of wafer orientation show that face-down or face-vertical processing gives the best particle results.

## SOURCES OF PARTICLES IN VACUUM PROCESSES

One principal source of particles is the product wafer itself, which may be broken, chipped, or abraded in handling or processing. Brittle photoresist films on the edge or backside of wafers may also generate particles. The use of some PR edge bead removal techniques is recommended.[12] Mechanical wear between moving mechanisms produces a broad distribution of particle sizes, including submicron particles, and these can be widely dispersed in vacuum.[1,17] Particles may be introduced from process gases or nitrogen used for venting up chambers or load locks. Hence, all those precautions recommended for preserving the purity of gases throughout the distribution system and equipment, such as utilizing point-of-use filters, must be observed.[18] Accumulations on chamber walls, boats, platens, or other fixtures which have been coated by deposited or back-sputtered material and reaction by-products may generate microscopic flakes due to film stress. Thermal cycling of these films may cause the release of copious quantities of particles, as in atmospheric furnaces.[19] Proper cleaning techniques and

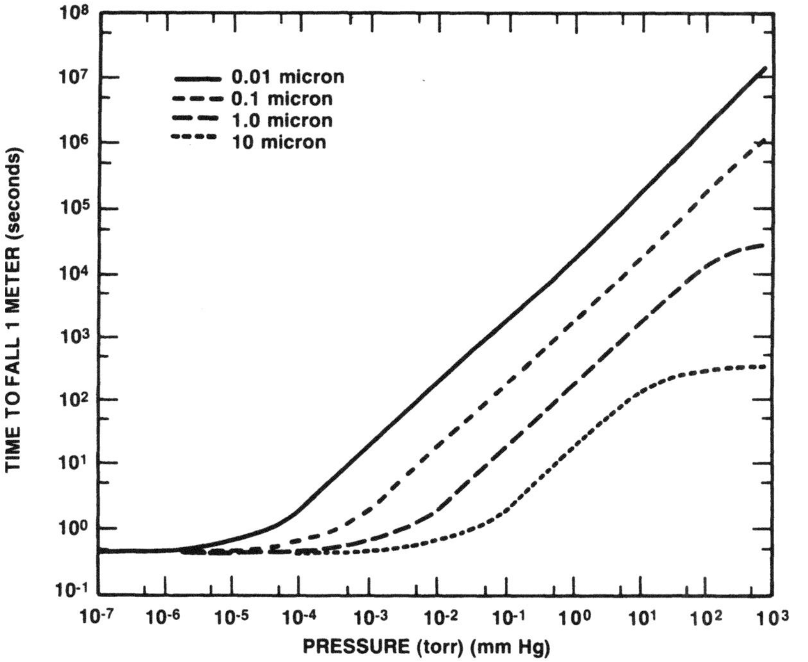


Figure 3. Gravitational settling of particles: time to fall one meter versus pressure for various diameter particles. (from Ref. 1).

preventive maintenance schedules must be established and adhered to, using particle monitors as an indicator of effectiveness. Identification of particles generated inside process equipment requires micro-analytical techniques such as scanning electron microscopy (SEM) with energy dispersive X-ray spectrometry (EDS) capability for determining elemental composition.[20] Fluorescence microscopy has been demonstrated to be beneficial for enhancing observation of organic particles, while micro-FTIR (Fourier-transform infrared spectroscopy) and micro-Raman spectroscopy techniques are severely constrained.[20]

## TRANSFER OF PARTICLES IN VACUUM

It is generally thought that particles are transported by aerodynamic forces and that, for particles less than 50 μm diameter, the primary forces of adhesion are van der Waals forces.[21,22] If turbulent gas flow conditions are present during the pumpdown or venting stages, substantial amounts of particles may be stirred up and added to the wafer surface. Pressure differentials between load locks and process chambers when opening connecting gate valves have been found to cause particle movement between chambers.[1] In processing wafers with photoresist, it may be necessary to select processing parameters to limit the incident power flux to avoid overheating which can cause catastrophic disintegration of the PR film.[12] Adjustments should be made so that the PR temperature does not exceed the manufacturer's recommended maximum value. Particles may adhere to the surface of wafers before, during, or after the key process in the vacuum cycle, with different consequences depending upon the process. For example, if a 1.0 μm diameter silicon particle is on the surface at the time that the implant step is performed, the implant would effectively be blocked from entering the substrate (for the normal range of energies used for any of the common dopants). Particles added after the key process itself is complete, while venting the chamber or load lock, may be fed into the next process step.

## EQUIPMENT MODIFICATIONS FOR PARTICLE REDUCTION

In recent years, steady improvements in the particle performance of vacuum processes have been realized as a result of much effort by both equipment manufacturers and users. These efforts were often expended in parallel by groups working in different companies, different locations of the same company, or in entirely different process disciplines! Some modifications and design changes can only be achieved in a totally new system configuration. Others are retro-fittable to existing systems and available as kits from the vendors. Simple changes in operating procedures or timing sequences controlled in software can often be effective in reducing particle counts. As mentioned earlier, certain process constraints may prohibit the implementation of a scheme which is beneficial in another process.

A universal recommendation seems to be to use "soft roughing" and slow venting schemes. The effect upon particle counts on daily monitors from such a modification[23] to a medium current implanter is seen in Figure 4. However, it is not clear that these modifications are universally effective. The principal consideration in designing a system might not be to limit the Reynolds number, but the fluid velocity, which may be the more important parameter with regard to dislodgement and deposition of particles.[24] In many existing systems, the timing sequence for switching valves from a low conductance to higher conductance mode is not optimized. When system throughput is dominated by pumpdown and venting times, a trade-off between particle performance and throughput may exist. It is beneficial to provide separate roughing and venting lines so that the flow of entrained particles will be unidirectional.

Automatic wafer handling techniques can eliminate particles introduced by operators. The use of "pick and place" wafer handlers with backside contact only is preferred. Ideally, no wafers or cassettes should be below the plane in which wafers are being handled. Any possibility of sliding contact and abrasion should be eliminated. If clamps must be used, the area of contact on the front sides or edges of wafers must be minimized. The use of low shedding Teflon or stainless steel tubing for both process and venting gases with point-of-use filters is also desirable. Only low particle generating components, such as O-ring seals, valves

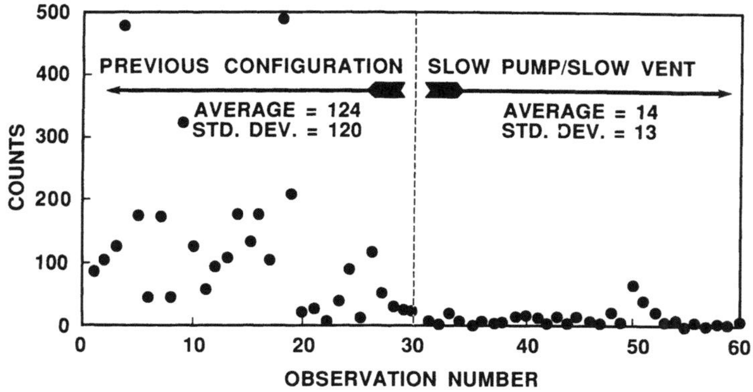


Figure 4. Particle counts on daily monitors on a Varian/Extrion 350D medium current implanter before and after modification, for particles greater than 0.5 μm diameter on 100 mm diameter wafers. (from Ref. 23).

and mass flow controllers (MFC) are recommended; shock waves which result from rapidly actuating valves or MFC's should be avoided.

In high current or batch implanters, used for high dose applications because of the need to distribute the high beam power over a large area, the wafers are arranged on a disk or wheel which spins at 1000 RPM or more during the implant. Figure 5 shows the impact of changing from a mode of operation where the rotation of the disk was started in room air as the chamber began to close to a mode in which the rotation was initiated after the chamber was in the high vacuum state.[23] The turbulence which was present in the former case can be used to advantage in an "auto-clean" mode of operation. With no wafers on the disk, the system can be programmed to run with the disk spinning while simultaneously venting and pumping. The resulting aerodynamic forces tend to dislodge particles inside the system, which are then entrained in the gas flow and pumped away. This feature is available on all high current systems on the market, and is effective in restoring specified levels of particle performance, especially after performing preventive maintenance on the end station of the implanter.

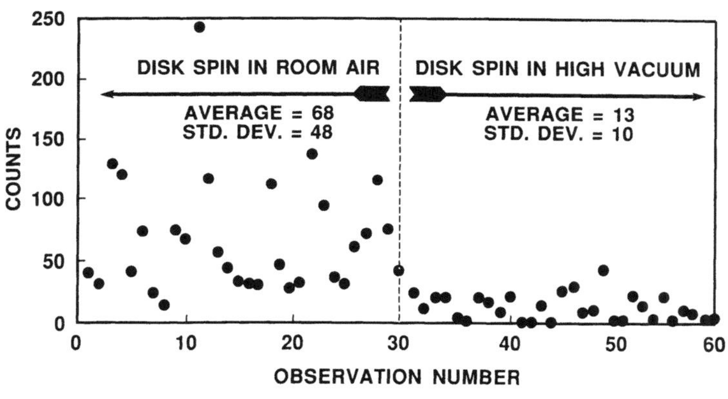


Figure 5. Particle counts on daily monitors on an Eaton/Nova NV10-80 high current implanter before and after modification, for particles greater than 0.5 μm diameter on 100 mm diameter wafers. (from Ref. 23).

In addition to some of the references cited above, there have been numerous studies published which deal with one or more of the processes listed in Table II. Some observations are unique to one process or particular equipment only, but some results are similar to those obtained in other systems or process disciplines. Specifics of the particle performance of the epitaxial Si growth process have been reported.[25] Efforts to characterize particle levels in LPCVD systems have been described for an oxide process,[25] and a tungsten process.[26] Equipment developments for particle reduction in a plasma enhanced chemical vapor deposition (PECVD) system were also published.[27,28] These point out the need to calibrate laser surface scanners for making measurements on wafers according to the specific film thickness which has been deposited. Studies done in plasma and reactive ion etching (RIE) processes have yielded additional insights into the particle performance of a variety of systems, etching different thin films.[13,20,29,30,31,32] In a load locked sputtering system, a particle flux monitor (PFM) was used for quality control purposes after a strong correlation was demonstrated between PFM counts and particle add-ons as measured on a laser surface scanner.[33]

Examples of results obtained from equipment modifications to implanters were given above.[12,23] A large amount of work has been done in this process technology, with many studies performed with in-situ particle flux monitors.[13,14,30,33,34] Several other reports stress the broader implications of dopant and heavy metal contamination and the effect of process induced crystal damage upon the apparent defect density.[35,36]

Two recent basic studies are noteworthy because they shed more light on particle transport in vacuum chambers during pumping and venting cycles. In the dynamic particulate characterization of a vacuum load lock system, particle counts from a particle flux monitor and pressure in the load lock were recorded as a function of time during the pumping and venting cycles.[37] Data were obtained for a range of pumping and venting conditions. As seen in Figure 6, a strong correlation between particle counts and the calculated Reynolds number was observed, indicating that high particle counts are encountered under turbulent pumping conditions. These researchers also observed water aerosol formation during pumpdown as had been reported earlier- see inset by P. Borden in reference 34. The mechanism is postulated to be one by which water droplets homogeneously condense or nucleate on very fine particles in the humid room air when the pressure drops at the beginning of the pumpdown.

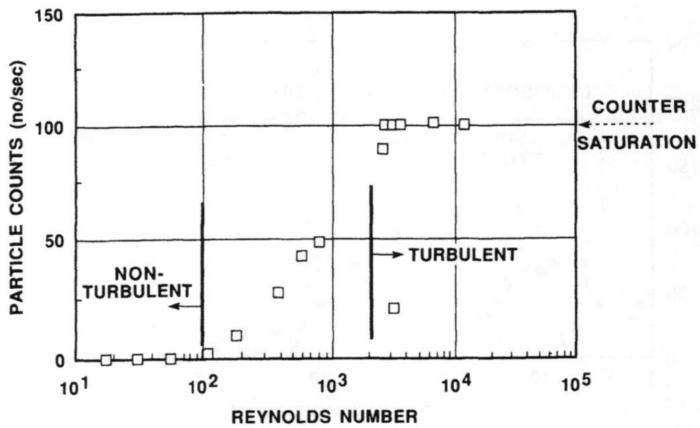


Figure 6. Particle flux counts in load lock versus calculated Reynolds number. (from Ref. 37).

An experimental study on the prevention of aerodynamic particle transport has recently been reported.[38] The minimum velocity required to detach particles from surfaces was measured by observing the particle's motion under a microscope while varying the pumping and venting conditions. A variety of different particles were used in a number of different sizes on insulating and conducting surfaces. Results for the measured minimum velocity to detach 5 μm diameter nickel particles from a platinum coated Lexan window are shown in Figure 7. Using the 5 μm data as an upper bound ensures that smaller particles will not be dislodged and transported during the roughing and venting cycles. As a result of their efforts, it is possible to tailor the venting and roughing cycles to dramatically reduce particle transport from process chamber surfaces without significantly lengthening those cycle times.

A third recent study discusses plasma generated particles, in which in-situ laser light scattering techniques were used to diagnose the temporal and spatial relationship of the particles.[39] Results show that certain chlorine- and fluorine-containing plasmas produce particles under silicon etching conditions. The data suggest that etch products react with gas phase species to form negatively charged clusters. The light scattering signals correlate directly with surface counts from laser scanners. When the rf field is turned off at the end of the etch process, the particles drop and contaminate the wafer.

## OPERATIONAL PROCEDURES FOR MONITORING PROCESSES

Although there are no standard operating procedures defined, the methodology for characterizing the particle performance of a process system has focussed in most practical instances to monitoring the "particles per wafer per pass" or "PWP" measurements.[40] Meaningful particle monitoring programs can only be implemented after verifying that the surface scanning system is correctly calibrated for the films of interest, again taking precautions with measurements made on wafers with deposited films from the full flow process.[25] In

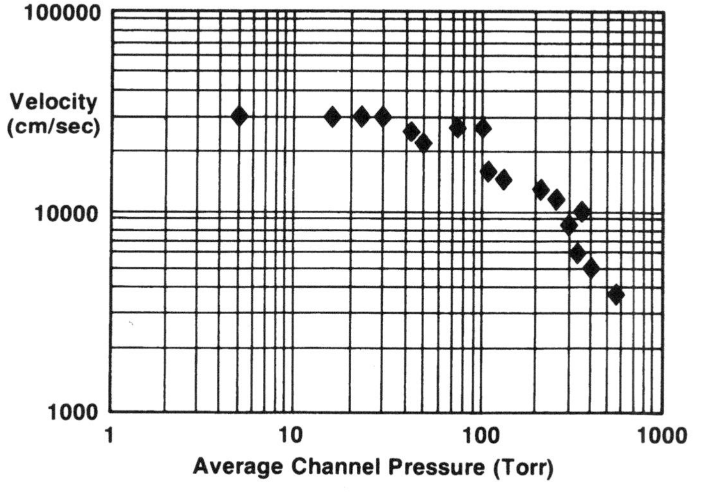


Figure 7. Experimental minimum velocity to detach 5 μm diameter nickel particles from a conductive surface versus pressure. (from Ref. 38).

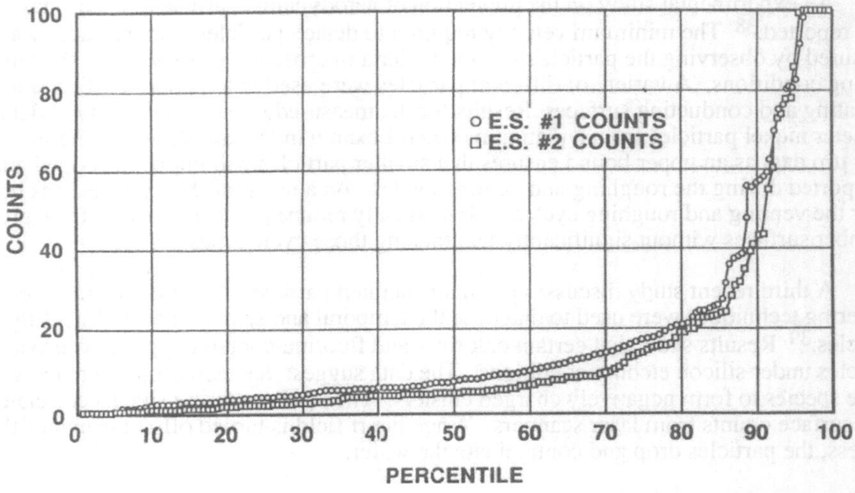


Figure 8. Percentile values of particle counts by end station (E.S.) on daily monitors over a two year period on a Varian/Extrion 120-10 high current implanter, for particles greater than 0.5 μm diameter on 100 mm diameter wafers. (from Ref. 23).

addition, one must ensure that careful planning has been used in experimental programs, and that adequate training has been given to all operators or technicians involved in handling wafers or making measurements. Results can be highly variable, with the standard deviation often greater than the average values, and proper statistical techniques must be employed in analyzing and reducing the data. It is worth mentioning that surface scanning instruments typically use generous edge exclusion, and it is sometimes useful to examine the edges and backsides of wafers to see what particles may be deposited there. Tullis has discussed many details of proper experimentation and characterization in further depth.[40] Monitoring the particle performance of the various processes is only one aspect of the total program of contamination control in the wafer fabrication area. However, because it gives real time feedback, particle monitoring is an important part of the overall culture of the clean room which is necessary to maintain high yields in the production environment.

When PWP's are plotted on trend charts or control charts, and production specifications for process qualification are established, realistic "shut-down" limits should be set and reaction strategies must be in place. Figure 8 shows the percentile values of particle counts on daily monitors run in a high current batch implanter operation over a two year period. It indicates that if the shut-down limit were set at 50 counts, the system performance would be acceptable about 90% of the time. Since then, further improvements have been made and the system currently has a shut-down limit of 20 counts, running below that limit about 95% of the time. Clean-up routines should be specified and performed on a periodic basis as part of the regular preventive maintenance schedule. All instances of wafer breakage should trigger an immediate shut-down for the process and efforts must be made to determine the specific reason for breaking wafers so that such catastrophic events can be minimized in the future. Comparisons between laser scanning instruments or between process equipment performance in different wafer fabrication lines (or from the vendor's factory to the user's processing area) are difficult because of the extreme sensitivity of resulting measurements to the threshold settings on different instruments.

## IN-SITU PARTICLE DETECTORS

The status and needs of in-situ real-time process particle detection have been reviewed.[41] It is generally felt that the available detection methods are inadequate to meet the requirements of the next generation of submicron devices.[4] Table III lists the desired features and gives status of the various detection methods with regard to these issues. The "wafer" detection method means surface scanner measurements on wafers run through the processing system. "Optical (in-situ)" means an in-situ optical detector mounted inside the vacuum chamber, and "Optical (external)" means measuring particles inside the vacuum system from outside through a window. It appears that the particle diameter detection limit for optical techniques will be in the 0.1 to 0.2 μm range due to the rapid decrease in the intensity of scattered light as particle size decreases. These authors conclude that new (non-optical) approaches need to be vigorously pursued to meet future needs. Many studies cited earlier were done with sensors placed inside chambers or in pumping lines. Although these sensors do not detect particles below about 0.3 μm diameter, it has been demonstrated that presently in many processes, particle flux monitors can be used to give warning of out-of-control situations, once a process baseline has been established.

## PRESENT LEVELS OF PERFORMANCE

Because of the wide range of processes and system configurations used, it is difficult to make comparisons and to state a benchmark figure for defect density or PWP which is "typical" of vacuum processes at any point in time. On new equipment, the particle performance guarantees, which are now an essential part of the purchase and acceptance specifications, have continued to move downward. In response to the user's requirement, both the guaranteed defect densities and the specified lower particle diameter have steadily decreased. Many systems are guaranteed by the manufacturer to add fewer than 0.05 particles/$cm^2$ of sizes greater than 0.5 μm diameter, and tighter specifications can be negotiated. In virtually every process discipline, rapid progress has been made in recent years in reducing equipment and process generated

Table III. In-situ Particle Detector Issues and Status (from Ref. 41).

| DESIRED FEATURE | DETECTION METHOD | | |
|---|---|---|---|
| | WAFER | OPTICAL (In-situ) | OPTICAL (External) |
| Large area detection | 75 $cm^2$ (100 mm wafer) | 0.1 $cm^2$ | 0.03 $cm^2$ |
| <0.1μm detectable size | 0.3 μm (bare wafer) 1.0 μm (patterned wafer) | >0.5 μm | 0.1 μm |
| Real-time detection | no, integrates | yes | yes |
| Sizing capability | yes | no (in vacuum) | yes (in room) |
| Inside chamber | yes | yes | no |
| Pumping line monitor | no | yes | yes |
| Cost effective | $65k-125k | $15k | $15k |
| In situ wafer measurement | no | no | no |

particles. The defect "learning curve" for one of the latest generation ion implantation systems shows that the particle performance has been driven down (by an order of magnitude over a period of two years) to meet the required defect density levels for advanced devices.[35] Similar trends have been reported by various equipment manufacturers and wafer fabrication users for other processes.

It is perhaps safe to say that the particle performance of many properly maintained vacuum processes compares favorably to other processes which operate in clean room ambients, such as PR tracks. However, the possibilities for failing a qualification test are often higher for the more complex vacuum equipment. It is important to note that it is not always necessary to change over to a whole new system in order to make progress.

## FUTURE REQUIREMENTS FOR PARTICLE CONTROL

The urgent need for further development in both metrology and processing capability are evident when one considers that the total particle budget goal for 16 or 64 MBit DRAM technology is less than 0.001 PWP/step.[4] This corresponds to about two killer defects per 8 inch wafer for the total process sequence! More comprehensive basic studies will be required to gain a better understanding of the equipment design and process interactions as they relate to particle generation and subsequent transfer to the critical device surfaces. Some processes, such as LPCVD, are regarded as inherently dirty and it will be desirable to provide for complete self-cleaning of the process chamber, possibly after every wafer.[3] Process innovation will be required to comply with the suggestion that we must avoid the use of PR in the future.[20] The emerging trend toward cluster tools, which will allow sequences of certain processes to be performed in one vacuum cycle, and will provide multiple chambers, is probably a step in the right direction.

## SUMMARY

The needs for detection and control of particles in vacuum processes of wafer fabrication are reviewed with respect to the trends in the semiconductor industry. Although the system configurations and operating conditions vary greatly throughout the processes used in advanced IC manufacturing, some common design and operational features can be implemented to further enhance the particle performance in each of these technologies. The references cited give more detailed information on specific studies made to date. In many cases, dramatic improvements have been made. The aggressive goals for the future require increasingly lower particle densities at still smaller particle sizes. Further cross-fertilization of ideas is needed to realize maximum transfer of experience from one discipline to another. Continued cooperation between the manufacturers and users of both processing and monitoring equipment is imperative to meet these goals in the wafer fabrication environment.

## REFERENCES


1. R. A. Bowling and G. B. Larrabee, Particle control for semiconductor processing in vacuum systems, Proceedings 1986 Microcontamination Conference, p. 161.
2. T. Hattori, Total contamination control for ULSI wafer fabrication, Microelectronic Manuf. and Test., 31 (April 1988).
3. T. Ohmi, What's the contamination control target in ULSI manufacturing?, Proceedings 1988 Microcontamination Conference, p. 55.
4. T. Seidel and K. Dillenbeck, Contamination in vacuum, paper presented at Semicon East, Boston, MA, Sept. 1989.
5. D. W. Cooper, Particulate contamination and microelectronics manufacturing: An introduction, Aerosol Sci. Technol., 5, 287 (1986).
6. C. M. Osburn, R. P. Donovan, H. Berger, and G. W. Jones, The effects of contamination on semiconductor manufacturing yield, J. Environ. Sci., 31(2), 45 (1988).
7. H. S. Nagaraj, B. L. Owens, and R. J. Miller, Particulate generation in devices used in clean manufacturing, in "Particles in Gases and Liquids 1: Detection, Characterization, and Control," K. L. Mittal, editor, pp. 283-293, Plenum Press, New York, 1989.

8. S. D. Cheung, In-situ monitoring of particulate contamination in integrated circuit process equipment, ibid, pp. 167-173.
9. P. G. Borden, J. Munson, and D. W. Bartelson, A real-time fallout monitor for 5-250 micrometer particles, ibid, pp. 175-184.
10. J. F. O'Hanlon, Advances in vacuum contamination control for electronic materials processing, J. Vac. Sci. Technol., A5(4), 2067 (1987).
11. J. F. O'Hanlon, Contamination reduction in vacuum processing systems, J. Vac. Sci. Technol., A7(3), 2500 (1989).
12. T. C. Smith, Photoresist problems and particle contamination, in "Ion Implantation: Science and Technology," 2nd edition, J. F. Ziegler, editor, p. 345, Academic Press, New York, 1988.
13. P. G. Borden, Y. Baron, and B. McGinley, Monitoring particles in vacuum-process equipment, Microcontamination, 5(10), 30 (1987).
14. P. G. Borden and L. A. Larson, Benefits of real-time, in-situ particle monitoring in production medium current implantation, Proc. of IEEE International Semiconductor Manufacturing Science Symposium, May 22, 1989, p.61.
15. R. Shasteen, H. Boisselle, M. Ishaq, and E. McGuire, Inline defect control on a 1-Mb DRAM production line, Proceedings 1988 Microcontamination Conference, p. 13.
16. J. F. O'Hanlon and D. B. Fraser, American Vacuum Society recommended practices for pumping hazardous gases, J. Vac. Sci. Technol., A6(3), 1226 (1988).
17. R. J. Miller, D. W. Cooper, H. S. Nagaraj, B. L. Owens, M. H. Peters, H. L. Wolfe, and J. J. Wu, Mechanisms of contaminant particle production, migration, and adhesion, J. Vac. Sci. Technol., A6(3), 2097 (1988).
18. W. Kroll, Contamination prevention and protection for process gases, Solid State Technol., 220 (March 1984).
19. S. D. Cheung and D. A. Hope, In-situ monitoring of particulate contamination in integrated circuit process equipment, Proc. 34th Annual Technical Meeting of Institute of Environmental Sciences, (1988), p. 524.
20. T. Hattori and S. Koyata, Detection and identification of process equipment generated particles for yield improvement, Proceedings 1988 Microcontamination Conference, p. 1.
21. R. A. Bowling, An analysis of particle adhesion on semiconductor surfaces, J. Electrochem. Soc., 132(9), 2208 (1985).
22A. K. L. Mittal, editor, "Particles on Surfaces 1: Detection, Adhesion, and Removal," Plenum Press, New York, 1988.
22B. K. L. Mittal, editor, "Particles on Surfaces 2: Detection, Adhesion, and Removal," Plenum Press, New York, 1989.
23. T. C. Smith, Particle contamination in ion implanters, Nuclear Instrum. Methods in Phys. Res., B37/38, 486 (1989).
24. M. H. Hablanian, If you rough slowly, do you get a clean vacuum chamber?, Res. Dev., 81 (April 1989).
25. B. Fishkin and E. J. Baker , Particle performance evaluation of CVD and epitaxial processes and equipment, Proc. 34th Annual Technical Meeting of Institute of Environmental Sciences, (1988), p. 517.
26. K. G. Hays, K. L. Bishop, and T. R. White, Particle reduction investigation of a LPCVD tungsten reactor, Proceedings 1987 Microcontamination Conference, p. 1.
27. D. G. Hemmes, Microcontamination reduction in PECVD systems - Part 1, Semiconductor Intl., 90 (May 1987).
28. D. G. Hemmes, Microcontamination reduction in PECVD Systems - Part 2, Semiconductor Intl., 98 (June 1987).
29. E. J. Baker and B. Fishkin, Particle control in process equipment: A case study, Microcontamination, 4(9), 52 (1986).
30. P. G. Borden and J. Gregg, Measurement and control of particle levels inside vacuum processing equipment, Proc. 35th Annual Technical Meeting of Institute of Environmental Sciences, (1989), p. 325.
31. S. DeOrnellas, A. Cofer, and M. BenDor, Particulate contamination in dry etch system design, Proceedings 1988 Microcontamination Conference, p. 66.
32. R. Lachenbruch, O. Gomez, and B. Chapman, Control of particulate emissions from plasma etching systems, Proceedings of 12th Annual TEGAL Plasma Seminar, (1986), p. 11.

33. P. G. Borden and W. Knodle, Process control through the measurement of particle flux to wafer surfaces in vacuum process equipment, Proc. 34th Annual Technical Meeting of Institute of Environmental Sciences, (1988), p. 429.
34. W. Weisenberger, Particle control in high-current ion implanters, Semiconductor Intl., 188 (May 1988).
35. J. Strain, S. Moffatt, and M. Current, Characterization and reduction of particle contamination in ion implantation processing, Proceedings 1988 Microcontamination Conference, p. 42.
36. M. I. Current and L. A. Larson, Ultra-pure processing: A key challenge for ion implantation processing for fabrication of ULSI devices, paper presented at the Materials Research Society Symposium, San Diego, CA, April, 1989.
37. D. Chen, T. Seidel, S. Belinski, and S. Hackwood, Dynamic particulate characterization of a vacuum load lock system, J. Vac. Sci. Technol., A7(5), 3105 (1989).
38. M. E. Mack, D. K. Stone, J. A. Jost, and P. Sferlazzo, Prevention of aerodynamic particle transport in air handling wafer processing equipment, paper presented at Semicon East, Boston, MA, Sept. 1989.
39. G. S. Selwyn, J. Singh, and R. S. Bennett, In-situ laser diagnostic studies of plasma -generated particulate contamination, J. Vac. Sci. Technol. A7(4), 2758 (1989).
40. B. J. Tullis, Particle contamination by process equipment, in "Handbook of Contamination Control in Microelectronics," D. L. Tolliver, editor, p. 410, Noyes Publications, Park Ridge, NJ, 1988.
41. R. A. Bowling, G. B. Larrabee and W. G. Fisher, Status and needs of in-situ real-time process particle detection, Proc. 34th Annual Technical Meeting of Institute of Environmental Sciences, (1988), p.508.

# OPERATOR-GENERATED PARTICLES: CHARACTERIZATION, MONITORING AND CONTROL

Eric S. Burnett

Araclean Services Inc.
One North Beacon Street, P.O.Box 458
La Grange, IL 60525

Most operator/cleanroom garment systems in service today can be characterized as broad spectrum particle generators, enclosed by inefficient mechanical filters which may also be sources of particulates. Excluding only the costly total-body containment systems fabricated from PTFE laminates, with "bubble" helmet and independent air supply, all cleanroom garment systems allow some particles, body detritus and effluvia to enter the airstream. The subsequent gravitational or airborne transportation and fate of these particles and aerosols, and their impact on product yield, is critically dependent on airflow patterns and the effectiveness of the physical and electrostatic screening between operator and product.

In this paper, sources and characteristics of operator-generated particles and aerosols are reviewed to aid in identifying requirements for improved monitoring and control. Following a brief discussion of current test methodology in the USA, which is generally limited to fabric filtration efficiency determination and non-destructive statistical sampling of garments for particle generation and release, "point-of-entry" and "point-of-use" particulate sampling is examined. This approach is believed to be appropriate for modern Class 1 and Class 10 cleanrooms equipped for automated particle monitoring and data logging. The ergonomic and economic trade-offs for alternate control strategies are also discussed.

## INTRODUCTION

When the Fine Particle Society was formed twenty years ago, it is doubtful that any of the founding members were thinking in terms of particles commensurate in size with human viruses (0.01 to 0.1 micrometer). Today, as we look at the developing technologies for the next decade, monitoring and control of particles below 0.1 micrometer is mandatory to achieve acceptable yields.

*Particles in Gases and Liquids 2,* Edited by K.L. Mittal
Plenum Press, New York, 1990

As one example, it has been projected that the flying head height in next-generation magnetic disc drives will decrease from 10.0 microinches (0.25 micrometer) to less than 5.0 microinches (0.125 micrometer)[1]. Assuming surface defects on the disk of 2.0 to 3.0 microinches (0.05 to 0.075 micrometer), the critical size to avoid data loss or a head crash is less than 2.0 microinches (0.05 micrometer).

For CMOS/ULSI devices such as a 16 Mbit DRAM, the minimum feature size is reported to be 0.7 micrometers.[2] Applying the "Rule-of-Ten" for allowable defect size due to particulate contamination, the critical particle size is 0.07 micrometers.

One other development with similar requirements is the ongoing DOD/DARPA-sponsored MMIC (millimeter wave and microwave monolithic integrated circuit) program, using GaAs devices with 0.5 micrometer feature size.[3] Using the "Rule-of-Ten" scaling factor, the critical particle size for acceptable yields is 0.05 micrometer.

One further factor to be considered is the allowable defect density, expressed in defects/cm$^2$, to achieve a profitable device yield. For VLS1 devices with 0.7 micrometer feature size and sixteen levels of integration, it has been estimated that allowable defect density cannot exceed 0.38 defects/cm$^2$ ; a tenfold reduction over current Class 100 facility performance[2].

Current technology does not permit direct characterization, monitoring and control of particles less than 0.1 micrometer in size, with most observations limited to particles greater than 0.5 micrometer in size. Whyte and Bailey reported releases between 3.5 x 10$^4$ and 8.2 x 10$^5$ particles/minute (0.5 micrometer and larger) from operators wearing hood, coverall, full-length overboots and latex gloves[4]. The same study also concluded that most particles (95%) were released from the operator through the fabric, based on dispersal chamber tests. However, with appropriate countermeasures, including the use of low-linting undergarments, Shiromaru et al. have shown that releases from the facial region may predominate[5]. These releases, which may include respiratory effluvia [6] and traces of cosmetics,[7] can certainly contain components below 0.1 micrometer in size.

These examples underscore the need for a re-examination of our approach to characterization, monitoring and control of aerosols emanating from the human operator/garment system.

PARTICLE SOURCE CHARACTERIZATION

The human operator is without equal as a potential source of particulate, chemical and biological contamination in the modern cleanroom environment. At rest, an ungowned operator may release 10$^5$ particles ($\geq$0.3 micrometer) to the atmosphere, increasing to more than 10$^6$ particles per minute with moderate activity. Conventionally gowned in a woven polyester coverall, hood, facemask, foot coverings and latex gloves the operator may still release 10$^3$ to 10$^4$ particles per minute, primarily from the upper torso, neck and face.[8] These particles are being released from a volume of less than 100 cubic feet (0.3 m$^3$,) with a corresponding aerosol density of 10 to 100 particles/cubic foot/minute (0.35 to 3.5 particles/liter/minute). This broad spectrum release (Figure 1) extends from the visible range, typified by human hair and skin cells to the molecular range for viable and non-viable organisms (bacteria, viruses and pyrogens).

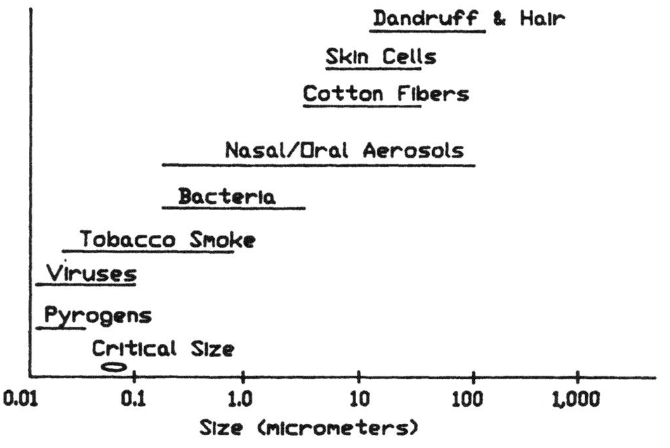


Figure 1. Human Aerosol Spectral Distribution

The cleanroom garments worn by the operator may also contribute to the aerosol burden through the release of occluded particles and fibers. At the 9th International Symposium on Contamination Control in Los Angeles, CA., in September 1988, Fukumoto, reporting on contamination control practices at Mitsubishi Electric Corporation, stated that 46% of the particulate contamination found on 16 kbit DRAMS was associated with dandruff and skin flakes from personnel, while 8% was associated with fibers from clothing.[9] Obviously, in the search for sub-micron excellence we cannot afford to ignore the more prevalent macroscopic particles and aerosols.

Often ignored in the past, the effects of nasal and oral effluvia received a great deal of attention following a flight computer failure five seconds before launch on the Space Shuttle program. An intensive investigation of this failure by Thomas and Calabrese identified the cause of failure as etching of the aluminum metallization by KCl crystals from human oral aerosols.[6] Subsequent investigations by Lowry, et al., at the Analytical Services Laboratory of Harris Semiconductor, Melbourne, Fl., further characterized the chemical composition of nasal and oral aerosols and other body effluvia, including perspiration, fingerprints, dandruff and facial cosmetics.[10] Some of the characteristics of nasal and oral aerosols are summarized in Table I. The size and detectability of human aerosols is shown in Table II.

Table I. Characteristics of Nasal and Oral Aerosols.

| | | |
|---|---|---|
| * | Size Range | - 0.2 to > 50 micrometers |
| * | Nasal Effluvia | - High Sodium & Chlorine<br>Low Potassium |
| * | Oral Effluvia | - High Potassium & Chlorine<br>Low Sodium |

Table II. Tabulation of Human Aerosols by Size and Detectability.

| Particle/Aerosol | Size Range (Micrometers) | Detection | | |
|---|---|---|---|---|
| | | Visual | Optical Microscope | SEM |
| Human Hair or Dandruff | 20 to 200 | X | X | X |
| Skin Cells | 10 to 50 | X | X | X |
| Fibers | 5 to 100 | X | X | X |
| Nasal and oral aerosols | 0.2 to >50 | X | X | X |
| Bacteria | 0.2 to 20 | | X | X |
| Tobacco smoke | 0.02 to 1.0 | | X | X |
| Viruses | 0.01 to 0.1 | | | X |
| Pyrogens | 0.002 to 0.02 | | | X |

More recently, Shiromaru, et al.,[5] have determined that operators wearing full-body woven polyester apparel (consisting of an open-face hood, coverall, calf-length boots and long-sleeve gloves) release almost 90% of particles over 0.17 micrometer from the face and front of the neck when in motion. This problem may be further exacerbated if the operator speaks, sneezes or coughs. The subsequent fate of these releases is dependent on the local aerodynamics, while their impact on product yield is governed by the size and mass of the particles or aerosols and the related attraction mechanisms (inertial, gravitational, diffusive and/or electrostatic).

MONITORING OF PARTICLES FROM GARMENTS AND OPERATORS

In the USA, test methodology to determine the contamination control capabilities of cleanroom garments and accessories has lagged far behind user requirements in most critical areas of the microelectronics industry. Specifically, filtration efficiency and air permeability measurements rely on vacuuming a very small area of new fabric, repeated after 50 processing cycles. Similarly, the most common test method for determining the release of particles from a garment, ASTM F51-68 (reissued in 1984), relies on small area sampling (0.1 $m^2$ or less) with membrane filtration of the airstream and microscopic examination of the membrane to detect and count particles ≥5.0 micrometers in size. More recently, some users and garment processors have started to use the Helmke Drum tumbling test for complete garments, with an automatic (light-scattering) counter to determine the average particle release rate in particles/minute for particles ≥0.3 micrometers.[11] Even more recently, there has been increased interest in the use of a small dispersal test chamber, commonly referred to as a "Body Box," to determine particle release rate and spectrum from the operator/garment subsystem in motion, following a predetermined mobility protocol. It is the feeling of the author that this approach should be refined and improved, through the addition of strategically located, multiplexed probes with aerosol concentrators and automated data acquisition. This would facilitate "real world" assessment of operator/garment system aerosol releases and the location of such releases at the point-of-entry to the cleanroom. Ideally, this monitoring system, if proven to be operable and cost-effective, could be used to routinely check critical operator/garment systems for containment integrity prior to cleanroom entry, possibly as an integral component of the air shower.

Another approach to particle monitoring, although not in real-time, is the technique widely used in the pharmaceutical industry to detect the presence of viable bacteria and spores at the sterile filling stations. In this approach, specially precoated witness plates are strategically located adjacent to the product and used to detect the presence of viable organisms on surfaces (RODAC Plates) or in the airstream (Fall-Out Plates). The use of witness wafers (or magnetic and optical disks) at critical work stations, and other areas where there is an operator/product interface, is an inexpensive technique to develop time profiles and history of operator-generated particle and aerosol interactions with the product.

At least one organization in the USA, VTC Incorporated, has used witness wafers to determine cleanliness levels in Class 1, Class 10 and Class 100 areas of a modern CMOS VLSI/ULSI wafer fabrication facility.[12] Interestingly, at this ultra-modern facility, where worker separation from the wafers is maximized through full implementation of SMIF and robotic intra-station technology, witness plates indicated that the major contributor to wafer contamination is now the robotic tools. However, in the majority of cleanrooms in operation to-day in the USA, personnel potentially remain a major source of particle contamination.

## ALTERNATE STRATEGIES FOR CONTROL OF OPERATOR-GENERATED CONTAMINATION

The key to successful contamination control is complete isolation of the product through all critical phases of fabrication. (Table III) Through the development of the Standard Mechanical Interface (SMIF), intra-station robotic wafer handling and sealed carriers for inter-station transfer, modern semiconductor facilities are approaching this ideal. Automated Guided Vehicles (AGV) replace human operators in at least one facility in Japan for transfer of wafer cassettes between process bays. Unfortunately, total product isolation remains an elusive goal in the USA.

Table III. Control of operator-generated particles in the next decade.

PRODUCT ISOLATION

- SMIF
- Robotics
- Sealed Transfer and Transport
- Changed Operator Roles

OPERATOR CONTAMINANT CONTAINMENT

- Improved barrier fabrics and test methodology
- Improved garment and facial closures
- Multilayer cleanroom apparel
- Limited access and mobility
- Operational contraints and trade-offs

AERODYNAMIC DESIGN

- Improved equipment aerodynamics
- Physical shielding (operator/product)
- Selective air flows (air curtains)
- Mathematical and physical modelling

Where human operators must interface with exposed product or equipment, particulates and effluvial aerosols from the operator must either be contained or deflected away from the product. Particle and external effluvial release from the body may be minimized through the use of lightweight 100% polyester woven fabrics for both undergarments and overgarments. The overgarment uses elasticized neck and wrist closures and a covered, zippered body closure.

In the most critical applications, release of particles and aerosols from the head and face may be totally contained through the use of the "Space Suit" concept, consisting of a bubble helmet with independently-supplied HEPA filtered air. This approach was adopted for the Class 1 cleanroom to produce RHIC's at Sandia Laboratories.[13] The ultimate in control technology available to-day is the combination of PTFE-laminated or similar fabrics for the overgarments (as typified by Gore-Tex™ laminates) and the bubble helmet. Theoretically, complete control of particles and aerosols down to <0.1 micrometer should be attainable with this technology. Where cost or ergonomic factors preclude the use of PTFE-laminates and bubble helmets, the correct design and selection of face-shield and face mask, with double-gowning and the use of the newer ultrafine filtration fabrics may be an acceptable alternative. To further minimize the impact of operator-generated particles on product yield, aerodynamics and other transport mechanisms need to be better characterized through mathematical and physical modelling and appropriate modifications made at critical locations. The need for realtime monitoring to validate the results of these analyses and hardware changes is very apparent.

## SUMMARY

The human operator is without equal as a potential source of particulate contamination in the cleanroom environment. Beyond the release of hair, dandruff and skin cells, the aerosol burden may be further increased by perspiration and facial cosmetics in trace amounts, and will certainly be increased by nasal and oral effluvia. The particle size spectrum of these releases extends from the molecular range for viable and non-viable organisms (pyrogens, viruses and bacteria) to the macrorange, typified by human hair and skin flakes. While near-total containment of human debris is now possible in the most critical applications, using PTFE-laminated fabrics and a bubble helmet with an independent HEPA filtered air supply, this approach to particulate control is not yet widely used in most cleanrooms in the USA due to cost and ergonomic considerations. Lacking near-total containment, there exists an urgent need to develop and implement improved methodology for the monitoring and control of particle and aerosol release and transportation from the operator/garment system prior to entry and within existing cleanroom facilities. On-going development in both woven and laminated fabrics, body closures for garments and improved facial closures can be expected to reduce the impact of operator-generated contamination on product yield. However, complete separation between product and operator remains the ultimate goal. The role of the operator will shift from direct to indirect interfacing with the product, with increasingly stringent control imposed on the operator/equipment interface for system installation and maintenance. Where human operators cannot be replaced by automation because of the need for superior adaptive learning and response capabilities, the effects of their interaction with the product must be minimized through the use of improved barrier materials, aerodynamic design at the interface and restrictions on the numbers and activities of operators.

REFERENCES

1. R.W. Martin, Defect density management -- IBM experience, 9th International Symposium on Contamination Control, Los Angeles, CA. p.1 (September 1988).

2. J. Burnett, World class contamination control practices, Semiconductor International, p.160 (April 1988).

3. Editorial in Research & Development, pp.26 - 31. (June 1989).

4. W. Whyte and P.V. Bailey : Particle dispersion in relation to clothing, J. Environmental Sci., p.43 (March/April 1989).

5. N. Shiromaru, K. Shimoda, S. Shibuya, T. Saiki, M. Morisaki, T. Yoneda, T.Takenami and T. Ohmi, Measurement of the number of particles on cleanroom garments, 9th International Symposium on Contamination Control, Los Angeles, CA. p.424 (September 1988).

6. R.W. Thomas and D.W. Calabrese, The identification and elimination of human contamination in the manufacture of IC's, 23rd International Reliability Physics Symposium, Orlando, FL. p.228 (March 26-28, 1985).

7. Q.T. Phillips, W.D. Auser, J.M.Baldwin and G.J. Washington, Cosmetics in cleanrooms, 29th Annual Technical Meeting of the Institute of Environmental Sciences, Los Angeles, CA. p.317 (April 1983).

8. T. Hirakawa and M. Tsumoto, Influence of fabric construction on the performance of cleanroom garments, 7th International Symposium on Contamination Control, Paris, France (September 1984).

9. T. Fukumoto, Total contamination control management in semiconductor processing, 9th International Symposium on Contamination Control, Los Angeles, CA, p.12 (September 1988).

10. R.K. Lowry, J.H. Linn, G.M. Grove, and C.A. Vicroy, Analysis of human contaminants pinpoint sources of IC defects, Semiconductor International, p.73 (July 1987).

11. G.E. Helmke, A tumble test for determining the level of detachable particles associated with clean room garments and clean room wipers, 28th Annual Technical Meeting of the Institute of Environmental Sciences (1982).

12. W.L. Workman, Contamination control for a submicron CMOS VLSI/ULSI factory, 9th International Symposium on Contamination Control, Los Angeles, CA, p.104 (September 1988).

13. D.W. Miller, The architectural ergonomics of Sandia's RHIC-II gowning facility, 9th International Symposium on Contamination Control, Los Angeles, CA. p.98 (September 1988).

Gore-Tex is a registered Trademark of W.L Gore & Associates.

# BUILDING AND USING AN APPLICATION-SPECIFIC PARTICLE ATLAS

Randy Greb and Michael Mckeag

Intel Corp.
5200 NE Elam Young Parkway
Hillsboro, OR 97123

Application-specific particle atlases can be assembled and successfully applied to solve manufacturing contamination problems. The Intel Particle Atlas is used at Intel as a general materials reference for identification and elimination of contaminant particles in liquids and gases and on semiconductor wafer surfaces.

The Atlas is briefly described. It contains the optical and SEM/EDX (Scanning electron microscope/energy dispersive x-ray) properties of over 100 materials collected from Intel integrated circuit manufacturing sites.

We will describe methods of assembling and using the atlas which may be applied to the production and application of other atlases. A brief description of the analytical methods used for our atlas is presented, and the importance of the sample collection and documentation steps is discussed. Applications of the atlas to specific contamination problems, mistakes to avoid, and the benefits of assembling an atlas are also included.

## INTRODUCTION

Particulate contamination limits yields in integrated circuit manufacturing. As circuits and IC manufacturing processes become more complex, the sensitivity to particulate defects increases and more time and money must be expended in identifying and eliminating contamination sources. For example [1], to achieve a 78% yield in a typical sub-micron process containing 250 steps, each step must contribute no more than 0.001 killer defects/$cm^2$. A particle atlas can be a useful tool in the process of particle identification and source elimination.

The McCrone Particle Atlas [2] is a valuable reference which describes particle analysis techniques and properties of a wide variety of particle types. To expedite interpretation of analytical results, an additional atlas was produced internally, which concentrates on materials and analytical techniques which are more specific to our environment.

*Particles in Gases and Liquids 2,* Edited by K.L. Mittal
Plenum Press, New York, 1990

We believe atlases of this type can be used for applications in pollution control, forensics, and other industries which require control of cleanroom manufacturing areas. Using the atlas provided several valuable lessons about assembling the atlas. We will describe some of those in this paper.

## DESCRIPTION OF THE INTEL PARTICLE ATLAS

### Format

The atlas includes three main sections:

Introduction. The introductory section describes the purpose and scope of the atlas, and then provides detailed descriptions of all terms and analysis techniques used in the atlas. Several unique terms were used for the particle descriptions, and these are defined in this section.

Indexes. The properties of each of the atlas entries are cross-referenced in a variety of indexes. The materials are indexed by reflected light properties, transmitted light properties, optical fluorescence, source, general materials type, elemental components (EDX), and descriptive name.

Main Body. Two pages are used for each material included in the atlas. One page is used to describe the optical properties, and the other is used for SEM/EDX properties. Notes describing the source of the material and potential for contamination are included. Figures 1 through 4 show two example entries (4 pages total) taken from the Intel Particle Atlas.

## ANALYTICAL INFORMATION INCLUDED

The types of analytical information included in the atlas were chosen because of their ease of acquisition, the wide availability of appropriate instruments within the company, and their ability to uniquely identify materials.

All optical properties were collected with a Zeiss UEM microscope in both transmitted polarized light and reflected light modes. Blue-violet excitation was used to determine fluorescence properties.

The reflected light properties include shape, surface texture, luster, color, and fluorescence. The transmitted light properties include color, refractive index (Becke line method), birefringence, and extinction. Descriptive terms for each of the properties are carefully defined in the introductory section of the atlas. The McCrone classification code [2] for each material is also included. A representative color, optical micrograph is included for each entry.

The SEM/EDX properties include a representative SEI (secondary electron image) micrograph and an EDX spectrum for each entry. The EDX spectrum is a compromise which provides the best overall representation of the elements in the sample. All spectra were collected under the same sample conditions (working distance, tilt, live time acquisition) at 20kV accelerating voltage. The spectra are qualitative representations of what elements may be present in the collected material. No attempt at quantitative analysis should be made with these spectra as references.

White Viton O-ring

OPTICAL DESCRIPTION

| | |
|---|---|
| CLASSIFICATION CODE: | 110000 |
| SHAPE: | Irregular fragments with rough edges. |
| REFLECTED COLOR: | Silver-white. |
| LUSTER: | Speckly. |
| SURFACE TEXTURE: | Rough, flaky, and pitted. |
| HOMOGENEITY: | Consistent throughout |
| TRANSMITTED COLOR: | Black (opaque) |
| FLUORESCENCE: | Medium yellow-green. |

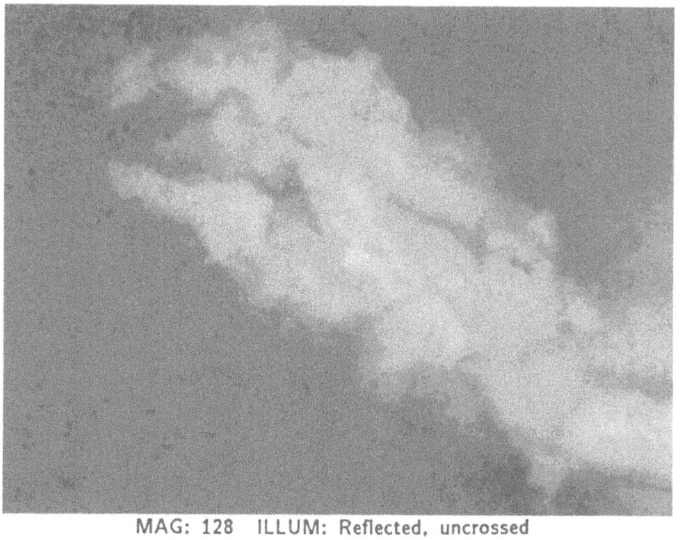

MAG: 128 ILLUM: Reflected, uncrossed

Notes: The rough surface and fluorescence are distinctive.

White Viton is used in a variety of O-ring and gasket applications.

**Figure 1. Sample page from the Intel Particle Atlas.**

White Viton O-ring

## SEM/EDX DESCRIPTION

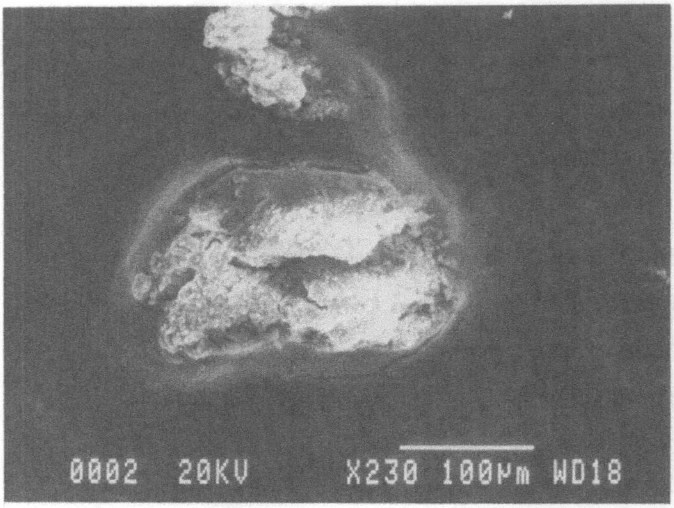


SEM IMAGE

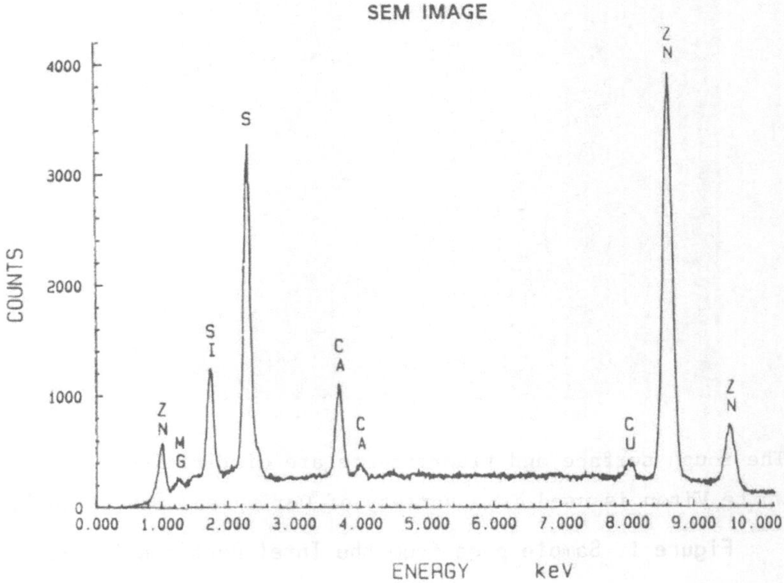


EDX SPECTRUM

**Figure 2. Sample page from the Intel Particle Atlas.**

Si Wafer Chips

## OPTICAL DESCRIPTION

| | |
|---|---|
| CLASSIFICATION CODE: | 110100 |
| SHAPE: | Grains and fragments with both sharp, smooth edges as well as rough edges. |
| REFLECTED COLOR: | Silver. |
| LUSTER: | Shiny, metallic. |
| SURFACE TEXTURE: | Mostly fractured with some smooth surfaces. Conchoidal fracture is obvious. |
| HOMOGENEITY: | Continuous. |
| TRANSMITTED COLOR: | Black (Opaque) |
| FLUORESCENCE: | None. |

MAG: 800 ILLUM: reflected, uncrossed

NOTES: The conchoidal fracture is the most distinguishing property of Si chips. They showed this type of fracture more than any other substance listed in this atlas (8/87).

This sample was prepared by grinding the edges of two broken Si wafers against each other.

Figure 3. Sample page from the Intel Particle Atlas.

SEM/EDX DESCRIPTION

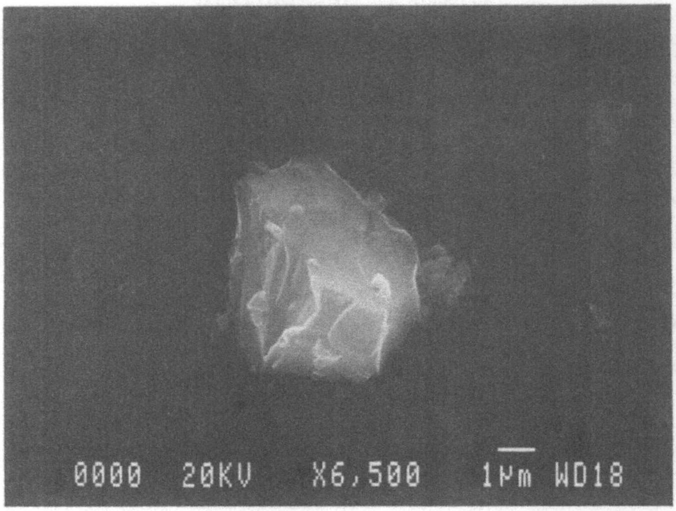


SEM IMAGE

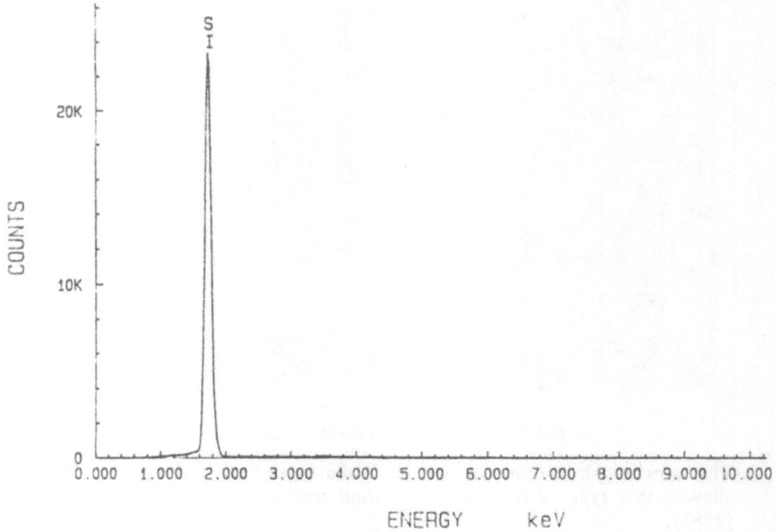


EDX SPECTRUM

**Figure 4. Sample page from the Intel Particle Atlas.**

MATERIALS INCLUDED

The materials in the atlas were picked to provide a representation of materials which may contaminate the process wafer environment. Samples were taken from the cleanroom, process equipment, DI water system, and spare parts supply. Emphasis was placed on materials which have the highest yield impact such as process equipment parts.

Samples of human-generated particles such as skin flakes, spittle, and hair were not included. These are already sufficently documented [3-5] and are not significant yield impactors given current stringent cleanroom practices.

The atlas collection is being expanded to include assembly area materials and descriptions of some materials that have been modified by thermal processing.

APPLICATIONS OF THE ATLAS

Short Term Benefits

The Intel Particle Atlas is a valuable tool that is used extensively for identification of contaminant particles. It is used to both positively identify particle sources and thus eliminate them, and to rule out suspected sources which may be incorrect.

In one case, white particles were accumulating at a cleanroom work station. The contamination was intermittent, and airborne particle counters were not sufficient for tracking down the source. Using the particle atlas as a reference, the white particles were quickly and easily identified as fiberglass. Based on this analysis, a faulty laminar flow hood installation was located and fixed. The white particles went away.

In another case, particle levels on ion implanter monitor wafers were trending high. The suspected source was the platen material. Analysis of the particles on a sample wafer indicated the presence of Si chips and stainless steel (again using the particle atlas as a quick reference). The platen material was in the particle atlas, and none of the contaminant particles had similar properties. Based on this analysis, the platen material was not changed, and a special cleaning of the implanter was completed. The particle levels were reduced and valuable resources were saved.

Long Term Benefits

Several long term benefits have resulted from internal publication of the atlas:

(1) The atlas is now also used as a general materials reference. It provides a useful reference of materials properties which many process engineers would not have easy access to otherwise.

(2) The atlas provides a standard for describing particles.

(3) The use of the atlas encourages process engineers and technicians to track down the sources of particles. If a match from the atlas is not found, then additional samples can be collected and characterized using the atlas descriptions as the standard.

Table 1. Initial method of indexing elemental compositions.

A2 ASSEMBLY AREA MATERIALS ELEMENTAL COMPOSITION - EDX

| PAGE NUMBER | F | Na | Mg | Al | Si | S | Cl | K | Ca | Ti | Cr | Fe | Zn | Ag | Pb |
|---|---|---|---|---|---|---|---|---|---|---|---|---|---|---|---|
| 32 | F | - | - | - | - | - | - | - | - | - | - | - | - | - | - |
| 12 | F | - | Mg | - | - | - | - | - | Ca | - | - | - | - | - | - |
| 6 | - | Na | Mg | | Si | S | - | - | Ca | - | - | - | - | - | - |
| 27 | - | Na | Mg | Al | Si | - | Cl | - | Ca | Ti | - | Fe | Zn | - | - |
| 2 | - | Na | Mg | Al | Si | S | Cl | K | - | Ti | - | Fe | - | - | - |
| 31 | - | Na | - | - | Si | S | - | - | Ca | Ti | - | - | Zn | - | - |
| 1 | - | Na | - | - | - | S | - | - | - | Ti | - | - | Zn | - | - |
| 13 | - | Na | - | Al | Si | S | Cl | - | - | - | - | - | - | - | - |
| 18 | - | - | - | Al | - | - | - | - | - | - | - | - | - | - | - |
| 33 | - | - | - | Al | Si | S | Cl | - | Ca | - | - | - | - | - | - |

Table 2. New Method of indexing elemental compositions.

INDEX 7

ELEMENTAL COMPOSITIONS SORTED BY PRIMARY ELEMENT 1

| PELEMENT 1 | PELEMENT 2 | SELEMENT 1 | SELEMENT 2 | TELEMENT 1 | TELEMENT 2 | TELEMENT 3 | NAME | PAGE |
|---|---|---|---|---|---|---|---|---|
| AL | CA | CU | FE | S | . | . | POUR ROCK | Ø05 |
| AL | SI | F | . | S | . | . | MILLIPORE FILTER MEMBRANE | Ø43 |
| AL | SI | TI | . | S | . | . | P/E SPUTTER CHAMBER PREC. | Ø51 |
| AL | TI | SI | . | . | . | . | SWAM DUST | ØØ9 |
| AL | TI | . | . | SI | . | . | MEISSNER TRAP PRECIPITATE | Ø61 |
| AL | . | TI | CA | SI | K | CL | GREEN SCOUR PADS | ØØ3 |
| AL | . | . | . | . | FE | . | IMPLANTER PLATEN | Ø87 |
| AL | . | . | . | . | . | . | ASM CERAMIC SLED PARTS | Ø57 |
| CA | . | SI | TI | AL | FE | CL | RUN CARD MATERIAL | Ø45 |
| CA | . | SI | . | . | . | S | THERMAL PRINT PAPER | Ø71 |
| CA | . | . | . | K | SI | AL | RED GASKET DIH20 | Ø97 |
| CL | AL | . | . | . | . | . | PLASMA-THERM POLYMER | 14Ø |
| CL | CA | SI | . | . | . | . | ANTI-FATIGUE MAT | ØØ7 |
| CL | F | SI | AL | . | . | S | LRC CHAMBER POLYMER | 137 |
| CL | S | CA | ZN | SI | . | K | ORANGE ACID GLOVE | Ø49 |
| CL | S | SI | F | AL | K | . | TEFLON SWAGELOCK CONNECTOR | Ø39 |
| CL | S | SI | . | . | . | . | CUNO 5u FILTER | 148 |
| CL | SI | MG | CA | . | . | . | LRC TOP PLATE POLYMER | 135 |
| CL | . | CA | . | . | . | . | PVC PIPING | 157 |
| CL | . | SI | . | AL | S | . | ANION E.CH. RESIN BEADS | Ø55 |
| CL | . | SI | . | . | . | . | NITRIDE PIGTAIL PREC. | 113 |
| CL | . | SI | . | . | . | . | FAB GLOVE | Ø75 |
| CL | . | . | . | CA | DI | CL | DIH2Ø TUBING | Ø99 |
| CU | ZN | PB | SI | . | . | . | BRASS SWAGELOCK | Ø41 |
| F | SI | . | . | . | . | . | PVDF TUBING | Ø53 |

Note: Pelement, selement, and telement represent primary element, secondary element, and trace element, respectively.

NOTES ON ASSEMBLING AN ATLAS

(1) Complete documentation of the sources of the atlas materials is critical. Documentation should include the physical source, vendor name, part number, date collected, and any unique information related to the application of the material. If the material is a process byproduct, then a complete description of the process and machine that the material came from should be included.

For example, in one case, plasma etch polymer was a suspected cause of a defect. The properties of the defect were compared with properties of etch polymer collected from the same machine type and cataloged in the atlas. A conclusion based on the atlas data could not be reached however, because the atlas documentation on the etch process (gases used, periodic maintenance details, process specifics) was not complete.

(2) Indexes for elemental compositions must emphasize the qualitative nature of the information. The indexing method used for the McCrone particle atlas [2] is an excellent model to follow. Our indexes were initially ordered alphabetically and by EDX spectrum peak height (see table 1). The indexes now simply indicate what elements were present in a material (see table 2). Work is in progress to determine an accurate method of defining "key indicator" elements and elements which "may be present".

(3) All terms used for descriptions should be carefully defined somewhere in the atlas. The differences between terms must be clear and distinct so that different users will describe unknowns as consistently as possible.

(4) Even when terms are clearly defined, variations in different descriptions of the same material may occur. These come from sample to sample variations, variations in the way users describe the same sample, and variations in equipment setup. Therefore, if there is more than one possible description for a material, include them all in the indexes.

(5) Selection of samples should depend on the application. We collected samples of materials that contacted the wafer environment (air, water, chemical, or vacuum chamber environments). Always collect more samples and a larger variety of samples than appears necessary. In several cases, atlas materials from one environment have identified an unknown in a completely different environment.

(6) The availability of good software is critical when assembling an atlas. A quality word processing program is needed to assemble the atlas material into a professional form. Data manipulation software which can easily handle text variables is helpful for sorting descriptive information and assembling it into indexes.

(7) In some cases, color photomicrographs are a useful addition to a particle description. Color copiers which produce high quality copies are now available and are extremely useful when a wide distribution of the atlas is required.

(8) The materials in the Intel Particle Atlas are indexed in variety of ways, as discussed earlier. One index that was not included in the initial revision was a general materials index. Such an index would be sorted by general materials class such as plastics, metal alloys, paper types, process byproducts, and any other general classification that would apply. These indexes are being included in the next revision of the atlas since it has become a widely used general materials reference of optical and SEM/EDX properties.

(9) Use of a "thin window" EDX detector to collect x-ray spectra can provide additional elemental information (low-z elements). Care must be taken to accurately describe analysis conditions (accel. voltage, working distances, beam current, detector description, and any other appropriate sample conditions), or misinterpretation of the spectrum may result.

## NOTES ON USING AN ATLAS

(1) We have been most successful in using the atlas when we rely on the indexes for the initial determination of potential matches, and then use the detailed descriptions to make the final determination of the source of the unknown. Merely browsing through the atlas pages is not efficient.

(2) Thorough training of any potential atlas user is always beneficial. An atlas can be used most effectively if the technical content is fully understood by the user.

(3) The person who assembles the atlas should also be the primary user and caretaker of the atlas. That person knows the atlas best and will continue to serve as a key resource.

(4) Save the samples collected for the atlas. The need may arise for additional analysis of the reference materials in the atlas.

## CONCLUSIONS

A particle atlas built for a specific application can be an extremely useful reference. Carefully collecting and documenting meaningful atlas materials is the critical step in assembling an atlas. An application-specific atlas can provide benefits both as a resource for identification of unknown particulate and as an educational source.

## REFERENCES


1. C.M. Osburn, R.P. Donovan, H. Berger, and G.W. Jones, J. Environmental Sci., 45-57, (March/April 1988).

2. W.C. McCrone, editor, "The Particle Atlas", Ann Arbor Science Publishers, Ann Arbor, MI, 1973.

3. Q.T. Phillips, W.D. Auser, J.M. Baldwin, and G.J. Washington, in "Proceedings of the 7th International Symposium on Contamination Control" held in Paris, 1984.

4. R.W. Thomas, D.W. Calabrese, in "Proceedings of the IEEE Reliability Physics Meeting" held in Orlando, FL, 1985.

5. R.K. Lowry, J.H. Linn, G.M. Grove, and C.A. Vicroy, Semiconductor International, 73-77, (July 1987).

# PARTICULATE CLEANLINESS TESTING OF FILTERS AND EQUIPMENT IN PROCESS FLUIDS (GAS OR LIQUID)

Susan Goldsmith, George Grundelman, Mark Smith
IBR,Inc.
727 W. Ellsworth Rd. Bldg #1
Ann Arbor, MI 48108

In many industries there is a strong correlation between particulate contamination and product yield or performance. For this reason particulate cleanliness tests have been performed by a wide variety of methods. The field of particulate cleanliness testing has been progressing very rapidly due to improved methodology,instrumention, and data analysis. This paper describes the selection of accurate test methods, apparatus, instrumentation, and data analysis.

## INTRODUCTION

In recent years, various standards groups, including the American Society for Testing and Materials (ASTM), Semiconductor Equipment and Materials International (SEMI),and the Association for the Advancement of Medical Instrumentation (AAMI) have attempted to standardize at least portions of cleanliness testing[1-4]. Table I shows a representative list of products which are commonly tested for particulate cleanliness.

Table I. Commonly Tested Products

<u>Active:</u>
- Pumps
- Flow Controllers
- Valves
- Pressure Regulators
- Heat Exchangers
- Clean Room Wipes and Swabs

<u>Passive:</u>
- Tubing
- Fittings
- Clean Room Chairs and Garments
- Reservoirs and tanks

Meaningful cleanliness testing requires rigorous attention in four major areas:

- Test Methods which are meaningful, reproducible, and accurate.
- Test Apparati which allow low particulate backgrounds. This includes not only the test system itself, but also the particle sampling system.
- Particle counting/monitoring instrumentation which is capable of detecting very low particulate levels at very small particle sizes.
- Analysis of data which is founded on an understanding of low populations and erratic arrival rates.

We will discuss each of these areas.

TEST METHODS OVERVIEW

Particulate cleanliness testing evaluates both particle generation and particle entrapment and release. By selecting the test method carefully, a user can differentiate between particle generation and particle entrapment and release.

The purpose of cleanliness tests is to evaluate relative levels of particulate generation/release from the test samples, as a function of time or volume. The methods basically consist of flowing ultraclean fluid through each test sample, and continuously monitoring effluent particle counts. During this test, process conditions can be simulated.

Particle generation testing is done on clean components, both in the as-manufactured condition, and after extensive use in a clean environment. Initial particulate levels can be from sources such as manufacturing debris and initial seal seating. Particulate levels can also rise during vibration, thermal or mechanical shock, chemical exposure, and elevated flow or pressure. Particles generated after extended use can be from filter saturation, part fatigue, and erosion of surfaces and seals. For extended use studies, test samples are actuated or used off the cleanliness stand, and are then re-installed and re-tested for cleanliness. A typical sequence, for a valve, is shown is Figure 1. The test control is done by removing a test sample from the stand, not actuating it, and then re-installing and re-testing it.

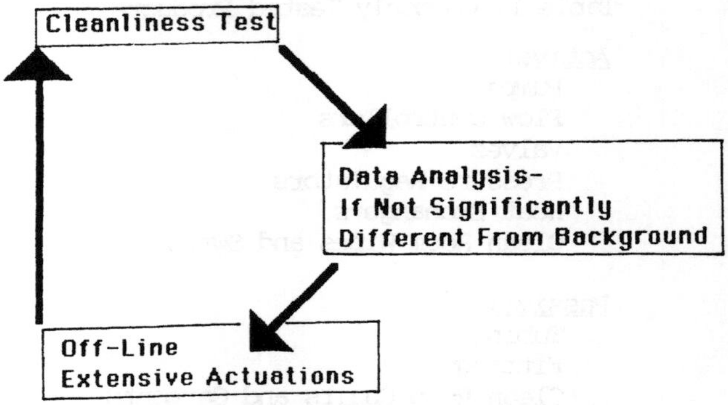


Figure 1. Testing sequence for valves

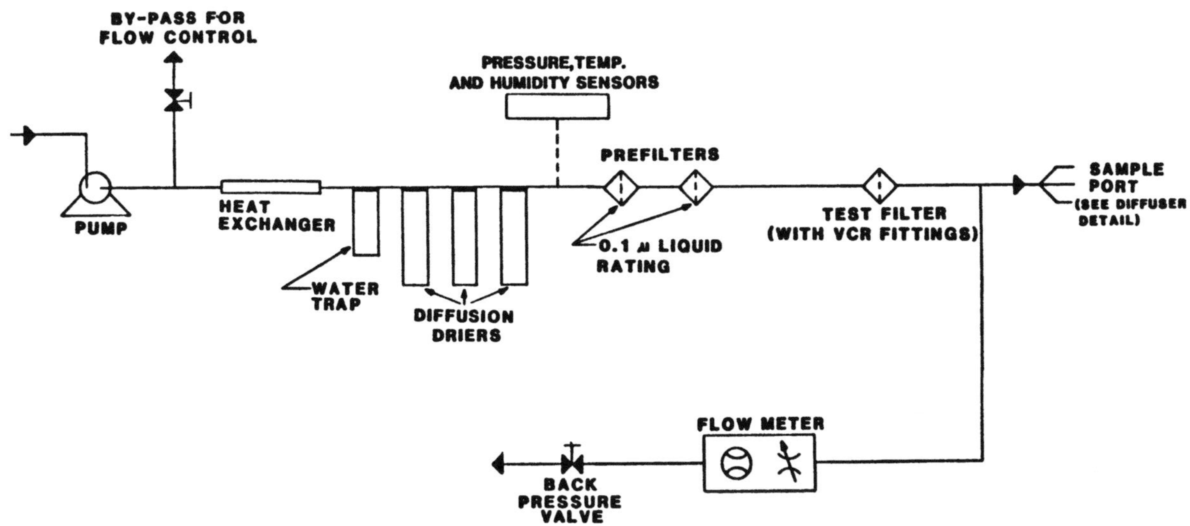


Figure 2. Test apparatus for gas-line filters

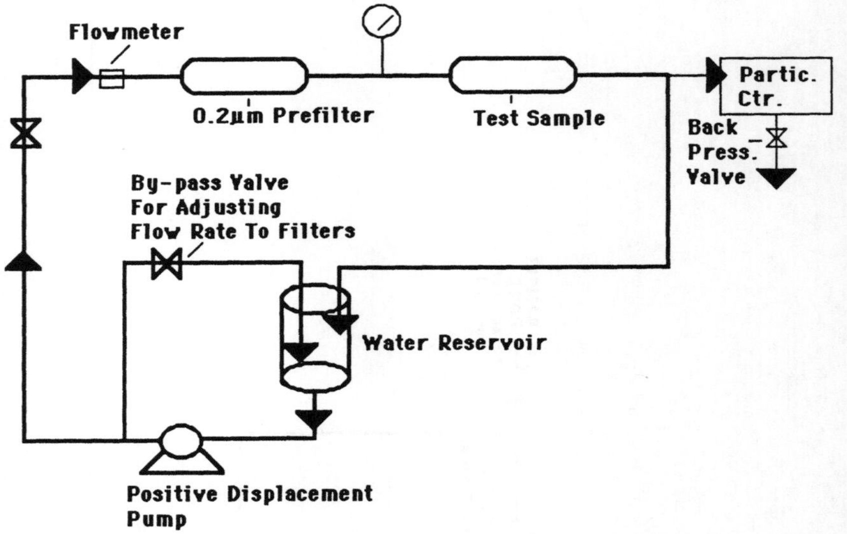


Figure 3. Test apparatus for liquid filters

Particle entrapment and release testing is done on components which have been deliberately exposed to particulate contaminate. Particulate entrapment/release occurs when filters or components contain "dead legs" or rough surfaces. Particulte cleanliness is measured after parts have been deliberately contaminated with particles.

TEST APPARATUS GUIDELINES

Figures 2-4 show typical testing set-ups for various components. Note that the common denominators for these apparati include ultraclean feed fluid and sampling for the particle counters.

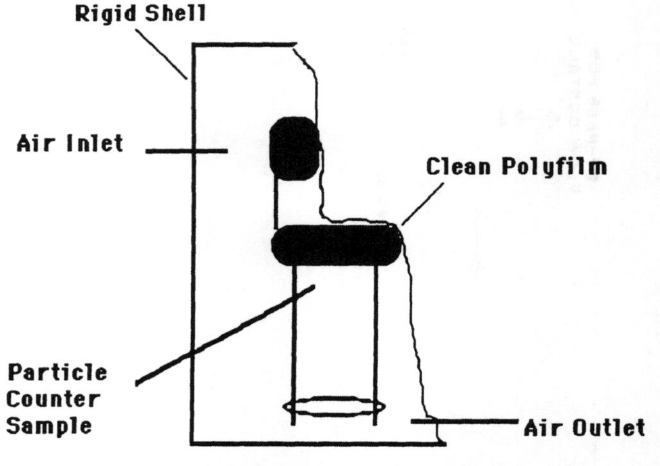


Figure 4. Test apparatus for cleanroom chairs

The heart of any cleanliness test apparatus is an accurate sampling system for the particle counter[5]. A good system must:

- Draw a representative sample, by not only preserving particle size distribution but also by minimizing particle loss due to impaction or diffusion. Figure 5 shows various sample port orientations.

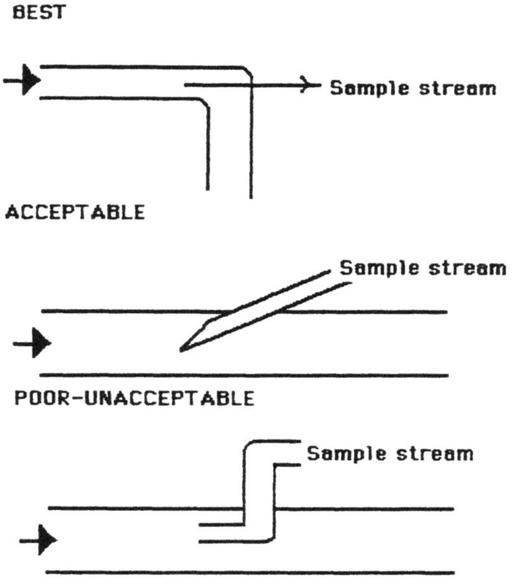


Figure 5. Sample port designs

Test systems must provide very low particle levels. This can be achieved by using the guidelines required for any ultrapure piping system:

- Stainless steel or PFA (perfluoroalkoxy) construction
- High purity, orbital welding for stainless tubing.
- Smooth interior fusion welding for PFA systems.
- When necessary, fittings should be gland seals with polymeric seals.
- No dead legs for particle entrapment/release.
- Continuous flowing systems, even when not in use.
- Redundant filtration - typically two 0.1 micrometer liquid rated filters in series before the test sample.
- Rigorous cleaning on a routine basis
- Used exclusively for cleanliness testing. All particle entrapment, filter saturation conditioning should be done on another test apparatus, to prevent contamination.

- Meet all requirements for ultrapure systems - including no dead legs and very smooth surfaces and seals.

- Minimize contaminant introduced by sampling. Ideally, this means there are no fittings, valves, or bottles between the test sample and the counting sensor. When this is not possible, designs should minimize these.

- Remove pressure and/or flow fluctuations. A smooth, steady flow rate is required by all standard particle counters.

- Reduce sample pressure to particle counter's pressure limit.

For liquid sampling, the specific areas of concern are:

- Avoid introduction of bubbles, either by vacuum draw or degassing in the sensor.

- Removal of entrained air or gas. This air may have been introduced by the pump or may be clinging/releasing from filter and tubing surfaces.

Probably the most difficult sampling requirement is downstream of the pumps used in viscous, concentrated acids. On-line sampling is still possible, but needs to be carefully done. Figure 6 shows a good system.

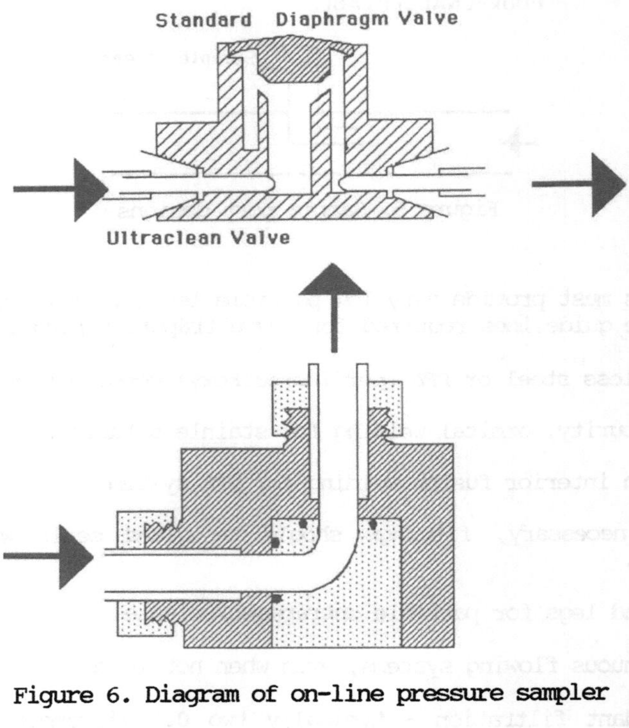


Figure 6. Diagram of on-line pressure sampler

The sampling sequence for the system shown in Figure 6 is as follows:

- Liquid flows into the first reservoir, through a machined sampling valve. This cannot be a standard solenoid valve. Standard valves contain "dead legs" which can act as particle entrapment/release points - see Figure 7.

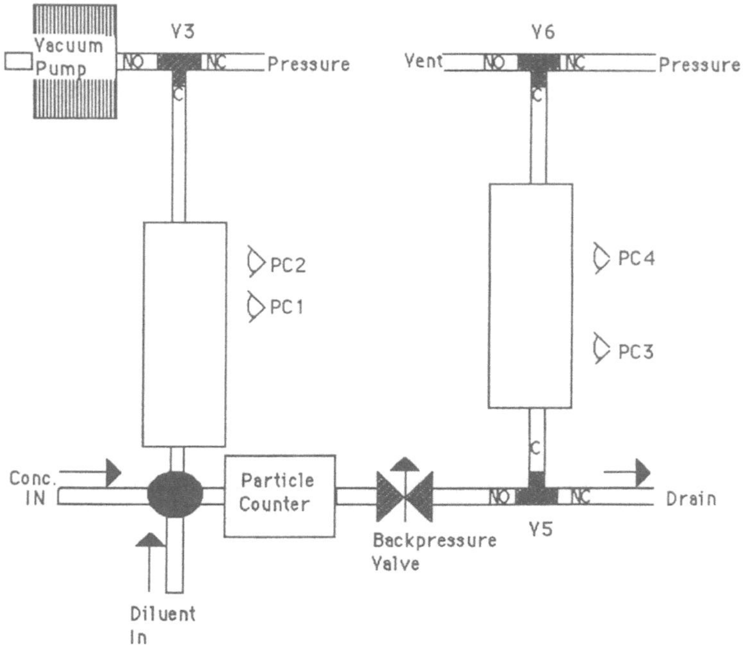


Figure 7. Comparison of clean valve to standard diaphragm valve

- In the first reservoir, the entrained air is removed by evacuation.
- Sample is smoothly pushed by a pressure which prevents degassing in the sensor, through the particle counter. Backpressure is regulated to prevent degassing.

Careful determination of sampling system background counts is critical for meaningful data. A good negative control can be performed by installing redundant 0.1micrometer filters on the sampling line. A good positive control can be performed by injecting a known quantity of latex spheres into the sample port.

## PARTICLE COUNTING INSTRUMENTATION

Instrumentation has improved greatly in both performance and in manufacturer supplied technical information. Any instrument used in cleanliness testing must be fully evaluated for:

- Counting efficiency
- Sizing accuracy
- Effect of varying fluid and particle indices of refraction
- Electrical noise
- Drift in calibration and noise level

Counting efficiency of a particle counter in any actual application is determined by a variety of parameters. Instrumentation size sensitivity, signal-to-noise ratio at the minimum size, sensor condition,and sensor design must all be understood. These include particle size, particle reflectivity (often metallics are not visible to laser scattering sensors),liquid index of refraction, contaminant index of refraction, laser wavelength,sensor condition, and sensor design.

In gases, optical particle counter (OPC) efficiencies can be easily checked by comparsion to condensation nuclei counter (CNC) data for the same sample stream. CNC counters can detect much smaller particles than optical counters. Therefore, it can be assumed that a CNC counter will have a counting efficiency close to 100% at the opticle counter's lowest detectable particle size. In gas applications, index of refraction differences are very small. For this reason, calibration in air or nitrogen will correlate very well to other gases. Care must be taken to verify metallic particle counting efficiency of OPC 's.

In liquids, counter efficiencies can be checked by comparison to various microscopic (either optical or SEM) methods[3]. Typically, these validations are done with latex sphere suspensions in pure water. Limited testing has been reported in other liquids and particles.

Sizing calibrations of liquid sensors is greatly affected by the relative index of refraction of liquid and particles[6]. Increasing index of refraction of the liquid decreases the amount of light scattered by a particle of known size, which makes it appear smaller. For example, a 0.5micrometer particle in HF (index = 1.4) will appear to be only 0.4micrometer to a counter which was calibrated in water (index = 1.33). Extensive mathematical correlations are available [3,5].

The effect of index of refraction on apparent counts can be minimized by:

- Accurate dilution with clean, low index of refraction solvent. For example, dilution of sulfuric acid (n=1.43) with water can be done. The relatively high particle counts of the sulfuric acid allow the diluted counts to still be statistically meaningful.

- Operating in a narrow particle size range, and determining the correction factor for the instrument experimentally. Correction factors are determined by comparison to microscopic sizing.

## STATISTICAL ANALYSIS

Particulate cleanliness data require careful application of statistical analysis [7-9]. The very low count rate, coupled with sporadic arrival rates from particle bursts, requires unique analysis. In general, Poisson distribution statistics is used. Data can be represented visually by graphing, or as summary analysis (mean and variance values). Careful selection of time intervals is required to balance the need for low variance, which requires large sample volumes, and the desire to detect spikes, which occur in very small volume increments.

## CONCLUSIONS

Meaningful cleanliness testing requires paying rigorous attention to test methods, apparatus, instrumentation, sampling, and data analysis. Evaluations of products for particulate cleanliness must be reported with complete descriptions of methods and instrumentation. Levels of instrumentation error and test system contamination must be reported and understood.

REFERENCES


1) ASTM F01.10.06E "Standard Practice for Evaluating Relative Levels of Particle Shedding from Filter Cartridges as a Function of Flush Volume, Using Electronics Grade Water".

2) ASTM F01.10.17 " Standard Practice for Evaluating Relative Levels of Particle Shedding from Gas-Line Filters as a Function of Time and Mechanical Shock.

3) SEMI Standard CI-88, "Specifications for Reagents,Sec. 3.9 Calibration and measurement Method for Particles in Liquids"

4) AAMI "American National Standard for Blood Transfusion Microfilters", section 4.2.3.1.

5) D.W. Cooper, Monitoring contaminant particles in gases and liquids: A Review, in "Particles in Gases and Liquids 1: Detection, Characterization, and Control" K.L. Mittal, editor, pp. 1-33, Plenum Press, New York, 1989

6) R.G. Knollenberg, The importance of media refractive index in estimating liquid and surface microcontamination measurements, J. Environmental Sci, 30, No. 2, 50-58 (March/April 1987)

7) R. A. Van Slooten, Statistical treatment of particle counts in clean gases, Microcontamination, 33-48 (Feb. 1986)

8) Thomas J. Bzik, Statistical management and analysis of particle count data in ultraclean enviroments: Part I, Microcontamination, 59-63,98-99(May 1986)

9) Thomas J. Bzik, Statistical management and analysis of particle count data in ultraclean environments: Part II, Microcontamination,35-41( June 1986)

# OPTICAL PARTICLE COUNTER PERFORMANCE DEFINITIONS EFFECTS ON SUBMICROMETER PARTICLE MEASUREMENT

A. Lieberman

Particle Measuring Systems, Inc.
46729 S. Fremont Blvd.
Fremont, CA 95438

Measurement of submicrometer particles, particularly in microelectronic process fluids, requires detection of sparse numbers of particles at the maximum sensitivity of a measurement device. Frequently, these measurements must be carried out in aggressive liquids. Under these conditions, the counting and sizing accuracy of the measurement device is strongly affected by some device performance parameters which are sometimes poorly defined. These include the particle properties observed by the device, its particle sizing resolution, counting accuracy at the point of maximum sensitivity, and its sizing accuracy. The effects of changes in these parameters on reported particle size and concentration data are discussed, along with their effects on inter- and intra-device correlations. Emphasis is placed on operation of optical particle counting systems.

## INTRODUCTION

Acceptable production of modern semiconductor devices requires that exposure to particulate contamination be minimized throughout the process. This includes a need for assurance that both the processing environment and any processing fluids used in the fabrication process be as clean as possible. Particle specifications at Class 1 or Class 10 for 0.1 or 0.2 µm are now being used for some process gases and are expected for many cleanroom air levels in the near future. Optical particle counters (OPCs) are used widely to monitor essentially all of the fluids used in semiconductor cleanrooms. These fluids include the ambient atmosphere where the process tools operate and where the assembly work is carried out, all the process liquids, compressed gases and low pressure gases, and vapor emissions from the process tools, the work enclosures and the workers. For all of these fluids, a primary necessity is knowledge of the particle content and of changes that occur in that content. For this reason, the OPC data should be as accurate as possible. Minor differences in OPC operation can result in significant differences in the data produced by that OPC.

*Particles in Gases and Liquids 2*, Edited by K.L. Mittal
Plenum Press, New York, 1990

The operating specifications for OPC performance are stated in terminologies which may not clearly define some of the phenomena that can affect OPC performance. Neither OPC design nor theory will be discussed here. There have been a number of publications[2], [3], [4] that define OPC operating principles clearly. This discussion is aimed at providing a more clear understanding of the meaning of some of the performance specification limits which are frequently used when OPC selections are made. It is hoped that this paper will aid the OPC purchaser and user in a better understanding of how the OPC performance descriptions affect the data which are produced by that OPC. In this way, the OPC user can obtain the most use from that data. The quality of the OPC data must be considered, as well as its quantity when the user wishes to obtain a useful indication of the true work environment cleanliness. This includes a definition of cleanliness for the fabrication area atmosphere and the interior of process tools, as well as for process gases and liquids.

Although there are differences in the OPCs used in each of these areas, there are some common performance requirements for good OPC operation. The OPC must define the number and size of particles randomly distributed in very clean fluid. The particle size distribution is usually a power law function; however, the particle concentration in a cleanroom environment is normally very low. Sufficient particle population data for a statistically valid definition of the true concentration value is difficult to obtain from a fluid sample that can be examined in a reasonably short time period. A typical particle size distribution function for ambient airborne particles[5] has a form similar to:

$$dN/dR = CR^{-k} \qquad (1)$$

N is the particle population per unit gas volume,
R is the particle radius,
C is a constant relating to the aerosol volume fraction,
k is an exponential constant that may vary from 2.2 to 4.

Particle population can decrease very rapidly with increasing particle size for many common particle size distributions. For these materials, the particle population becomes very small with increasing particle size.

## OPC MEASUREMENT REQUIREMENTS

In considering the problem of sparse data, we assume first that the data are derived solely from a particle source. Most specifications for defining particle loading in clean fluids call for a particle concentration statement that includes definition of a confidence level. That indicates how much faith one should have in a stated value. The confidence level states how certain one can be that the true unknown parameter lies between lower and upper confidence limits. For a measured sample average, knowledge of the standard deviation (the breadth of the data) compared to the measured average value allows one to state the limits within which the true average will fall with a specific confidence level. Uncertainty in the value of the mean is determined by the standard error, which is equal to the standard deviation divided by the square root of the number of measurements. As the quantity of data increases, the relative standard deviation and the standard error will decrease. Therefore, the reliability of the data will be better as the quantity of data increases.

The quantity of data from an OPC can be increased by increasing the sample size or the OPC sensitivity. In either case, it is also necessary to make sure that the quality of the data is not degraded. The data quality can be degraded if large portions of the data are produced by noise sources, rather than by particles. OPC operating requirements for good data are summarized in Table I.

Table I. Particle Counter Operating Specification Requirements.

| |
|---|
| Adequate Particle Sizing Sensitivity |
| Particle Sizing Accuracy |
| Particle Sizing Resolution |
| Particle Counting Accuracy |
| Maximum Sample Volumetric Flow Rate |
| Maximum Particle Concentration Capability |
| Minimum Data Processing Dead Time |
| Correlation Capability with Other Instruments |
| Adequate Particle Size Definition Range |
| Component Compatibility with Particles and Fluids |
| Documented Calibration Record |
| Reliability, Ease of Service |

## Particle Size Sensitivity

The need for adequate particle size sensitivity arises from the fact that clean fluid particle concentrations are low, but increase inversely with a power of the particle size, as shown in equation 1. Since the statistical validity of the data increases with data quantity, the maximum data quantity is desired. More data are obtained if the OPC can detect the smallest possible size particle. This requirement for maximum sensitivity means that the OPC is operating with particle signals close to the OPC noise level. At that point, the signal-to-noise ratio is at its minimum. If particle data are close to zero, OPC noise may be an unacceptably large portion of the data. We can either quantify the data sources and factor out OPC noise data or we can simply reduce OPC noise to an insignificant level. Obviously, the latter procedure is more desirable. However, it is not always possible. For very clean gas, very long measurement periods may be required even for an OPC with sub-tenth micrometer sensitivity. Most electronic device noise levels are defined in terms of RMS noise levels. A useful OPC noise level is the point where no more than one noise count occurs during a standard sample period. Usually, this point will correspond to the RMS noise level plus three to four standard deviations. For <u>any</u> OPC, <u>given sufficient time,</u> some random noise events can exceed a minimum threshold level. If the noise level is defined as the point where noise pulses are an insignificant fraction of the signal pulses, then a signal-to-noise ratio of at least two-to-one should suffice. At this time, the most sensitive OPCs for either liquid or air measurement are capable of counting and sizing individual 0.05 μm particles.

## Accurate Particle Sizing

Particle sizing accuracy is needed since most desired information is expressed in terms of particle concentration at and above a specific particle size. For the power law size distributions, frequently seen, any error in particle size measurement results in a concentration error larger than the sizing error by the power of the distribution function. Figure 1 shows that a concentration error of some 15% will result for a third power distribution if the sizing error is 5%.

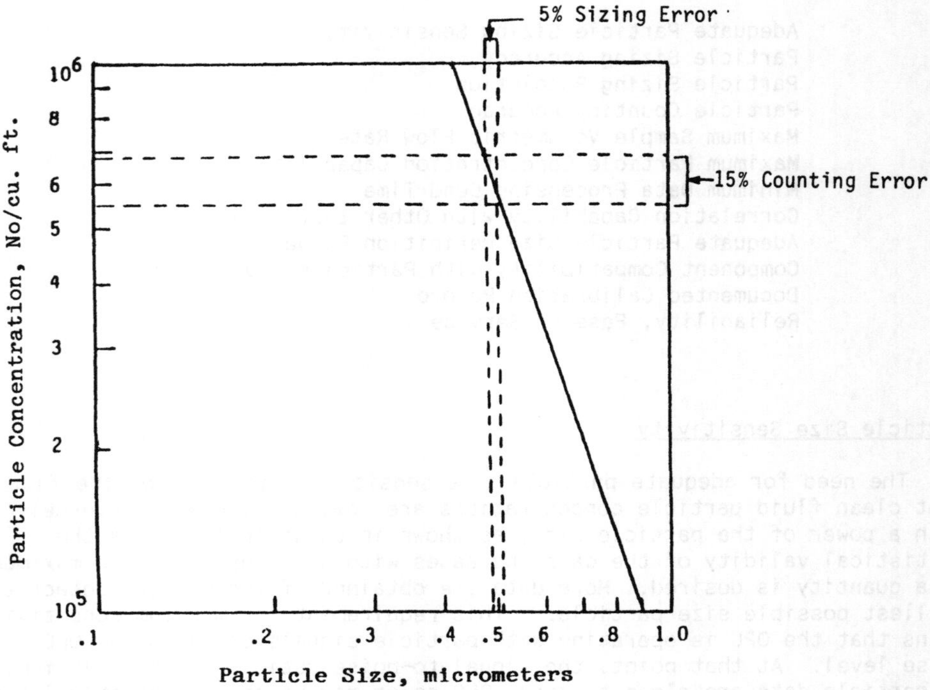


Figure 1. Counting Error as an Effect of Sizing Error for a Logarithmic Particle Size Distribution.

## Particle Size Resolution

Particle size resolution is defined as the ability to differentiate between particles that are closely sized. It can be quantified as the increase in standard deviation caused by the OPC to the true standard deviation of a suspension of monosized particles. The measurement is made at a particle size two to five times larger than the minimum sensitivity of the OPC[6]. Particle concentration data for suspensions of polydisperse particles may be significantly higher from an OPC with poor resolution than those obtained using an OPC with better resolution. This effect occurs since the power function particle size distribution results in more particles just under the smallest detection threshold than the number of particles just over that threshold by the same amount. The number of counted "subcountable" particles will be greater than the number of uncounted particles that are just over the threshold level. Figure 2 shows how poor resolution sensors can affect the range over which a particle size distribution can be reported. The upper part of Figure 2 shows the pulse height distributions that will be obtained from a series of monosized particles observed by an OPC with resolution that varies with particle size. The worst case is for the smallest particle size. The lower part of the figure shows the range of particle size distributions that may be derived as resolution improves. The center line shows the actual particle size distribution.

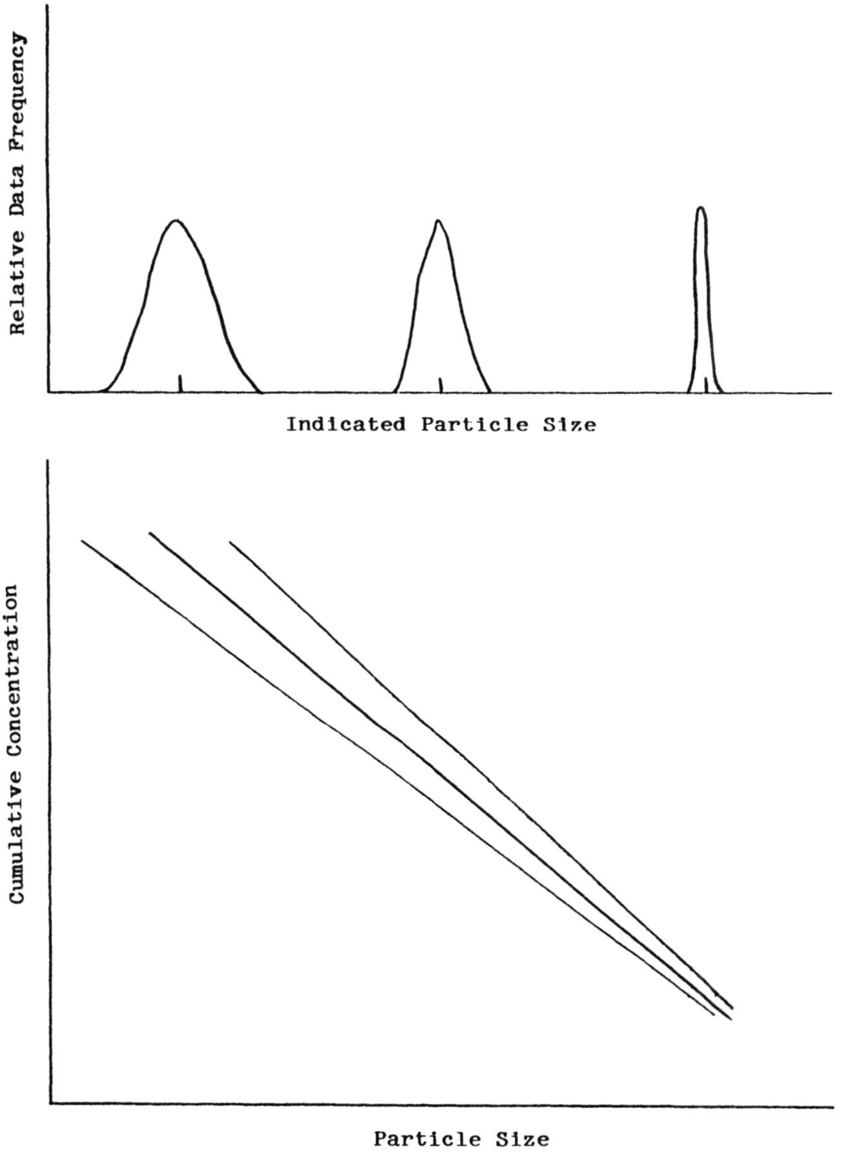


Figure 2. Instrument Resolutions and Potential Size Distribution Ranges.

## Particle Counting Accuracy

Accurate data are expected for all instrumental measurements. Counting efficiency for an OPC, as a percentage, is defined as the ratio of the reported particle population to the true population in any fluid sample, multiplied by 100. Even though it is normally assumed that any OPC counting efficiency is 100%, this assumption is not always justified. In some situations, an OPC optical system is misaligned so that part of the sampled fluid bypasses the OPC sensing zone. This situation can usually be remedied by sensor realignment.

Particularly for OPCs used for airborne particles, inlet sampling inefficiency and transport losses can cause production of erroneous data for particles larger than 1-2 µm in diameter[7]. For liquid-borne particle OPCs, the same effects can occur, but the differences in the viscosity, density and resulting shear forces in the fluid result in errors not becoming important for particles smaller than 10-15 µm in diameter. Airborne OPC inlets should be constructed so that the inlet nozzle is not blunt, but not so sharp that a hazard exists for personnel. Sample inlet dimensions should permit isokinetic sampling if at all possible. For air moving at 0.51 m/sec (100 ft per minute), an OPC sampling at 0.00047 $m^3$/sec (1 CFM) should have a sampler inlet with an area of 35.5 $mm^2$ (1.3-1.4 $in^2$). The transport line from the inlet nozzle to the detection area of the OPC should be as short as practical for particles smaller than 0.1 µm and larger than 1-2 µm. If direction changes in the transit line are required, the radius of curvature should be at least 8 inches so as to minimize centrifugal effects on larger particles.

When an OPC is calibrated, its inherent limited resolution will result in production of a pulse amplitude distribution from a suspension of monosized calibration particles. Standard practice is to establish the reference pulse amplitude level at the median of the pulse amplitude distribution produced by these particles. Thus, half of the calibration particles will be sized in the size range for particles ≥ the median particle size and half will be sized in the range ≤ the median particle size. The effects of cross-channel sensitivity can, of course, result in sizing error[8]; however counting accuracy will not be affected except in the smallest size range. If the calibration particles are sized at the minimum measurable size for the OPC, then the counting efficiency for calibration at the smallest size range will never be more than 50%. When polydisperse particles are counted by the OPC, particles which produce pulses with amplitudes ≥ the smallest particle size threshold level will be counted with 100% efficiency. Therefore, it can be stated that OPC <u>operational</u> counting efficiency is 100% for all particles with an equivalent optical size ≥ the smallest particle size threshold, even though the calibration counting efficiency at that size is 50%. Counting accuracy for all particles larger than the smallest measurable size should be 100%, even though excessive cross-channel sensitivity may result in some incorrect particle distributions for those particles which are sized close to the channel bounds. The relationship between counting accuracy for monosized calibration particles and for polydisperse ambient particles is shown in Figure 3.

## Maximum Sample Volumetric Flow Rate

Particle concentration data for the very clean semiconductor process fluids in use at this time may result in extremely sparse data. For example, the proposed SEMI specification for certain process gases[1] would allow no more than 10 particles ≥ 0.1 µm per standard cubic foot in acceptably clean gas. The particle spatial distribution in any sample is normal-

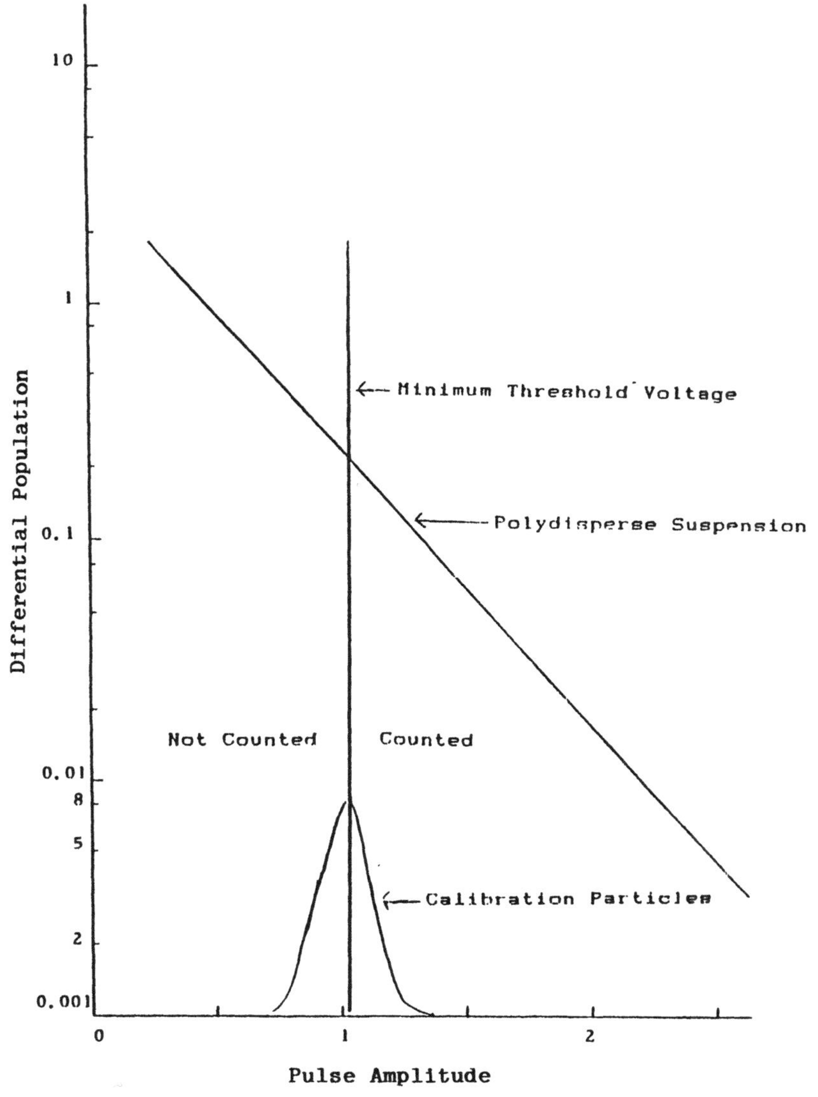


Figure 3. Particle Counting Basis for Monodisperse Calibration Particles and for Polydisperse Suspensions.

ly non-uniform. Sufficient data should be acquired from a sample so that at least an indication of that distribution and its form can be stated with some assurance that the particle concentration in the entire batch of gas will not exceed the measured median value. The SEMI specification, based on an OPC sample flow rate of either 0.01 or 0.1 CFM, requires that samples of eight cubic feet be measured. The advantage in sampling time and personnel labor savings for a high flow rate OPC is obvious.

## Maximum Particle Concentration Capability

In examining submicrometer particle suspensions, the effect of excessive particle concentration seldom occurs. However, there are many times when it is necessary to characterize suspensions with high concentrations of particles. The OPC observes the total amount of light emitted from the sensing volume; it does not image individual particles. If a single large particle or several small particles are present in the sensing volume, the OPC response can be the same for both. Since particles in a suspension are always randomly distributed, local inhomogeneities in the particle spatial distribution can cause coincident particles to be present in the sensing volume at some time, no matter how low the concentration and how small the sensing volume. In addition to particle count loss due to physical coincidence within the sensing volume, count data can be lost due to inability of a data processing system to measure pulse amplitudes and record pulses accurately when the pulse frequency becomes very high. Modern OPC systems are limited in concentration by physical coincidence rather than by electronic system capabilities. Particle pulse duration times of 10-40 μsec with electronic system dead time requirements of 2-5 μsec are typical of post-1988 instruments. For these parameters, a uniformly spaced particle frequency of some 20,000/sec can be handled by the electronic system. Thus, the limiting factor should be the physical coincidence error.

The physical coincidence error can be calculated from basic probability considerations[9]. Assuming a Poisson distribution in the sample volume, the ratio of measured to true particle concentration is:

$$N_i/N = (1 - e^{Nv})/Nv \qquad (2)$$

where $N_i$ is the indicated concentration
$N$ is the true concentration
$v$ is the sensing volume

Most OPC manufacturers specify a maximum recommended concentration where coincidence occurs less than 10% of the time. This level allows use of an approximation for coincidence error, shown in Equation 3.

$$N_i = 1 - Nv/2 \qquad (3)$$

For a condition where the sensing volume contains one or more particles 10% of the time, the indicated concentration will be approximately 95% of the true concentration.

Figure 4 illustrates how physical coincidence error increases with particle concentration for a specific OPC with fixed sensing volume. Note that these comments apply to measurement of particles whose volume is not a significant portion of the sensing volume. If a large particle is present in the suspension, then the trailing edge signal can still mask a small particle signal.

## Minimum Data Processing Dead Time

The OPC data processing electronic systems is based upon pulse height analysis. This means that the electronic system must count each pulse and define its amplitude in relation to a "zero level" voltage. The zero level point must be reached between successive pulses. A finite residence time, or dead time, at that point is required to ensure that an accurate value for pulse amplitude can be established. If another particle arrives in the sensing volume during the dead time, then the pulse from that particle is not detected. Older, analog comparator circuit pulse counting systems may require as much as 90 μsec dead time to recover from pulses produced by large particles. WHen this time interval is added to the 10-30 μsec pulse duration for most particles, then a maximum pulse counting rate of some 10,000 per second can be accommodated if uniformly spaced pulses arrive. Since the particles are randomly spaced, the pulse counting system must be able to handle the short-term concentrations as well as the long-term average. A rule of thumb observation states that the short-term rate can be up to a factor of 10 greater than the long-term average rate. This means that the 10 kHz counter should not be used for concentrations which will cause long-term average pulse rates greater than 1000/sec.

Correlation with Other Measurements

Ideally, any OPC should produce particle size and concentration data that can be compared well with data from any other particle characterization device. Since OPC data are affected by particle properties that may differ from those that may control other devices, then direct correlation

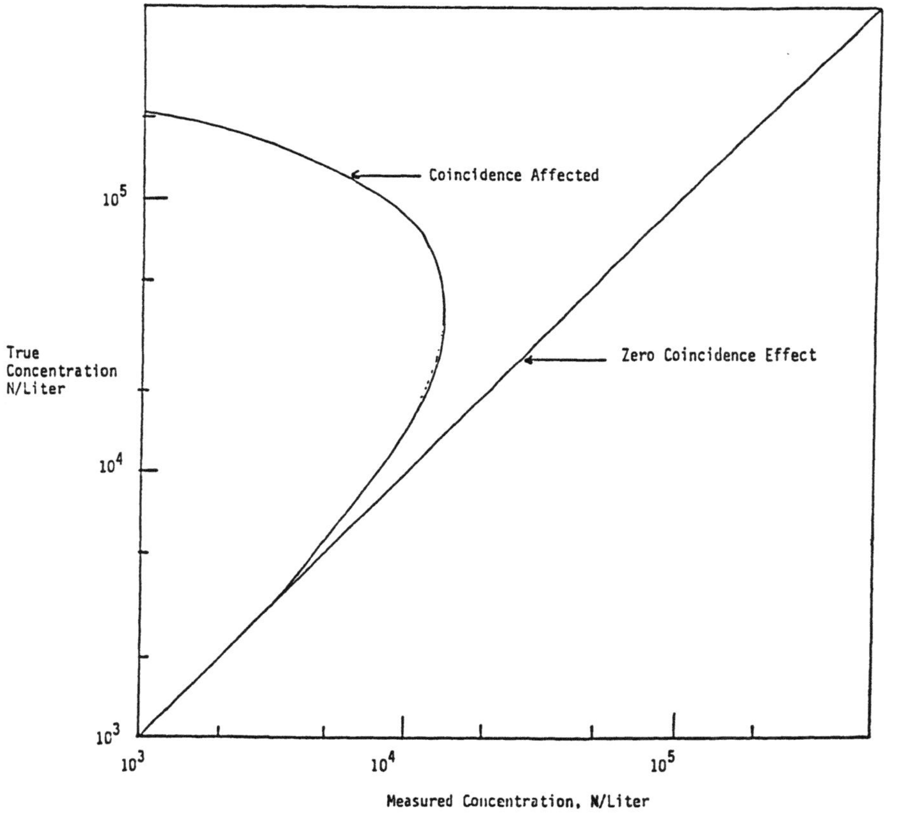


Figure 4. Coincidence Error Effects on Some Concentration Data.

to any device without a means of converting data to match the controlling particle properties is not possible. At best, the slopes of the particle size distribution functions from the OPC and another device may correspond. An empirically derived conversion factor may then be developed to aid in correlating particle size data. In this situation, experience has shown that the particle size distribution data may correspond acceptably well over 80-90% of the particle size range, but the tails of the distributions will vary to some extent. Therefore, correlation with other devices is possible, but verification is required for the specific instrument type and particle composition.

Inter- and intra-facility particle data correlation in the semiconductor industry is required particularly for verification of particle concentrations in process fluids as supplied by a vendor and as received or in use in process lines. For this application, most vendors and their customers use OPCs that have similar sensitivity and design features. The major variation in these OPCs may be in optical design, with lesser variations in illumination wavelength. For these variations, response changes can be predicted from physical theory well enough so that correlation can be accomplished between two different OPCs. The OPCs must be in good operating order and calibrated in accordance with ASTM procedures. The ASTM methods provide for initial primary calibration with monodisperse latex spheres, followed by matching test OPC output for actual particles with that from a standard, calibrated OPC. The methods for liquid-borne OPCs are:

F658-87, "Standard Practice for Defining Size Calibration, Resolution, and Counting Accuracy of a Liquid-Borne Particle Counter using Near-Monodisperse Spherical Particulate Material".

F1226-89, "Test Method for Calibration of Liquid-Borne Particle Counters for Sub-Micrometer Particle Sizing"

The methods for air-borne OPCs are:

F328-80, "Standard Practice for Determining Counting and Sizing Accuracy of an Airborne Particle Counter using Near-Monodisperse Spherical Particulate Materials"

F649-80, "Standard Practice for Secondary Calibration of Airborne Particle Counters using Comparison Procedures"

## Adequate Particle Size Measurement Range

Most semiconductor process fluid particle measurement requires an OPC to indicate trends, rather than to provide detailed size and concentration data at any specific time. For this purpose, an OPC which states only the population of particles with a single size threshold is adequate. An OPC which can characterize particles over a broad range of sizes is convenient to aid in defining particle sources where the particle size distribution varies with the source. An OPC which is capable of sizing particles smaller than approximately 0.5 µm produces signals which can vary with particle size. The OPC response varies with the square of particle radius for particles larger than approximately 0.7 µm and the variation increases to a sixth power relationship for particles of approximately 0.1 µm. If one wishes to define a broad range of particle sizes, particularly in the submicrometer range, then a very large signal voltage range must be measured accurately. This may result in larger design and material costs for such an OPC. Therefore, adequate particle size measurement range is recommended, rather than maximum range. Adequacy means sufficient size range to characterize those particles which are expected to occur in significant quantity. For example, assume a particle size distribution with a -3 power relationship between population and particle size. If 10,000 0.1 µm particles are counted, ten 1 µm particles and only one or two 2 µm particles should be present. A particle size definition dynamic range of 20:1 is adequate for this situation; the particles which will be present in any statistically significant quantity will be recorded. OPC selection should be based on requirements of the system to be measured, rather than on maximum OPC capabilities.

### Other OPC Requirements

Detailed discussion of the remaining points in Table I is not required. Some fluids may damage some OPC components. Knowledge of the limitations of the system should be at hand before passing aggressive liquids or corrosive gases through susceptible materials. The importance of the last two requirements in Table I, documented calibration records and reliability and serviceability, is obvious. The OPC owner and user should not need to be reminded that knowledge of these points is necessary for production of useful and valid data.

## CONCLUSIONS

The performance definitions for optical particle counter operation have been pointed out in detail for those OPC operating parameters which will have the most effect on the validity of the data reported by the OPC. Many of the OPC parameters which are discussed are not usually described in OPC descriptive literature. However, the performance of the OPC in providing acceptable data is strongly affected by operating parameters which are usually defined only in an operating or service manual. The effects of interactions between the properties of the particle population, the conditions in the environment where those particles are present and the important operating specifications for the OPC making the measurement are shown. It is suggested further that maximum OPC performance may not always be required for all measurements.

## REFERENCES


1. SEMI Document No. 1831, "Particle Specification for Grade 10/0.1 Nitrogen and Argon Delivered as Pipeline Gas", 1989, Semiconductor Equipment and Materials International, 805 E. Middlefield Rd., Mountain View, CA 94043

2. J. R. Hodkinson, and J. R. Greenfield, Appl. Optics, 4, 1463-1474 (1965).

3. J. Gebhart, P. Blankenburg, and C. Roth, in "Aerosols, Science, Technology and Industrial Applications of Airborne Particles", B. Y. H. Liu, D. Y. H. Pui and H. J. Fissan, Editors, pp. 7-10, Elsevier Science Publishing Co., New York, NY, 1984.

4. A. Lieberman, in "Semiconductor Fabrication, Technology and Metrology", D. C. Gupta, Editor, ASTM STP 990, American Society for Testing and Materials, Philadelphia, PA, 1989.

5. K. T. Whitby, R. B. Husar and B. Y. H. Liu, in "Aerosols and Atmospheric Chemistry", G. M. Hidy, Editor, Academic Press, New York, 1972.

6. ASTM F658-87, "Standard Practice for Defining Size Calibration, Resolution, and Counting Accuracy of a Liquid-Borne Particle Counter using Spherical Particulate Material", Amer. Society for Testing and Materials, Philadelphia, PA 19103

7. S. P. Belyaev, and L. M. Levin, J. Aerosol Sci., 5, 325 (1974).

8. R. Jaenicke, Aerosol Sci., 30, 95-111 (1972).

9. N. K. Nikiforova, and Y. S. Sedunov, Aerosol Science, 30, 441-453 (1972).

# PART II. PARTICLE GENERATION AND DEPOSITION

# ASSESSING WIPING MATERIALS FOR THEIR POTENTIAL TO CONTRIBUTE PARTICLES TO CLEAN ENVIRONMENTS: A NOVEL APPROACH

Charles F. Mattina

GDE Analytical
PO Box 779
Lenox, MA 01240

Steven J. Paley

The Texwipe Company
PO Box 575
Upper Saddle River, NJ 07458

A method is described for determining the number of particles that are readily releasable from the surface of a wiper, wetted with a liquid, under conditions of zero or near-zero mechanical stress. Excellent reproducibility is attainable with this method; furthermore, the test is quickly and easily performed. In the unique model underlying this test, the number of particles readily releasable from wipers is interpreted as a measure of the y-intercepts of stress-strain curves which are hypothesized to exist for wipers. It is proposed that the determination of this quantity is the first step toward establishing valid empirical criteria for estimating the potential of wipers to contribute particles to clean environments.

## INTRODUCTION

In the science of contamination control--some would characterize it as a black art--there is an obsession, understandably, with knowing the number of particles that are contributed to clean environments by people, by processes and, certainly, by the various items that are used in those clean environments. As a natural corollary of this preoccupation, there is a corresponding obsession on the part of those who are responsible for testing, developing, using, and offering those items for sale to characterize them in some way, before they are ever used in a clean environment, for their potential to contribute particles to that environment.

The subject of testing wipers for particles is itself fraught with difficulty. Indeed, what is actually meant when reference is made to the "number of particles" that a wiper can contribute to an environment? By itself, the phrase is in fact meaningless without further clarification. A wiper can both release particles which are already present on its surface, and it can also generate particles which currently do not exist but, in response to an applied stress, will be created in some proportion to that stress. The combined phenomena of release and generation of particles from wipers result from a plethora of complex physical interactions among the particles, the substrate, and the medium used to administer the stress. Almost certainly, if one were clever enough to understand, quantitate, and unify these complex interactions, they would take expression as complex mathematical functions. In this paper, a description is given of the authors' attempts to develop these functions--or, rather, the results of these functions--empirically.

*Particles in Gases and Liquids 2*, Edited by K.L. Mittal
Plenum Press, New York, 1990

Many tests[1-6] are extant which profess to measure the contribution of particles from wipers to clean environments, and these span a wide range of sophistication and scientific content. There are dry tests, tests done in the presence of a liquid, tests which attempt to duplicate the actual process of wiping, and tests which seek to apply stress to a wiper in some other form. The shortcoming of all of these tests is twofold: none relates the number of particles obtained to a range of known and realistic stresses delivered to the wiper, and none seeks to draw the important distinction between particle release and particle generation. The authors of this paper contend that there is a need for a test method for wipers which provides quantitative information concerning these two distinct phenomena as a function of known and reasonable stresses. Such a test ought to be the most useful index of the propensity of wipers to contribute particles to a clean environment and ought to provide information about the behavior of wipers over a realistic range of conditions of use.

The model used in developing this test rests on the hypothesis that the mathematical functions describing the conditions of release and generation of particles might be analogous to the stress-strain curves of classical physics which depict the behavior of many different kinds of materials. Just as the strain exhibited by a bar of steel or piece of plastic is proportional to the stress delivered to that item, so it is proposed that the number of particles contributed by a wiper to an environment ought to be proportional to the mechanical stress delivered to that wiper.

This concept is more easily comprehended by examining a plot of some hypothetical curves, curves which one might reasonably expect to exist for wiping materials, and curves which perhaps could be derived from first principles if the complex physics and mathematics underlying the phenomena responsible for the release and generation of particles were completely understood. Figure 1 depicts such a hypothetical representation. On the abscissa as the independent variable is some as-yet unspecified mechanical stress per unit area delivered to the wipers. The ordinate, the dependent variable, is the strain--defined here as the number of particles generated and released per unit area of wiper--resulting from the mechanical stress applied thereto. Before individually discussing these curves and

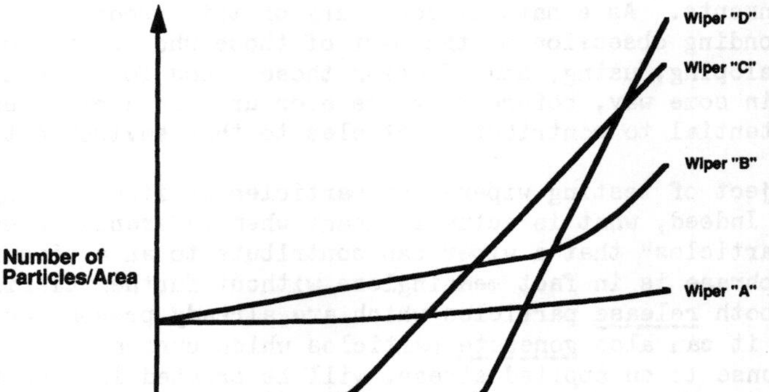


Figure 1. Hypothetical stress-strain curves for four wipers.

what they mean, two other aspects of the hypothesis should be noted. These are:

1. All curves exhibit some positive finite value at the condition of "zero mechanical stress," i.e., some y-intercept which, from the model, is interpretable as the number of particles which already exist on the surface of the wiper before any mechanical stress is delivered thereto. These are the "readily releasable" particles. Some complex release function describes their (relatively rapid) removal from the surface of a wiper, but it is only the integral of that function, the total number of readily releasable particles, that is of concern in developing this model.

2. All wipers exhibit curves which, upon continued application of mechanical stress, rise monotonically--with either shallow or precipitous slopes--as new particles, which formerly did not exist, are created in proportion to that stress according to some generation function. These are the "generated" particles. It is necessary to determine the shape of the curves which describes this generation of particles.

Philosophical questions may be raised concerning precisely what degree of intimacy must be undermined before a particle can properly be said to have an existence independent of the substrate from which it was created. For that matter, conversely, what degree of intimacy between a bona fide particle and the substrate on which it rests is required before it is in fact one with the substrate? Consequently, among the "generated particles" will be included some particles which are merely less readily releasable.

Returning to the hypothetical curves in Figure 1, consider the trace representing Wiper A. This curve might be characteristic of a wiper which, initially, has some modest level of readily releasable particles on its surface and, because of the resiliency of its construction, displays a weak generation function and thereby is resistant to the creation of very many more particles in response to the application of increased mechanical stress. Therefore, its stress-strain curve remains quite shallow. Another curve, for Wiper B, describes the behavior of a wiper with a level of readily releasable particles equal to that of wiper A but, because of the construction of this wiper, its generation function is quite strong. Thus Wiper B responds to increased stresses by creating particles to a greater extent and with a different proportionality than does Wiper A. Similarly, Curve C describes the behavior of a wiper which has a substantially lower level of readily releasable particles than either A or B but, because of the peculiarities of its construction, the generation function rises precipitously in response to the stress. Curve D is characteristic of a wiper which initially contains an intermediate level of readily releasable particles and has a generation function which responds only weakly to moderate stress but, after some critical amount of stress, begins to generate particles quite rapidly.

These hypothetical curves emphasize the importance of distinguishing between and quantifying the number of particles that are already present on the surface of a wiper and those particles which are not yet in existence but which will be created in proportion to a known and relevant mechanical stress delivered to the wiper. All other tests[1-6] with which the authors are familiar deliver a single, usually unknown, probably unrealistic, generally not reproducible stress to wipers--by twisting, abrading, folding, shaking or wetting them. The numbers obtained from these tests describe the generation of particles from the wipers at a single, usually unknown, probably unrealistic, generally not reproducible point along one

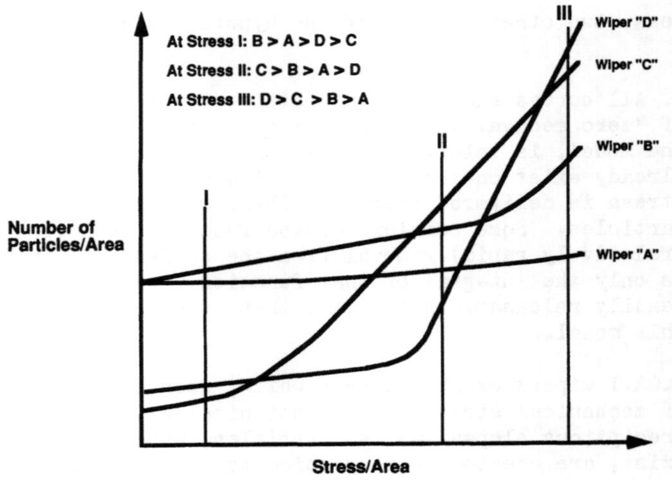


Figure 2. Hypothetical stress-strain curves for four wipers compared at three different applied stresses.

of these curves. The authors propose that the whole curve should be known for a wiper, or at least the portion of the curve extending from the condition of zero mechanical stress outward to some magnitude of applied stress which spans the range of reasonable mechanical stresses that a wiper is likely to encounter during use.

Plotted in Figure 2 are the same four hypothetical curves as in Figure 1 but with the addition of three reference lines--designated by Roman numerals I, II, III--to represent the delivery of three different stresses administered, perhaps, by three different test methods. It is immediately apparent that the conclusion one draws concerning which wiper will contribute the fewest particles will depend upon where one looks along the abscissa: it will depend upon the magnitude of the mechanical stress delivered. At a stress equal to that designated by Roman numeral I, one concludes that B A D C is the ranking of the wipers from worst to best. But at a stress equal to that of Roman numeral II, the ranking C B A D is obtained; at Roman numeral III, the ranking is D C B A. The fact that such disparate conclusions can be drawn is precisely the point. Any test which asks naively "how many particles are there" on a wiper ignores both the distinction between those particles which already exist and those which are yet to be created, and furthermore ignores the dependence of the generated particles upon mechanical stresses administered to the wiper. Unlike the analysis for a particular metal ion or for the fraction of a wiper soluble in some solvent, there is no unique answer to the number of particles a wiper can contribute to an environment. That value depends both upon the number of particles initially present on the wiper and also upon the magnitude of the mechanical stress delivered: if one hits the wiper harder, one obtains more particles. The authors contend that quantifying the number of particles contributed by a wiper as a function of the mechanical stress administered to that wiper will be the most useful way to characterize the different wiping materials that are used in clean environments.

At the outset of this work, it was decided that the first step in obtaining these curves ought to be to develop an experimental technique for determining their y-intercepts, that is, the number of particles readily releasable from wipers under conditions of zero, or near-zero, mechanical stress. In the next part of this paper, the methodology that has been developed for determining these y-intercepts is described and the data that have been generated are presented. Toward the end of this paper, a description is given of the authors' initial approaches for developing a technique for delivering known stresses to wipers in order to establish empirically the shapes of the postulated stress-strain curves.

EXPERIMENTAL

It is appropriate to discuss here the rationale behind the testing of wipers and other materials in a wetted condition. The reason for wet testing has very little to do with the fact that wipers sometimes become wet when they are used. Rather, it has to do with the physics of the attachment of particles, with the large number of particles that are already present on most wipers, with the fact that tests for particles should reflect the risks to which one is exposing his process and environment, and with the fact that dry testing is not equipped to give users a reflection of those risks. Further amplification of these points is in order.

As wet testing has shown,[3,6] and as the data to be presented in this paper will corroborate, the number of particles (equal to or greater than 0.5 μm) already present on the cleanest of cleanroom wipers is in the range of several million per square meter. Other wipers that are routinely used in cleanrooms can be shown by wet testing to have particle burdens (equal to or greater than 0.5 μm) in the range of tens and even hundreds of millions per square meter. Yet the dry testing of these same items yields numbers that are much lower, understating the true burden by orders of magnitude even before any mechanical stress is applied.[4]

The reason for this is that the forces which hold particles to wipers--coulombic forces, capillary forces, London-van der Waals forces--are enormous.[7,8] Attempting to dislodge these particles with a stream of air cannot remove more than a fraction of the actual number present. But in the presence of a liquid, these same forces are readily overcome,[9,10] the particles are released into the body of the fluid and are easily sluiced away for counting.

> [This last point, concerning the collection and counting of particles, is another reason why dry testing is to be eschewed. With dry testing, it is extremely difficult to collect but a fraction--and a size-selected fraction at that--of the number of particles released and generated, because the phenomena of fallout, reattachment, deposition, etc. are all taking place.[11] In most dry tests with which the authors are familiar, a large number of the particles released and generated never make it to the counting stage.[1,2,6]]

Thus, wet testing provides information about the true burden of releasable particles and therefore offers the most useful data about the item under consideration.

Proponents of dry testing offer two objections to wet testing. First, that wet testing gives numbers that are unacceptably large and, second, that the total number of particles present is not important if most of them are so tightly bound that they are not easily released.

Concerning the first objection, the choice of a test method should never rest upon one's preconceptions of the truth. Concerning the second objection, several studies[11,12] have shown that the observed deposition of particles from wipers onto surfaces in cleanrooms is quantitatively correlated with the results from testing these same items in a wetted condition. Wet testing provides data which correlate with the reality of contamination; dry testing does not. Further to the second objection, unpredictably large mechanical contact stresses invariably occur at the surfaces of wipers during normal operations; only through the application of equivalent (e.g., liquid) shear forces can these mechanical stresses be easily duplicated.[13,14] Still further to the second objection, changes in the moisture and humidity of the environment or in electrostatic conditions also affect the magnitude of the forces holding particles to wipers.[8] Such variations in local conditions are a major source of the great variation observed with dry testing whereas with wet testing these forces are unilaterally overcome.

The impact of all of the above factors upon the true releasable burden of particles is not captured in dry testing. With wet testing the entire picture is seen. Therefore, in seeking to develop an experimental design to assess the release of particles from a wiper under conditions of zero or near-zero mechanical stress, it was decided to perform the test in the presence of a liquid on the assumption that a valid measure of the number of readily releasable particles existing on the surface of the wiper will thereby be obtained.

Experimentally, the problem is equivalent to determining a release function, or at least the integral of a release function, without permitting the generation of particles to interfere with the experiment. Practically speaking, this is sometimes difficult to accomplish since the two physical phenomena are operating at the same time. It is the relative magnitude of release and generation which determines whether or not one can easily extract the desired results from the experimental data. For some wipers, where the generation of particles is known or suspected to be weak (such as knitted polyester fabric), it is relatively simple to design experiments to determine the number of readily releasable particles. For other wipers, where the generation of particles is likely to be very large (such as commonplace cellulosic wipers), determining the number of releasable particles under those same conditions is difficult, if not impossible. Therefore, the experimental technique selected has to accommodate both situations or it will not be of any value.

What was finally decided was to wet wipers by immersing them completely in a liquid. The liquid is poured off into a separate vessel and the wiper rewetted immediately twice again; the three washings are combined and analyzed for particles using an automatic particle counter or by optical microscopy. This method minimizes the difficulties described above yet still meets the criteria for the test:

a. Virtually no mechanical stress is applied to the wiper. No agitation occurs; the liquid merely overcomes the coulombic, capillary, and London-van der Waals forces holding the readily releasable particles to the wiper.

b. Since the wiper is wetted three times, rapidly, and since nearly all of the liquid is recovered for analysis, little opportunity is provided for a renegade generation function to contribute heavily to the number of particles removed. The authors recognize, however, that this method provides only an operationally defined measure of the number of particles present on the wiper, and therefore includes some particles that have been prematurely generated from wipers having strong generation functions as well as omitting some releasable particles which are more tightly bound to the wiper.

Table I. Particles $\geqslant 0.5\ \mu m/m^2$ (No. x $10^{-6}$) Released During Replicate Testing of Two Different Wiping Materials.

| | polyethylene terephthalate filament, knitted, laundered | 50% polyethyleneterephthalate, 50% cellulosic fiber |
|---|---|---|
| | 14 | 39 |
| | 14 | 50 |
| | 19 | 47 |
| | 17 | 51 |
| | 20 | 48 |
| | 17 | 50 |
| | 16 | 49 |
| | 13 | 37 |
| | 10 | 42 |
| | 10 | 41 |
| | 13 | 52* |
| | 11* | 51* |
| | 14* | 34* |
| | 17* | |
| | 14* | |
| Average | 15 | 46 |
| Standard Deviation | 3 | 6 |
| Relative Standard Deviation | 20% | 13% |
| Number of tests | 15 | 13 |

c. Once the particles have been removed from the wiper and are in a separate vessel, one has the luxury of determining their concentration in a relatively leisurely fashion without fear of the generation function contributing to the count.

d. Since the liquid is poured off, virtually in its entirety, problems with settling or incomplete mixing are easily remedied in the second vessel by gentle swirling or by analyzing the entire volume of liquid.

e. Three washes were judged sufficient since it was observed that--for a wiper for which generation was anticipated to be small (knitted polyester, for example)--after the third wash, the water in contact with the wiper had virtually the same concentration of particles as did fresh deionized water. It was concluded from this that three washes are enough to sluice the readily releasable particles from the surface and to rinse away traces of particle-bearing water from earlier washes.

f. The amount of water used, roughly a liter per wiper, is--from the point of view of the readily releasable particles--an enormous excess, an infinite source of fluid, for removing those particles from the surface.

Although it is proposed that this test at a zero or near-zero mechanical stress is the starting point for establishing complete stress-strain curves for wipers, the results obtained are also indicative of the risk to which one's environment and process will be exposed if a wiper is used in the gentlest possible fashion.

Table II. Particles released from wiping materials.

| generic description | particles ≥ 0.5 μm/m² (No. x $10^{-6}$) |
|---|---|
| polyethylene terephthalate, knitted, laundered | 30 |
| polyethylene terephthalate, knitted, laundered, heat-sealed edge | 10 |
| cotton cloth, woven | 340 |
| nylon, knitted, laundered | 26 |
| nylon, nonwoven, treated with surfactant | 410 |
| polyethylene terephthalate, nonwoven | 30 |
| 50% polypropylene, 50% cellulosic fiber, nonwoven | 160 |
| 50% polyethylene terephthalate, 50% cellulosic fiber, nonwoven | 46 |
| 100% cellulosic fiber, nonwoven, wet strengthened | 46 |
| polyurethane foam | 270 |

## RESULTS AND DISCUSSION

One of the problems with all tests for particles with which the authors are familiar is the extraordinary difficulty of obtaining reproducible results. Frequently when data are scattered, it is unknown whether the fault lies with the experimenter, the test method, or the variability of the item being tested. An attractive feature of this procedure is that it is capable of the excellent precision shown by the data in Table I. As can be seen from the standard deviations, the repeatability is quite good. The entries marked with asterisks are from analyses that were separated in time by six months from the other data points.

In generating these data, two packages of wiping materials--one a knitted and laundered polyester wiper, the other a nonwoven blend of polyester and cellulosic fiber--were removed from regular production. The packages were opened in a Class 100 clean bench and a stack of wipers removed from each. Individual wipers were then tested following the procedure described above. The apparatus used in the analyses consisted of a Hiac-Royco Model 4100 Particle Counter equipped with a Model 346-BCL Particle Sensor and a Model 3200 Tank Sampler.

The results of testing ten different wiping materials, each described generically, are presented in Table II. As is evident from the data, many millions of particles are present per square meter of wiper, and these are only the readily releasable particles, those corresponding to the *y*-intercepts of the curves; no appreciable generation of particles has yet taken place. It is the existence of numbers of this magnitude that caused the authors to be disturbed by other tests[4,6] for wipers for which the results were reckoned in terms of several hundred, or several thousand, or even tens of thousands of particles per square meter. Far more particles are present on wipers, even before any mechanical stresses have been delivered.

Another attractive feature of the test is that it is extremely easy to perform. With minimal experience, an individual can test four or five wipers in an hour.

DEVELOPING THE STRESS-STRAIN CURVES

It was mentioned above that a description would be given of the authors' first attempts to develop stress-strain curves for wipers. In order to do this, at least three details must be resolved at the outset:

1. The nature of the stress,
2. The magnitude of the stress, and
3. How the stress should be delivered to the wiper.

With respect to the nature of the stress, the reader may have noted that, thus far in this paper, the word "stress" has been used only euphemistically as a term for an as-yet unspecified lambasting of the wiper with a liquid. It is necessary to decide what form this independent variable, this "stress" per unit area, should take in order to be relevant.

Initially, the notion of administering some kind of force per unit area to the wiper was considered. Certainly when wipers are used, force is applied by pushing with the wiper upon the surface being wiped. Upon further consideration, however, the idea of using energy per unit area seemed more relevant, for the simple reason that when people use wipers, they do not merely apply a stationary force to the wiper but they translate that force through a distance--that is, they impart energy--during the process of wiping. However, while the administration of a force through a distance is indeed quite reasonable, it makes a distinct difference to the wiper whether the input of energy is fast or slow. In other words, the time rate of change of energy, or power--per unit area--ought to be taken into consideration in applying stress to the wiper.

Since power is an intensive variable, the question then arose whether different power settings should be integrated over different time intervals to the same constant energy, or whether the power setting should be held constant and integrated continuously over time to different total energies, or whether some combination of both should be employed. It was decided that the last option made the most sense, increasing both power and energy, as will be explained below.

The second point requiring resolution has to do with the magnitude of the stresses to be delivered. A reasonable range must be chosen both for the power density and for the integral of the power density, the energy, to span. A condition of zero mechanical stress has been said to exist during the careful wetting of the wiper to overcome the forces holding the readily releasable particles to the surface. It is necessary to decide what magnitude of mechanical stress (that is, at what point along the abscissa in Figure 1) should be administered to the wiper in order to establish an arbitrary minimum and thereby initiate the generation of particles.

After some consideration it was decided that the application of a net resultant normal force of 20 g applied over a circular area of wiper 10 cm in diameter ($7850 \times 10^{-6}$ $m^2$) which is then translated horizontally through a distance of 5 cm in a time of 10 sec might represent the administration of such a minimum mechanical stress to a wiper in a controlled environment. Assuming a smooth surface, and if a reasonable choice for the coefficient of dynamic friction is 0.1, then calculating the power density, the power applied per unit area, for this set of conditions is a simple matter:

$$\frac{(0.1 \times 20 \text{ g} \times 10^{-3} \text{ kg/g})(9.8 \text{ m/sec}^2)(5.0 \text{ cm} \times 10^{-2} \text{ m/cm})/10 \text{ sec}}{(7850 \times 10^{-6} \text{ m}^2)} =$$

$$0.0125 \text{ W/m}^2$$

If power of this magnitude were administered to a wiper for, say, 10 sec, 0.125 $J/m^2$ would be delivered to the wiper--a reasonably small energy density. The authors do not, of course, imply that this particular power or energy density represents some magical or unique minimum used in wiping. It is, however, a reasonable place to begin in developing the stress-strain curves since the intention is first to fix a point at the low end of the curve and then to increase this by two or three orders of magnitude in the process of establishing the entire curve.

As an upper limit of energy, it was decided to consider an overall increase of two orders of magnitude, achieved by applying one order of magnitude more power, 0.125 $W/m^2$, for one order of magnitude more time, 100 sec. Under those conditions, 12.5 $J/m^2$ will be delivered to the wiper. This is two orders of magnitude more energy than the minimum value, and it is delivered via a factor of ten in power and a factor of ten in time.

The most problematical question of all is how the stress should be delivered to the wiper in a fashion that will permit: a) its administration in a quantitative, reproducible and controllable fashion, and b) the convenient collection and counting of the particles released at each level of administration. In attempting to address this question, the authors considered and rejected a number of different options.

Owing to its simplicity, the use of an ultrasonic generator initially seemed very appealing. One might merely place the flask or beaker containing the wiper, after the releasable particles have been removed, into a sonic cleaner for some amount of time. It is certainly possible to know how much power is being generated by a sonic cleaner as a function of the time, amplitude and frequency of the sonic signal, and it should be a simple matter to calculate the input of power.

Although the power output of a sonic generator can indeed be calculated, unfortunately it is difficult to know what fraction of that power will actually be absorbed by the wiper. Much of the energy of a sonic

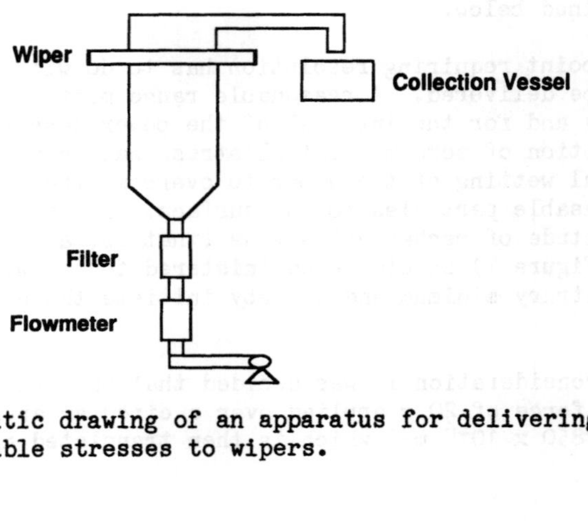


Figure 3. Schematic drawing of an apparatus for delivering known and controllable stresses to wipers.

generator is converted into noise, into heat, and into moving the water around in the sonic bath and bumping it into the walls. The portion of that energy that the wiper will absorb is some unknown and probably small fraction of the total power output. And since the intention is to apply known energy and power densities which span the range of reasonable stresses that a wiper might encounter in use, the notion of doing this with a sonic generator seemed ill-suited to the task. A further limitation of this approach became apparent when the authors noted that when an automatic particle counter is used in conjunction with a sonic generator, a large number of false counts are obtained which are ascribed to the presence of bubbles. These bubbles are produced either through cavitation caused by the sonic generator or else by virtue of the liquid's being degassed by the sonic generator. To be sure, one can certainly get around this problem by filtering the suspension of particles through a membrane filter, optically clearing the filter, and counting the particles using darkfield illumination and optical microscopy.[1] But owing to the time-consuming nature of this exercise, the approach was abandoned.

Other ideas for delivering the necessary power and energy to the wiper, none of them bearing much fruit, were considered before settling on something that, although not yet tested completely, seems to be reasonable. Shown schematically in Figure 3, the apparatus consists of two vertical glass tubes which clamp horizontally a circular cross section of wiper. Water is forced through the wiper via a constant velocity pump and is taken off at the top and subsequently analyzed for its content of particles. Knowing the mass of water pumped through the wiper, the velocity with which it is pumped, and the cross-sectional area of the wiper, it is a simple matter to calculate the power and energy delivered to the wiper per unit area. By changing the mass of water and/or the velocity and/or the cross-sectional area one can cover two or three orders of magnitude of energy.

From simple calculations, it can be shown that an apparatus of this design can be used to deliver the suggested range (0.0125 $W/m^2$ for 10 sec to 0.125 $W/m^2$ for 100 sec) of reasonable power and energy densities. If a cross-sectional area of 10 cm (7850 x $10^{-6}$ $m^2$) is chosen, then the lower limit of energy, 0.125 $J/m^2$, is achieved at a power input of 0.0125 $W/m^2$ by driving water through the wiper at a velocity of 3.7 x $10^{-2}$ m/sec for 10 sec. Similarly, the upper limit of energy, 12.5 $J/m^2$, can be administered at a power input of 0.125 $W/m^2$ using a velocity of 6.3 x $10^{-2}$ m/sec for 100 sec. Obviously, this apparatus can be used not only to deliver the proposed upper and lower limits of energy density, but intermediate levels as well--all at reasonable flow rates.

It is recognized that with the experimental design that has been described, only the input of energy from the water to the wiper will be known, not how much of that energy will be absorbed by the wiper. To determine the energy absorbed by the wiper, the pore volume of the wiper and the resultant pressure drop as the water transverses the wiper will have to be measured.

The authors have just recently settled on the experimental protocols for this part of the experiment and have acquired only fragmentary data. In the next phase of this work, this apparatus will be used in an attempt to generate empirically the stress-strain curves.

## SUMMARY

The procedure described in this paper provides a rapid and precise technique for determining the number of particles that are readily releasable from the surface of wiping materials. In the conceptual model upon which this test is based, stress-strain curves with positive y-intercepts are hypothesized to exist for all wiping materials. The readily releasable particles are interpreted as the y-intercepts of these curves under conditions of zero, or near-zero, mechanical stress. Determining the shape of the entire curves for wiping materials, at least over reasonable ranges of applied mechanical stress, will provide the most useful information for characterizing these wiping materials for their propensity to contribute particles to clean environments.

## REFERENCES


1. "Wipers Used in Clean Rooms and Other Controlled Environments," IES Recommended Practice RP-CC-004-87T, Institute of Environmental Sciences, Mount Prospect, Illinois 60056, October 1987.

2. G. E. Helmke, A Tumble Test for determining the level of detachable particles associated with clean room garments and clean room wipers, Proceedings, 28th Annual Technical Meeting, Institute of Environmental Sciences, 1982, p. 218.

3. W. B. Harding, Clean room wiping cloths, Proceedings, 27th Annual Technical Meeting, Institute of Environmental Sciences, 1981, p. 4.

4. W. J. Havel and C. Sheridan, Modified flex test for particulate analysis of dry wipers, Proceedings, 31st Annual Technical Meeting, Institute of Environmental Sciences, 1985, p. 80.

5. W. J. Soules, Testing wipers for abrasion resistance, Proceedings, 9th International Symposium on Contamination Control, 1988, p. 431.

6. C. H. Sheridan and S. Fahy, Wet and dry testing methods for evaluation of clean room wipers, Proceedings, 32nd Annual Technical Meeting, Institute of Environmental Sciences, 1986, p. 530.

7. M. Corn, The adhesion of solid particles to solid surfaces, I. A review, J. Air Pollution Control Assoc., 11(11), 523 (1961).

8. M. Corn, The adhesion of solid particles to solid surfaces, II, J. Air Pollution Control Assoc., 11(12), 566 (1961).

9. K. L. Mittal, Editor, "Particles on Surfaces 1: Detection, Adhesion and Removal," Plenum Press, New York, 1988.

10. K. L. Mittal, Editor, "Particles on Surfaces 2: Detection, Adhesion and Removal," Plenum Press, New York, 1989.

11. E. R. Crutcher, Boeing Aerospace Company, Seattle, Washington, personal communication, 1982.

12. R. D. Beeson, Lockheed Missiles and Space Company, Sunnyvale, California, personal communication, 1985.

13. A. Lieberman, Particle Measuring Systems, Fremont, California, personal communication, 1989.

14. D. C. Grant, B. Y. H. Liu, W. G. Fisher, and R. A. Bowling, Particle capture mechanisms in gases and liquids: An analysis of operative mechanisms in membrane/fibrous filters, J. Environmental Sci., 32(4), 43 (1989).

# ROBOTS AS A POTENTIAL SOURCE OF PARTICULATE CONTAMINATION

K. May, T. Pfeifer, and R. J. Schilling[1]

Electrical and Computer Engineering Department

Clarkson University, Potsdam, NY 13676

The replacement of human clean room operators by robots provides an opportunity to reduce a major source of airborne particulate contamination. However, robotic manipulators also generate particulates, albeit at a reduced rate. The detrimental effects of robot-induced contamination on product yield will become increasingly important as the minimum feature size of integrated circuits continues to shrink. This paper reports the results of an investigation of the airborne submicron particulate contamination generated by an Intelledex 660 robotic arm. Particles generated by the joints of the manipulator under various operating modes are examined. In addition, a comparison of the particulate contamination generated by several custom end-effectors designed for handling integrated circuit photomasks is presented.

## INTRODUCTION

The use of robotic manipulators for material handling in clean rooms is rapidly becoming commonplace in the semiconductor industry[1,2]. Existing clean rooms and specialized pieces of process control equipment are being retrofitted with robotic arms at key processing stations, and new clean room facilities are being designed with integrated robotic work cells[3,4,5,6]. The primary motivation for introducing robots in the clean room lies in the need to reduce contamination of materials due to airborne particles. Human operators are known to be a major source of particulate contamination[2,7]. By replacing human handling with robotic manipulation, this important source of contamination can be significantly reduced.

The intense competition in the semiconductor industry has led to a startling growth in the requirements and capabilities of integrated circuits. As the densities of circuits increase and the minimum feature sizes shrink, smaller and smaller particulates play a significant role in device performance. Several robots designed specifically for clean room operations have appeared in the marketplace[2,8]. Manufacturers have gone to considerable lengths to minimize the particulates generated by robotic manipulators. For example, special materials designed to resist shedding and flaking are used for

[1]This work was supported by I.B.M. Corp., Essex Jct., VT, Contract No. 140378JE548

the fabrication of exposed surfaces. Bearings are sealed using magnetic washers to hold ferrofluid lubricant in place. In some cases, the interior of the arm is kept at a vacuum to ensure that particulates that are generated by moving surfaces are pulled through the arm and purged outside the clean area or vented through a filtered return duct. These measures have been quite effective in eliminating the relatively "large" particles with a diameter of 1 $\mu$m and larger. However, there is increasing evidence that suggests that robots[9] as well as other devices used in clean room manufacturing[10] generate substantial numbers of particulates in the 0.1 $\mu$m range. Predictions for semiconductor industry in the year 1998 call for fully-automated Class 0.1 fabrication facilities with few, if any, human operators present, and all equipment accessible by robots[11]. Consequently, robot-generated particles are likely to become increasingly troublesome to semiconductor manufacturers in the near future.

## CONTAMINATION AND DEVICE YIELD

The overall device yield for an integrated circuit is affected by defect density[12] as well as by several fundamental design parameters such as die size and the number of critical mask levels[13]. In an effort to predict the yield of VLSI and ULSI circuits, the following Bose-Einstein model has been proposed[14].

$$\text{yield} = \frac{100}{(1 + D(x)A)^n}\ \% \tag{1}$$

Here $n$ is the number of critical mask levels, $A$ is the chip area in cm$^2$, $x$ is the minimum feature size in $\mu$m, and $D(x)$ is the effective defect density in defects/cm$^2$/level. Clearly, as future integrated circuits become larger ($A$) and more complex ($n$), the yield will drop if the effective defect density $D(x)$ remains constant. The problem of decreasing yield with increasing defect density is compounded by the fact that the density of particulates appears to increase with the inverse square of the particle size. Thus as the minimum feature size $x$ decreases, the effective defect density $D(x)$ increases as follows[13].

$$D(x) = D_0 \left(\frac{x_0}{x}\right)^2 \tag{2}$$

Here $x_0$ is a base size in $\mu$m, whereas $D_0$ is the corresponding base defect density in defects/cm$^2$/level when $x = x_0$. For example, for a 64K DRAM manufactured in 1984 using 3 $\mu$m line widths, the defect density for a "world class" manufacturer was 0.5 defects/cm$^2$/level[13]. Thus, if the model is initialized around this operating point, then:

$$x_0 = 3\ \mu\text{m} \tag{3}$$

$$D_0 = 0.5\ \text{defects/cm}^2\text{/level} \tag{4}$$

Shrinking the minimum feature size $x$ by a factor of two then quadruples the effective defect density. Hence for future integrated circuit fabrication, the decreasing minimum feature size, the increasing number of critical mask levels, and the increasing die size all combine to amplify the effects of contamination on yield. Indeed, it has been estimated that the defect density will have to be reduced by a full order of magnitude if commercially meaningful yields are to be maintained over the next several years[13].

## ROBOT-GENERATED PARTICLES

The robot used in this study was an Intelledex 660 robot shown in Figure 1. This is a six-axis light assembly robotic arm designed specifically for clean room operation. The robot was housed in a Class 10 clean room chamber. Airborne particulate contamination was measured using a Royco model 245 particle counter. The particle counter sampled the air at a rate of 1 CFM ($ft^3$/min) through an isokinetic sampling probe mounted horizontally on a 10″ tripod that sat on the 4′ × 4′ robot work surface. Five particle ranges could be detected based on their light scattering properties. The sensitivities were: 0.3 $\mu$m, 0.5 $\mu$m, 1.4 $\mu$m, 3.0 $\mu$m, and 5.0 $\mu$m.

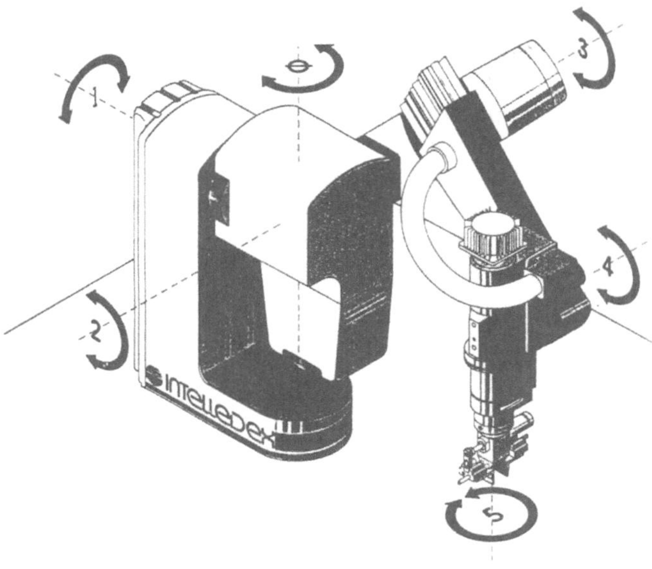


Figure 1. An Intelledex 660 robotic arm designed for clean room operation.

### Test Methods

For each test with the particle counter, the counter was first allowed to "warm up" for at least 10 minutes to ensure accurate measurements. The sampling probe was then placed at the location of interest. This involved a person entering the clean room chamber and manually placing the probe in the desired position. Once in place, the clean room chamber was closed to the outside environment and allowed to return to its steady state condition, which generally took 1 to 2 minutes. The particle counter was set to monitor samples at 1 CFM, and all samples were at least 15 minutes in duration to ensure statistical validity.

In each case, a "background" particle count was taken prior to the actual contamination test to determine the level of background contamination present. Upon conclusion of the background test, the contamination test of interest was performed in the same manner. Comparison of the two particle counts then gave an indication of the level of contamination, above the background level, generated by the test of interest. The robot and its end-effectors were carefully cleaned with special clean room wipes and an isopropyl alcohol solution prior to performing contamination tests.

## Clean Room Chamber

Prior to analyzing the robot and end-effectors, it was necessary to test the clean room chamber itself to determine the level of background contamination present. The clean room chamber housing the Intelledex 660 robot is a vertical laminar flow design with a flow rate of 100 ft/min. The inside dimensions of the room are $8' \times 5' \times 7'$. The robot sits on a $4' \times 4'$ metal table whose surface was partitioned into a $5 \times 5$ sampling grid as shown in Figure 2. The air sampling probe was mounted horizontally on a tripod 10″ above the table. Each grid location shown in Figure 2 was monitored for 15 minutes to provide a valid statistical sample. A total of 23 locations were sampled to produce the contamination profiles shown in Figure 3 and Figure 4. During the measurement process, the robotic arm was kept at rest in a forward-facing horizontal position with the tool or end-effector axis pointing straight down. For the purposes of plotting the surfaces, the locations with coordinates 2D and 2E, which lie inside the robot, represent data interpolated from neighboring measurements. The values plotted represent 95% upper confidence limits (UCL), a statistical limit that indicates that 95% of the samples should fall at or below this level. It is clear from Figure 3, which shows particle counts at 0.5 $\mu$m and larger, and Figure 4, which shows particle counts at 0.3 $\mu$m at larger, that the clean room chamber exhibits background contamination that lies well within Class 10 levels[15].

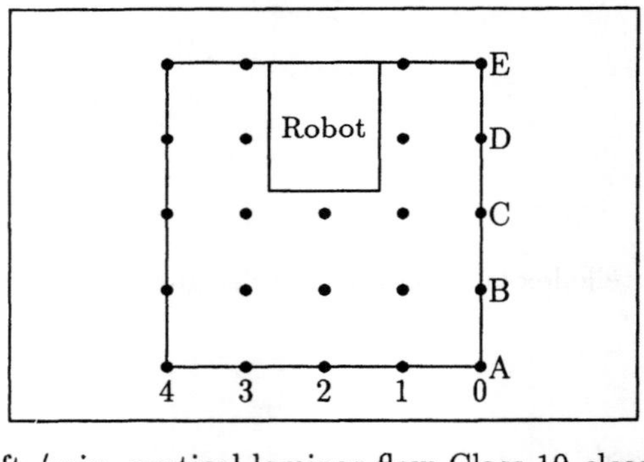


Figure 2. A 100 ft./min. vertical laminar flow Class 10 clean room chamber.

## Robot Joints

The Intelledex 660 robot was tested to determine the number of particles generated while the arm was in motion by examining each joint of the robot separately. For each joint, the location sampled was chosen to provide a worst case situation for particle generation. That is, when possible, samples where taken immediately below the joint of interest, down stream from the air flow. However, joint 0 and 1 could not be sampled in this manner due to the robot design. In these cases suitable locations near the joint were chosen. To isolate the contamination generated by each joint, only one joint was

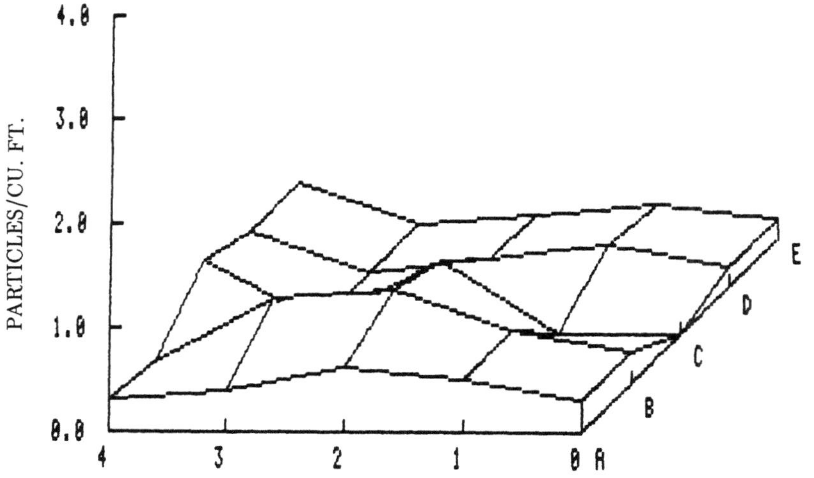


Figure 3. Background particles counts, 0.5 $\mu$m and larger, measured 10″ above the work surface with the robot horizontal.

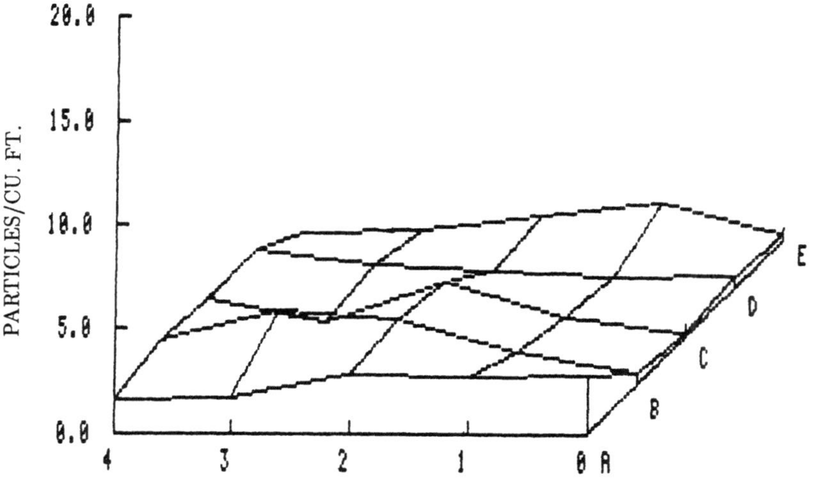


Figure 4. Background particles counts, 0.3 $\mu$m and larger, measured 10″ above the work surface with the robot horizontal.

activated at a time, and the range of travel was made as large as possible. The default operating speed (80% of maximum speed) was used for each joint.

The test configuration used for the robot was one with the arm in a horizontal position with the tool or end-effector axis pointing straight down. A background count was taken immediately prior to the actual test. The background and actual contamination tests were each 15 minutes in duration. A summary of the particles generated at the six joints of the robot is shown in Figure 5 for particles 0.5 $\mu$m and larger and in Figure 6 for particles 0.3 $\mu$m and larger. Note that the higher location numbers correspond to joints further out toward the end of the arm. The ordinate in each case represents particles/ft$^3$ obtained by averaging 15 one minute samples at 1 CFM. The difference between the moving count and the background count represents the average particles added. It appears to be insignificant at 0.5 $\mu$m and larger, but there is a contribution from joint 2 in the more sensitive range from 0.3 $\mu$m to 0.5 $\mu$m. These observations are consistent with the results of a similar contamination test made on a four-axis Adept One robot in a Class 1 facility. In that case there were no particles generated at 0.5 $\mu$m and larger, 1 particle/ft$^3$ at 0.3 $\mu$m and larger, and 5 particles/ft$^3$ at 0.1 $\mu$m and larger[9].

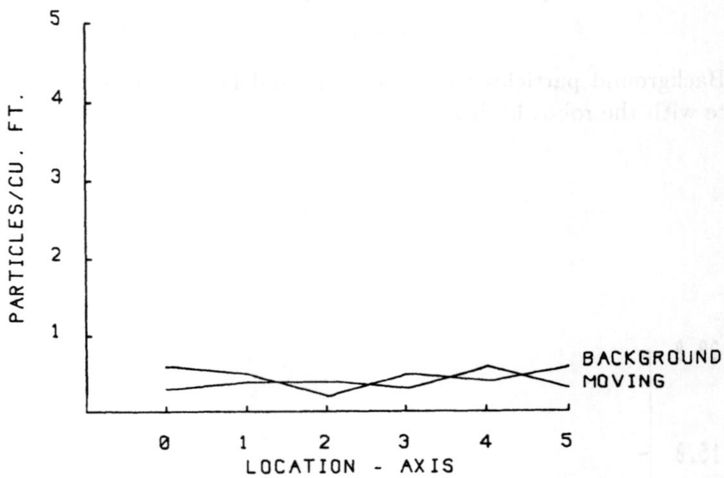


Figure 5. Particles generated by the joints of the robotic arm, 0.5 $\mu$m and larger, with only one axis moving at a time.

The difference between the particles detected in the background measurements and the particles detected when the robot was in motion does not appear to be significant except for joint 2, the shoulder pitch axis where more than 2 particles/ft$^3$ were generated in the size range from 0.3 $\mu$m to 0.5 $\mu$m. Further tests with this joint using different starting configurations revealed that these particles appeared to be shed from the upper arm housing, when the arm was pointing straight up, and deposited in a small recirculation zone of eddy currents beneath the shoulder where the probe was located. Therefore, they did not appear to be coming from the joint itself, but rather from the upper arm housing when it was located upstream of the sampling probe. This was particularly evident when background measurements were taken with the arm straight up where counts as high as 15 particles/ft$^3$ in the 0.3 $\mu$m to 0.5 $\mu$m range were observed.

An alternative, and more traditional, plot of the contamination produced by the joints of the robot is shown in Figure 7 which displays particle density vs. particle size

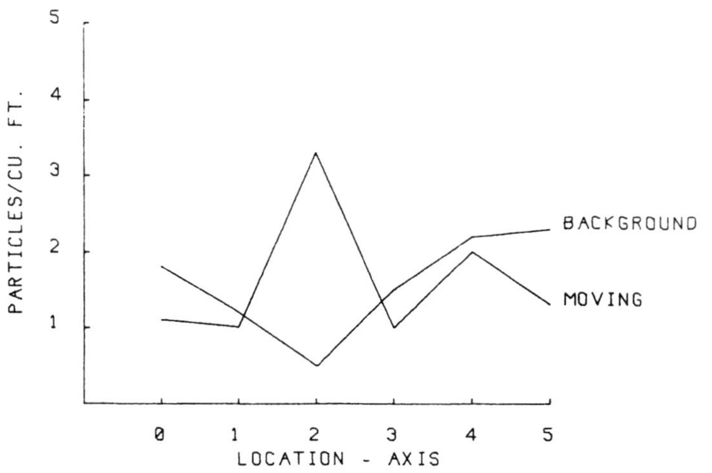


Figure 6. Particles generated by the joints of the robotic arm, 0.3 $\mu$m and larger, with only one axis moving at a time.

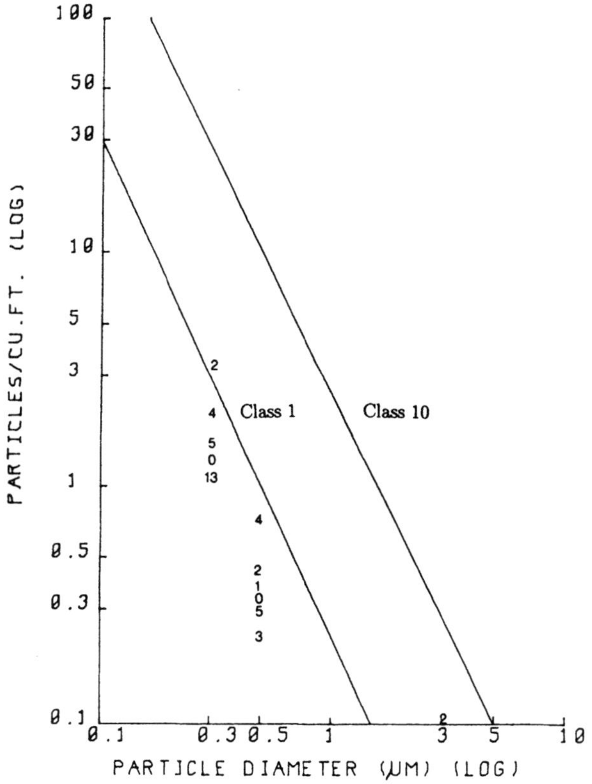


Figure 7. Distribution of particles generated by the motion of the joints of the robot: 0 denotes particles generated by joint 0, 1 denotes particles generated by joint 1, etc.

on a log-log scale. The ordinate in this case is the total particles/$ft^3$ observed when the robot was moving including contributions present in the background. It is apparent that the robot falls within the Class 10 level in this case.

## ROBOT END-EFFECTORS

The part of the robot that actually comes into contact with the object being manipulated, the robot tool or end-effector, is of crucial importance when it comes to the generation of particles because of its proximity to the material being manipulated. Although end-effectors vary widely depending upon the application, they all have some form of actuator which is energized when the object is grasped and de-energized when it is released. This actuation mechanism, plus the body of the end-effector itself, are potential sources of contamination in the immediate vicinity of the object being manipulated. A contamination analysis was performed on three end-effectors, each designed to manipulate $6'' \times 6''$ glass photomasks used in the manufacture of integrated circuits. Each end-effector was thoroughly cleaned with an isopropyl alcohol solution and special wipes prior to the test.

### Mechanical Gripper

The first end-effector tested was a traditional mechanical gripper design. The body of the gripper was made of Delrin, a commercial acetal resin. It had two stainless steel arms with small Delrin gripper heads designed to contact the edge of the photomask. Each gripper head was mounted to its arm with a Bendix Flexural Pivot, a special bearing made of flat crossed springs supporting concentric rotating sleeves. The flexural pivots are frictionless and do not require lubrication, features which make them highly suitable for clean room material handling. The flexural pivots support an angular deflection of up to $\pm 12$ degrees. Consequently, in the presence of small uncertainties in the position and orientation of the photomask, they provide the necessary compliance for reliable automated handling. In order to develop a fail-safe design, one in which the gripping force is not lost in the event of a power failure, linear springs were used to close the stainless steel gripper arms. Each arm was opened by a stepper motor driven linear actuator whose position was monitored with a linear potentiometer. Finally, an optical reflective sensor was mounted on each gripper head to verify the presence of a properly grasped photomask.

The mechanical gripper was tested for particulate generation by placing the sampling probe of the particle counter directly below the gripper housing, closest to the linear actuators. A 15 minute background count was taken with the gripper inoperative, and this was followed by a test with the end-effector cycling between open and closed positions. The results of the contamination test for the mechanical gripper are shown in Figure 8. It is clear that the average particle count (M), which was obtained from 15 one minute samples, was very close to the background count (BK), and well within Class 1 limits. The maximum particle count over the 15 samples was 1 particle/$ft^3$.

### Bi-metal Gripper

The second end-effector investigated was a less traditional, highly experimental design whose basic actuation mechanism was a pair of bi-metallic strips. The design philosophy here was to develop a very clean end-effector by eliminating the need for an electro-mechanical actuator with frictional surfaces, surfaces which generate particles and require lubrication. Consequently, the stainless steel arms were replaced by bi-metallic strips. Delrin heads were mounted on the bi-metallic strips using flexural pivots as with the mechanical gripper design. A 1″ × 4.5″ thin film thermofoil heater was attached to the high expansion side of each bi-metallic strip. A fail-safe design

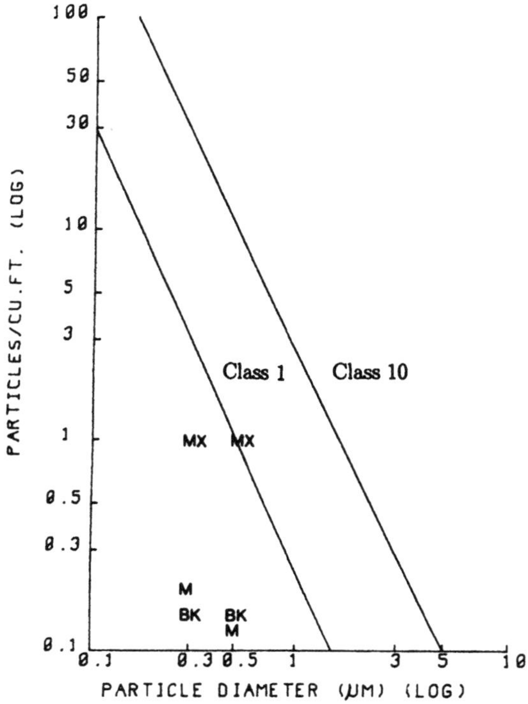


Figure 8. Distribution of particles generated by the actuation of the mechanical gripper: BK = background, M = mean, MX = maximum.

was achieved by mounting the bi-metallic strips in such a way that the metal with the lowest coefficient of thermal expansion was on the outside. Consequently, when the heaters were activated, the bi-metallic arms opened up and in the absence of power they closed. The gripper position was monitored by attaching a thermal sensor strip to the low expansion side of each bi-metallic strip. Since the resistance increased with temperature, the voltage drop across the sensor was used to monitor the deflection of the bi-metallic strip. A calibration curve which shows the opening between the two Delrin heads in inches as a function of the thermal sensor resistance in ohms is shown in Figure 9. Of course one of the drawbacks with a bi-metal gripper is the fact that the actuation mechanism for opening and closing the gripper is quite slow. To open the gripper by 1″ takes almost 90 seconds, and to close it by the same amount takes approximately 120 seconds. However, if the gripper is adjusted properly, a stroke of considerably less than 1″ may be adequate for grasping objects.

Perhaps a more serious problem with the bi-metal gripper is the fact that the bi-metallic strips get quite hot when the gripper is opened. The heaters used are rated at 40 watts, and when the bi-metallic strips get hot they may well generate gaseous emissions that go undetected by the particle counter. When the bi-metal gripper was operated for an extended period, an odor was detected but there was no noticeable change in the particle count. The maximum temperature of the bi-metallic strips

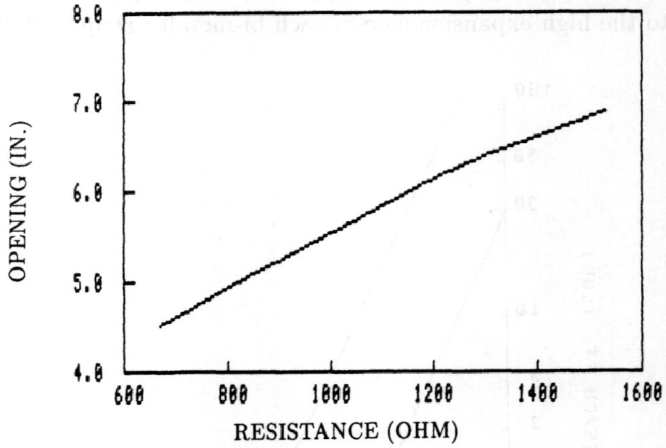


Figure 9. Calibration curve for the bi-metal gripper.

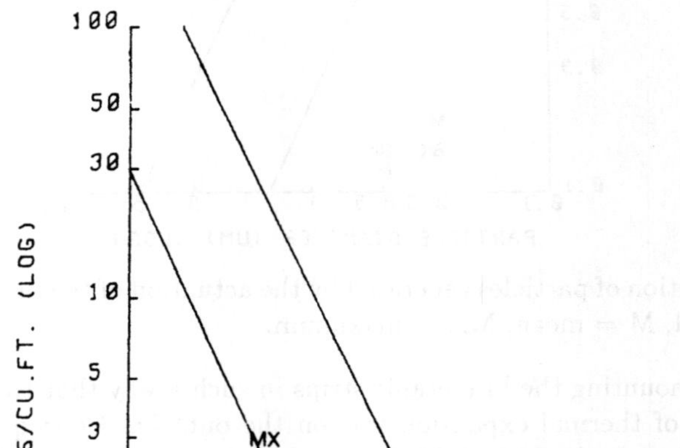


Figure 10. Distribution of particles generated by the actuation of the bi-metal gripper: BK = background, BM = mean, MX = maximum.

was not directly measured, but it was observed that they were too hot to touch and therefore presented a potential safety problem.

For the bi-metal gripper, particulate contamination testing focused on areas of the foil heaters and the bi-metallic strips with the sampling probe located directly below one the of the bi-metallic strips. A 15 minute background count was taken with the end-effector inoperative (heater turned off). The particles generated by the bi-metal gripper were then measured with the gripper cycling between open and closed positions. Due to the relatively slow nature of this actuation mechanism, this test was extended to 45 minutes to ensure an adequate number of cycles. The results of the tests are shown in Figure 10. Again the average particle count (labeled BM) and the background level (BK) are very small and within Class 1 limits. The maximum particle count over the 45 one minute samples was 3 particles/ft$^3$ at 0.3 $\mu$m and greater and 1 particle/ft$^3$ at 0.5 $\mu$m and greater. Interestingly enough, this is slightly more than the mechanical gripper, but is still quite clean.

### Small Servo Tool

The last end-effector tested was the small servo tool, a commercial gripper that is supplied as an option with the Intelledex 660 robot. This is a versatile general purpose parallel-jaw device that is not designed exclusively for clean room operation. It has an accuracy of $\pm 0.008''$, a stroke of $1.8''$, and a gripping force of 25 lbs., controllable in 0.1 lbs. increments. The actuator is a DC servo motor driving a rack-and-pinion mechanism. Each rack drives a slider block which follows a guide shaft. The slider blocks serve as mounting fixtures for attaching custom jaws. In this case two stainless steel arms with Delrin heads attached with flexural pivots were used as with the mechanical gripper. Optical proximity sensors were also used in the gripper heads to detect the presence of an object.

For the small servo tool the exposed rack-and-pinion drive mechanism was the natural focus for the investigation of particle generation. The sampling probe was placed directly below the centering racks, nearest the actuator housing. A 15 minute background particle count was taken with the small servo tool fully open, exposing the stainless steel centering racks. The end-effector was then cycled between its open and closed positions for 15 minutes while particle measurements were taken. During the cycling operation, the small servo tool was neither fully opened nor fully closed, because this would cause the slider blocks to hit mechanical limits, a potential source of additional particles. The results of the contamination measurements are shown in Figure 11. The results suggest that the small servo tool appears to add a significant number of particles to the environment, in comparison with the mechanical gripper and the bi-metal gripper. In this case the average particle count (SS) was considerably higher than the background count (BK). At both the 0.3 $\mu$m sensitivity and the 0.5 $\mu$m sensitivity, the average particle count is within Class 10 limits, but not within Class 1. Furthermore, several large particles were detected at the 3.0 $\mu$m and 5.0 $\mu$m sensitivities which lie in the Class 100 range. Finally, the maximum particle count (MX) shows that the particles generated appear to exceed Class 100 limits.

The most likely cause of the increased number of particles appears to be the friction between the spur gear and the exposed centering rack of the rack-and-pinion drive mechanism. Although these components were cleaned beforehand with alcohol and wipes, they were not disassembled for this purpose and it was therefore not possible to remove all of the lubricant which, in any event, is required for long term operation. It is anticipated that some sort of bellows or hood covering the exposed mechanism might reduce this source of contamination.

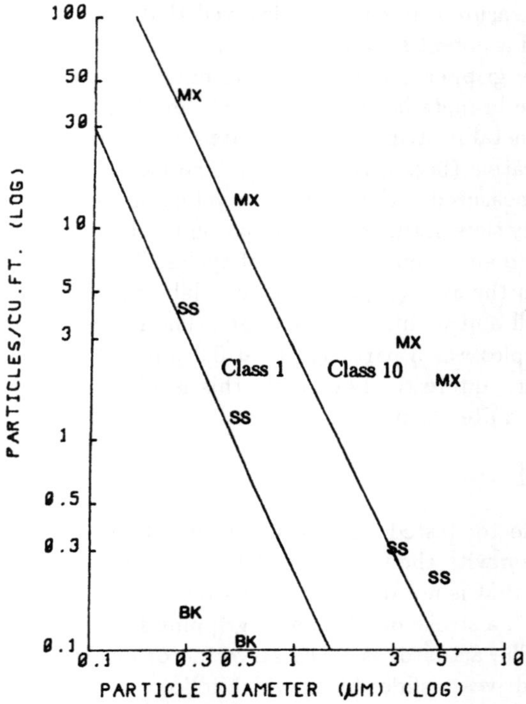


Figure 11. Distribution of particles generated by the actuation of the modified small servo tool: BK = background, SS = mean, MX = maximum.

## CONCLUSIONS

Airborne submicron particulate contamination generated by an Intelledex 660 robot and three special purpose end-effectors was investigated. For the range of particle sizes measured (0.3 $\mu$m and larger), the difference between the background levels with the robot at rest and levels generated by the motion of the robot was not found to be statistically significant. The three end-effectors investigated were all designed, or modified, to handle $6'' \times 6''$ glass photomasks used in the manufacture of integrated circuits. The mechanical gripper and the bi-metallic gripper were found to be the cleanest of the three. In each case, no statistically significant difference was observed between the background particle levels and the particles levels generated by gripper actuation. However, the modified small servo tool did generate significantly higher particle counts, including the presence of a few large particles which probably came from the exposed rack-and-pinion drive mechanism. Although most of the particle counts associated with the robot and end-effectors were near background levels, it is anticipated that as smaller particles (0.1 $\mu$m and smaller) begin to play a more significant role in generating defects and lowering device yield, the contamination produced by robotic manipulators may become a significant manufacturing issue. Due to the limitations of the instrumentation used, particles sizes smaller than 0.3 $\mu m$ were not examined in this study.

## REFERENCES


1. R. Iscoff, Robots in the clean room, Semiconductor Intl., 50-57 (Nov. 1984).
2. A. Chestnut, P.H. Singer, and K. Kearney, Robotic systems enhance manufacturing efficiency, Semiconductor Intl., 58-62 (October 1988)
3. P. S. Burggraaf, Semiconductor factory automation: Current theories, Semiconductor Intl., 88-97 (October 1985).
4. P. H. Singer, Automating inter-equipment transport, Semiconductor Intl., 56-61 (March 1985).
5. J. Dooley, Automated processing equipment, Semiconductor Intl., 120 (Nov. 1986).
6. P. Silberring, Put automation to work in assembly, Semiconductor Intl., 166-169 (April 1989).
7. S. A. Hoenig and S. Daniel, Improved contamination control in semiconductor manufacturing facilities, Solid State Technol., 119-128 (March, 1986).
8. T. Peterson, Anatomy of a clean room robot, Semiconductor Intl., 58-62 (Nov. 1984).
9. Special Report: Process Equipment, Materials, and Automated Systems, Microcontamination, 42-56 (October 1987).
10. H. S. Nagaraj, B. L. Owens and R. J. Miller, Particulate generation in devices used in clean manufacturing, in, "Particles in Gases and Liquids 1: Detection, Characterization and Control," K. L. Mittal, editor, pp. 283-293, Plenum Press, New York, 1989.
11. K. Skidmore, A look at the past and a glimpse into the future, Semiconductor Intl., 68-75 (Dec. 1988).
12. J. R. Monkowski, Particulate surface contamination and device failures, in, "Treatise on Clean Surface Technology," K. L. Mittal, editor, Vol. 1, pp. 123-148, Plenum Press, New York, 1987.
13. W. L. Morgan and J. R. Burnett, Concepts for world-class VLSI manufacturing plants, Semiconductor Intl., 136-147 (June 1984).
14. M. R. Gulett, A practical method for predicting IC yields, Semiconductor Intl., 87-94 (March 1981).
15. R. D. Helander, Certifying a Class 10 clean room using Federal Standard 209C, Microcontamination, 45-49, 76-78 (Sept. 1987).

# PARTICLE RELEASE FROM SURFACES BY MECHANICAL SHOCKS

Hwa-Chi Wang, Horng-Yuan Wen, and Gerhard Kasper

Chicago Research Center, American Air Liquide
5230 S. East Avenue, Countryside, IL 60525

The influence of mechanical shocks on a gas cylinder, one of the critical components in a gas distribution system, was evaluated with respect to particle generation. An experimental technique was developed for applying impact shocks and for simultaneous measurements of the shock intensity, the corresponding particle concentration and size distribution. It was found that a significant amount of submicron particles could be generated by impact shocks on the order of $10^3$ g. The particle size distribution based on number concentration indicated a peak around 0.1 $\mu$m. It is further shown that data reported here might correspond to ppb-ppm resuspension. It implies that allowable shock intensity for today's high-purity gas distribution systems should be defined based on an extremely low level of resuspension, whereas classical adhesion theory and experiments are based on 50% removal.

## INTRODUCTION

The cleanliness requirement in today's microelectronic manufacturing environment has been tightened by several orders of magnitude in the last decade. The allowable particle generation from a gas distribution system has decreased accordingly, and, in today's ultra-clean environment, is at least 7 orders of magnitude lower than ambient conditions. Since many prevailing concepts and conventions regarding particle release from surfaces were developed from air pollution studies, it is of critical importance to understand the basic assumptions behind the available theory and the limitations of extrapolating over several orders of magnitude.

A brief review on the process of particle release from surfaces is given in the next section. The need for understanding particle generation by mechanical processes, especially for components in a gas distribution system, is indicated. An experimental technique designed to meet this need is described in the following section. Experimental results, comparison with common perceptions, and implications for defining allowable shock intensity are discussed.

*Particles in Gases and Liquids 2,* Edited by K.L. Mittal
Plenum Press, New York, 1990

## PARTICLE RELEASE FROM SURFACES

Particle release from surfaces is governed by two distinct, but often co-existent, processes: a deterministic process characterized by a sudden rise and a rapid decay in particle concentration, and a stochastic process more or less following a power law decay[1,2].

Figure 1 is a good example of these two processes[3]. A contaminated stainless steel 1" tubing 10 feet in length was purged with filtered nitrogen and the hourly average particle concentration was monitored over 3 days. In the first two days, a gradual decay in particle concentration was observed. After a fairly low particle concentration was reached, a mechanical shock was applied to the tubing. Immediately after the shock, a spike of over 3 orders of magnitude increase in particle concentration was observed. A similar phenomenon was observed at the end of the third day when the flow velocity was doubled. It is clear that the gradual decay occurring under steady-state conditions follows the stochastic process, while the spikes corresponding to transient conditions are deterministic.

Many studies have been focused on the stochastic process of particle re-entrainment. Two most recent studies offered the kinetic[2] and the dynamic[1] pictures of this process. One of the highlights of the kinetic model is that measurable quantities such as particle concentration n(t) at time t and flow rate Q can be correlated with a simple equation.

$$n(t)\, Q / N = [1/F_{max}]\ [1/t]\ [1-\exp(-t/t_1)]\ [\exp(-t/t_2)]$$

where

$$N = \int n(t)\, dt.$$

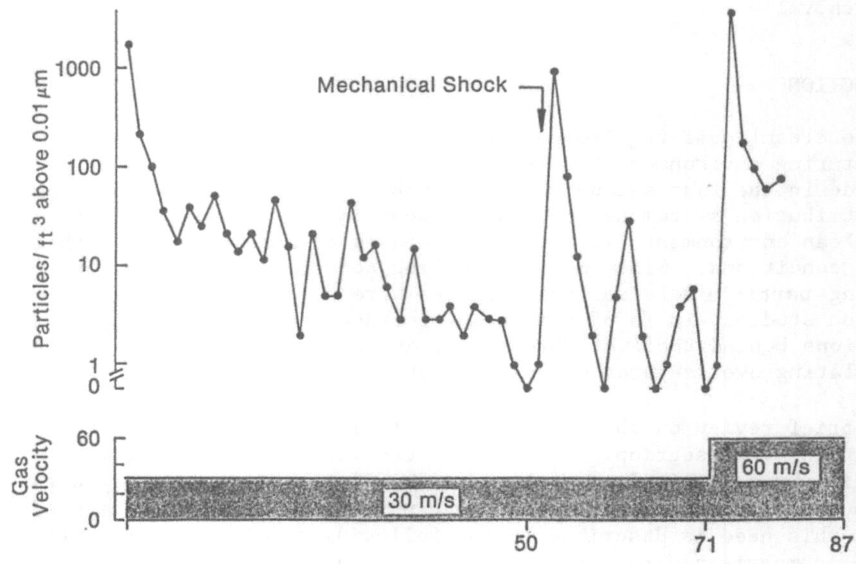


Figure 1. Particle release from a contaminated stainless steel 1" tubing. Both deterministic and stochastic processes were responsible for particle release.

$F_{max}$ is the ratio of adhesion force to removal force for removable particles, and $t_1$ and $t_2$ are the characteristic times for short term and long term behavior[2]. This equation shows a general 1/t dependence but deviates from such dependence when t is shorter than $t_1$ or longer than $t_2$. The first 50 hours data in Figure 1 are replotted in Figure 2 together with predictions from the above equation with $F_{max}$ = 4, 5 and 6 (as shown by dashed lines). The prediction of 1/t dependence is shown by solid line. Fairly good agreement between experimental results and theory is obtained.

In comparison, the deterministic behavior usually caused by transient conditions bears no time correlation (response time of the system is smaller than that of the instrument) but shows strong dependence on the intensity of the disturbance to the system. Therefore, the magnitude of the spike (Figure 1) should be correlated with the intensity of the disturbance. Such intensity is in general difficult to characterize and to quantify. Mechanical shocks may be one of the most common disturbances to a flow system and can be quantified. Hence, it is logical to investigate particle generation by mechanical shocks to a flow system.

## REVIEW OF PARTICLE RELEASE BY MECHANICAL PROCESSES

It is of interest to review what we know about particle release from surfaces by mechanical means. When designing a mechanical cleaning device or setting a maximum allowable impact shock for a high-purity

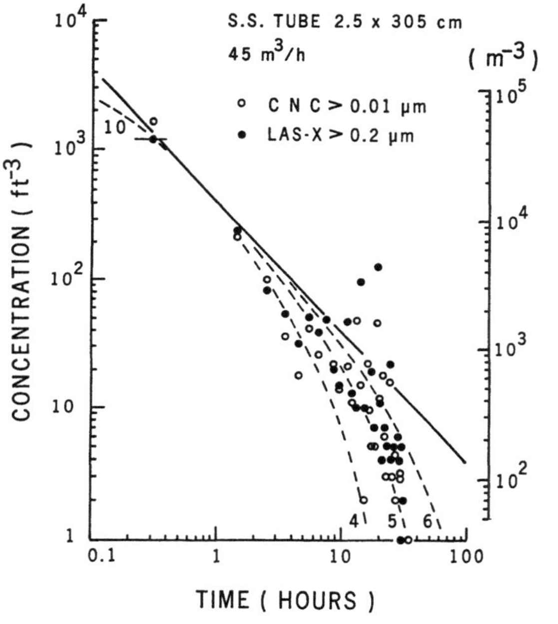


Figure 2. Comparison of particle release from a stainless steel tube purged at a flow rate of 45 $m^3/h$ with the kinetic model of Wen and Kasper[2]. The solid line shows a 1/t dependence while the dashed lines show the model prediction with $F_{max}$ = 4, 5, and 6.

system, there are two frequently asked questions: "What is the acceleration required to remove a particle from a surface?" and "What is the size distribution of particles generated by a mechanical process?"

To answer the first question requires the knowledge of adhesion force and removal force for the given particle-surface system. The classical adhesion theory[4] was proposed more than 50 years ago. Since then many advances have been made by including the elastic deformation resulting from the adhesion force[5,6] at the particle-surface interface. Extensive efforts[7,8] have been devoted in this field; however, the adhesion force for any given particle-surface system still can not be accurately calculated. A key reason is the uncertainties in material data for micron-sized particles, such as the Hamaker constant, the mechanical constant, the surface energy, etc.

To obtain an order-of-magnitude estimation, the adhesion force can be calculated from the classical Bradley-Hamaker theory,[4] provided that the Hamaker constant is known for the materials in contact at the interface. (The contact materials at the interface are not necessarily those of the particle and surface; layers of contaminants are often present unless in atomically clean, high vacuum conditions.) The removal force is usually assumed to be the same as or greater than the adhesion force. Based on these two common approaches, Ranade[9] calculated the acceleration required to remove aluminum oxide particles from an aluminum oxide surface as shown in Table I. Acceleration over $10^6$ g is required to remove particles smaller than 1$\mu$m. Since the Hamaker constant for typical materials varies less than one order of magnitude, similar acceleration is required to break the particle-surface contact for other materials. The above calculation can be considered as the lower limit. An even larger acceleration is needed if other adhesion mechanisms such as capillary forces and electrostatic forces are present, or if the JKR[5] or DMT[6] model is used to account for the adhesion-induced deformation.

To verify the classical theory, many experiments have been performed. The common techniques involve depositing known particles on a substrate, applying removal force of variable magnitude, usually in the form of centrifugal force or vibration, and evaluating the fraction of removal. It is recognized that particles of the same size can be removed over a wide range of removal forces, reflecting a distribution in adhesion force even for monodisperse particles. It was proposed, and followed by most researchers, to use the force corresponding to 50% removal as the adhesion force[10,11]. Order-of-magnitude agreement was generally obtained between theory and experimental results for 50% removal.

Table I. Acceleration Required to Remove $Al_2O_3$ particles from $Al_2O_3$ Surfaces[9].

| **Particle Size** ($\mu$m) | **Adhesion Force** (dyn) | **Acceleration Required** (g) |
|---|---|---|
| 10 | 9E-2 | 4.5E4 |
| 1 | 9E-3 | 4.5E6 |
| 0.1 | 9E-4 | 4.5E8 |

To answer the second question, a schematic diagram such as Figure 3[12] can be used as a guideline. Particles generated by mechanical means, i.e. friction, abrasion, etc., have a mass mean diameter on the order of 10 μm. This picture was confirmed by ambient measurement as shown in Figure 4[12], in which the size distribution was recorded without (triangle) and with (circle) the particle emission of mechanical origin. Hence, the majority of mechanically generated particles are greater than 1 μm.

In light of the above discussions, a naive extrapolation would lead to the conclusion: Mechanical shocks are of no concern to a gas distribution system since (1) shock intensity for normal operation seldom exceeds $10^3$ g[13] and (2) few particles greater than 1 μm are present in the system. In reality, however, impact shocks do cause significant problems in a gas distribution system as will be shown below. Thus, experiments are needed to quantify the relationship between shock intensity and particle generation.

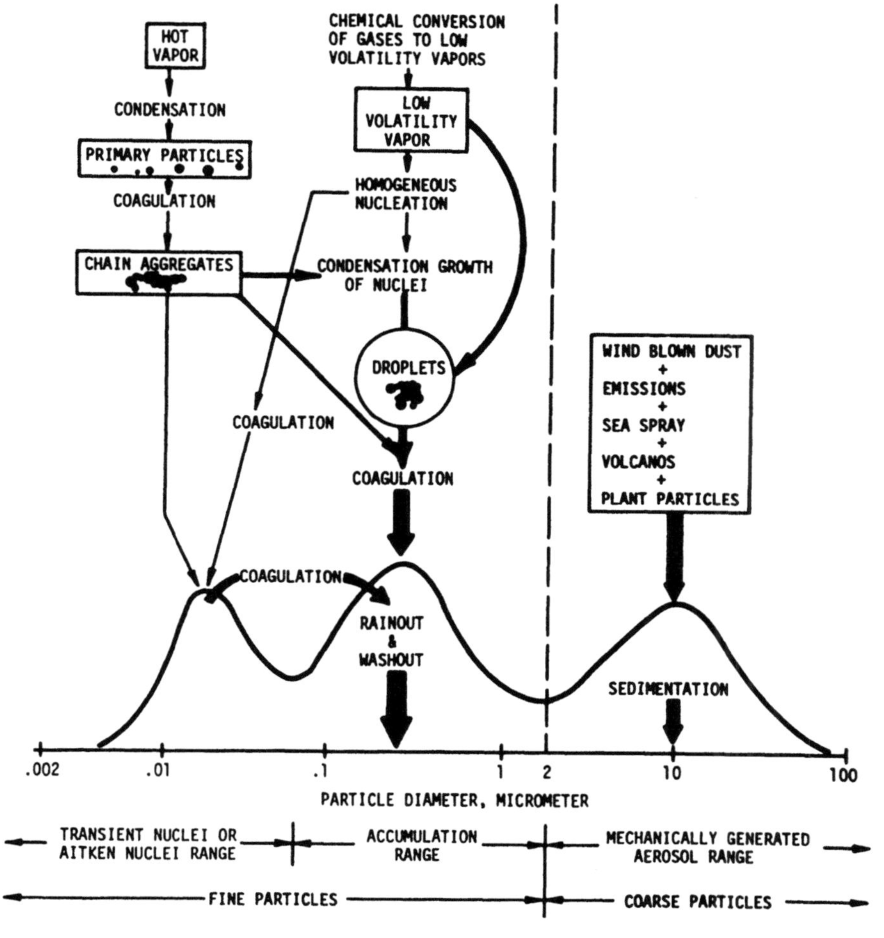


Figure 3. Source and size distribution for ambient aerosol. Mechanically generated particles are primarily coarse particles[12].

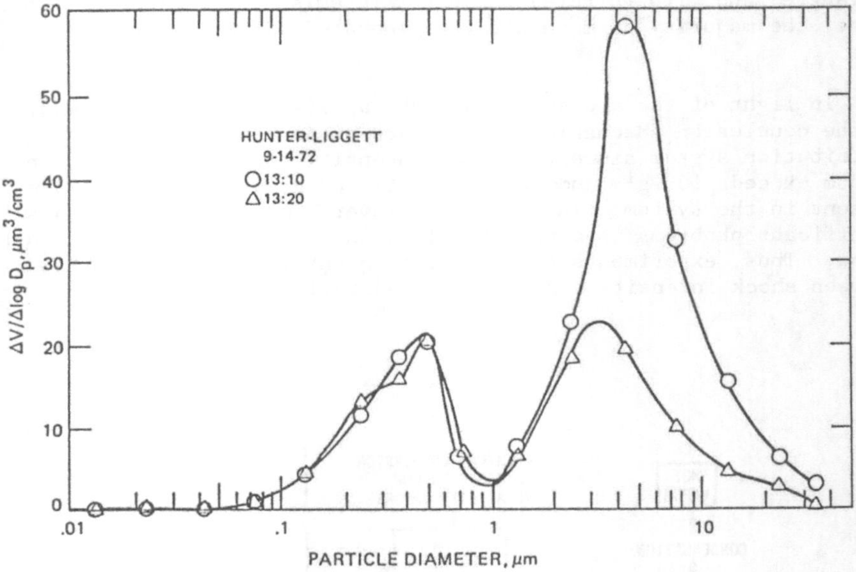


Figure 4. Sudden growth of the coarse particle mode due to local dust sources measured at Hunter-Liggett Military Reservation by Whitby and Sverdrup during the 1972 ACHEX experiment[12]. Circle: measured at 1:10 pm with local dust sources due to anthropogenic activity. Triangle: measured at 1:20 pm. Note only the coarse mode varies with local dust sources.

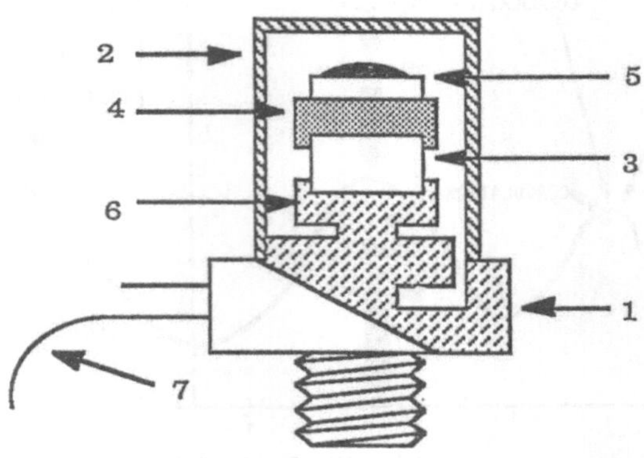


Figure 5. Schematic diagram of a piezoelectric accelerometer. 1: accelerometer base, 2: accelerometer cap, 3: Crystal, 4: mass, 5: circuit, 6: internal isolation base, and 7: cable.

## EXPERIMENTAL

To investigate the impact of mechanical shocks on a gas delivery system, we selected a gas cylinder for our experiments because (1) it is one of the components most susceptible to shocks, (2) shock intensity can be quantified more accurately, and (3) it can be pressurized to hold a large gas volume for satisfactory sampling using particle counters.

To measure the magnitude of the impact shocks, a piezoelectric accelerometer can be used to obtain an electrical signal proportional to acceleration associated with the shock. A low-impedance accelerometer as shown in Figure 5 is convenient since the piezo-electric crystal has a built-in impedance converter to transform the high-impedance voltage signals sensed by the crystal to low-impedance signals. Thereby, the output signals can be easily detected by an oscilloscope without appreciable signal loss or distortion.

Controllable impact shocks on the cylinders were obtained by a steel ball striking the cylinder. The ball was attached to a 60 cm chain with the other end fastened to the neck of the test cylinder and was lifted to form an angle with the cylinder before release. Shock intensity increases with the angle. The impact locations were roughly at the midpoints for most cylinders. The shock intensities measured at various locations are listed in Table II. The shock intensity remains relatively constant at various locations.

Figure 6 shows the schematic design of the experimental system used to measure the particle generation as a function of shock intensity. Pressurized nitrogen of fixed flow rate was filtered by an absolute filter and introduced into the test cylinder. The filling process was stopped when a pre-determined pressure was reached. The compressed nitrogen in the test cylinder was then released through a critical orifice pressure reducer to allow particle measurements. The performance of this orifice-type pressure reducer[14,15] is well documented. Two particle counters were used in parallel; the condensation nuclei counter (TSI 3760) registered the total number of particles greater than 0.01 $\mu$m, while the laser particle counter (PMS LAS-X) sized particles greater than 0.1 $\mu$m into 16 channels.

TABLE II. Distribution of Impact Shocks for Cylinders Subject to the Standardized Test*.

A. **AXIAL DISTRIBUTION**

| Position | POI** | 0.5m Above | 0.5m Below |
|---|---|---|---|
| Magnitude | 880g | 860g | 890g |

B. **RADIAL DISTRIBUTION**

| Position | POI | 45° | 90° | 135° | 180° |
|---|---|---|---|---|---|
| Magnitude | 1100g | 103 g | 950g | 1150g | 950g |

* Typical standard deviation for the measurements is about 200g.

** Point of impact

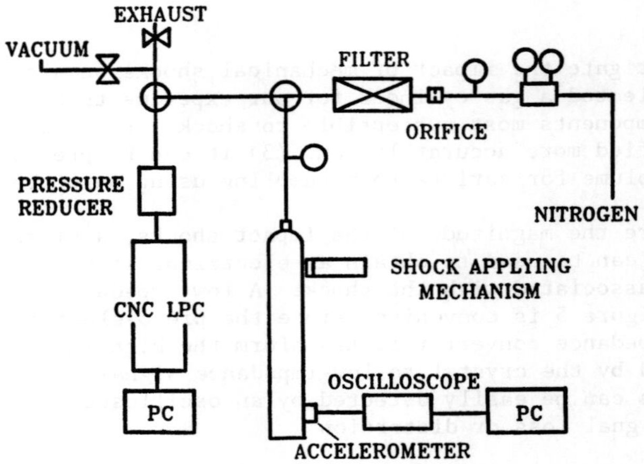


Figure 6. Experimental setup for cylinder shock test.

Mechanical shock was introduced and its magnitude determined while particles concentration was continuously monitored. The difference in particle counts before and after the shock yields the number of particles generated by the applied mechanical shock.

**RESULTS**

Figure 7 shows particle generation as a function of the magnitude of mechanical shocks. The vertical axis represents the total number of particles generated by the corresponding impact shock. Particle generation increases with shock intensity and seems to level off for shocks greater than $10^3$ g. It should be noted that the characteristics of particle generation may vary for other cylinders.

About $10^5$ particles were generated by an impact shock on the order of $10^3$ g. Such high particle generation is definitely not acceptable for gas delivery systems commonly used in microelectronics industry. There is no detectable particle generation for shocks less than $10^2$ g.

Also indicated in Figure 7 are shock intensities for various cylinder handling activities simulated in our laboratory. Normal handling includes transporting cylinders on a cart, securing cylinders to bench, and mounting pressure regulators. Typical shocks associated with normal handling are less than 100 g. However, accidental falling of gas cylinders can cause shocks on the order of 1000 g.

Figure 8 shows the size spectrum of the shock-generated particles recorded by particle counters. The geometric mean of channel boundary diameters was used and the concentration was normalized by the channel width. The peak is at 0.1 $\mu$m. Very few particles greater than 1 $\mu$m were found.

Since the critical orifice used for pressure reduction has a cut-off diameter around 8 $\mu$m[16], there is no significant distortion in size spectrum. As a precaution, an experiment was performed at ambient pressure bypassing the critical orifice. The cylinder was continuously purged with filtered nitrogen and monitored with particle counters.

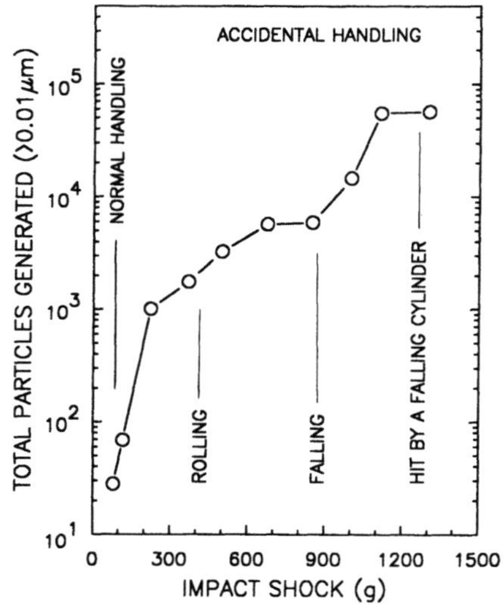


Figure 7. Total particle generation as a function of shock intensity. A contaminated cylinder was used to show the effects of impact shocks.

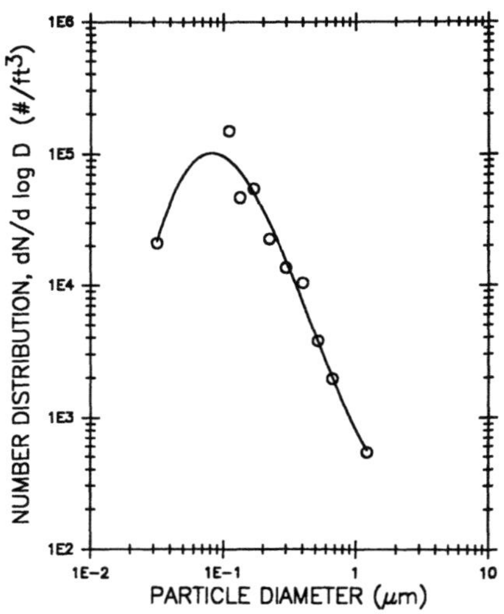


Figure 8. Size distribution of shock-generated particles from the test cylinder. The solid line is eye-fitted curve to the data.

Impact shocks were introduced and particle counts were integrated until the original background level was restored. Similar results to that shown in Figure 8 were obtained.

**DISCUSSION**

Relevant Removal Force

The force corresponding to 50% removal of deposited particles is frequently used to report experimental results and is often cited as the adhesion force for the particle-surface system. Figure 7 clearly demonstrates that a force several orders of magnitude smaller than the adhesion force can generate an intolerable amount of particles for most gas distribution systems. Hence, the adhesion force is not a good indicator in this case for allowable removable force (or allowable shock intensity). A better indicator should be defined according to the tolerable fraction of resuspension, which in turn depends upon the cleanliness requirement of the flow system.

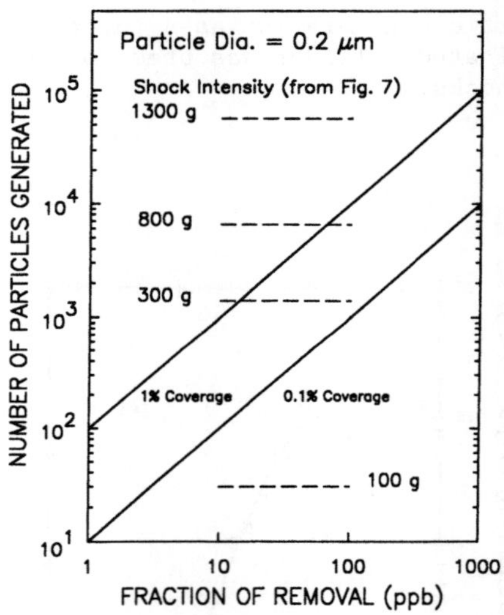


Figure 9. Total particle generation as a function of fraction of removal for two surface coverages, 1% and 0.1%. The fraction of removal is calculated assuming a uniform particle size of 0.2 μm and a smooth cylinder surface.

It is of interest to obtain order-of-magnitude correlation between shock intensity and the fraction of removal from data presented in Figure 7. Assuming a uniform particle diameter of 0.2 μm and a smooth internal surface of the cylinder, the total suspended particles of this size can be related to the fraction of removal for a given surface coverage. Based on the SEM photographs obtained from the inner surface of the cylinder, we assumed a surface coverage between 1% and 0.1%. The results are plotted in Figure 9. Also included are data from Figure 7, relating particle generation with shock intensity. Two observations are made: (1) The shock on the order of $10^3$ g can only resuspend 0.1 ppm of total deposited particles, and (2) Fraction of removal on the order of sub-ppm levels is not acceptable in today's ultra-clean environment.

Number Distribution vs. Mass Distribution

The statement "Particles generated by mechanical processes are primarily coarse particles" is true only for mass or volume distribution. In a field where mass is relevant, such as inhalation toxicology, the mass-based size distribution provides useful information. However, number concentration is monitored and correlated with yield in the microelectronics industry, and thus the number distribution should be used.

Figure 10 shows the number distribution and volume distribution for mechanically generated particles. The volume distribution clearly indicates a major fraction of coarse particles, while the number distribution shows a peak of submicron particles. The submicron peak would be totally overlooked if plotted in volume distribution.

Uncertainties

There are also a few uncertainties regarding the mechanisms of particle release. The inceptive motion of the resuspended particle can strongly influence the minimum removal force. A recent study[17] indicated that the removal force may be two orders of magnitude smaller than the adhesion force if particle removal is initiated by rotation. Shock generated particles may be foreign deposits or result from material failure due to high local stress concentration.

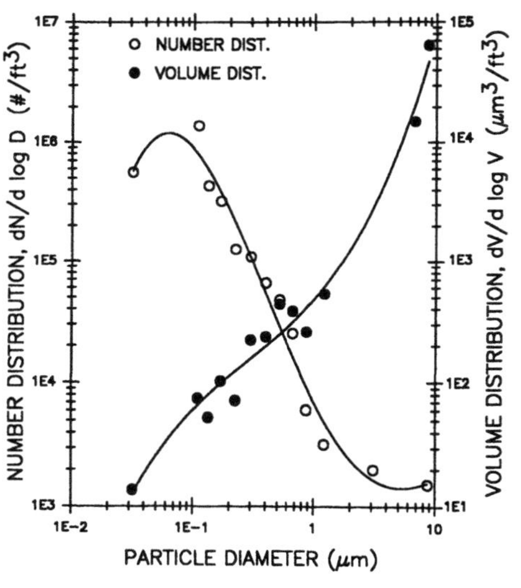


Figure 10. Number and volume distribution for mechanically generated particles. Solid lines are eye-fitted curves to show the general trend.

## CONCLUSION

The adhesion force corresponding to 50% removal can not be used to define the allowable shock intensity for a clean gas distribution system, where a much lower fraction of resuspension is encountered. An experimental technique, capable of simultaneous measurements of shock intensity and particle generation, is required to measure the response of a gas distribution system. The allowable shock intensity can then be defined based on the cleanliness requirement.

## ACKNOWLEDGEMENTS

We thank R. Udischas and D. Montgomery for valuable assistance in experimental setup and data collection.

## REFERENCES

1. M. W. Reeks, J. Reed, and D. Hall, J. Phys. D: Appl. Phys., 21, 574 (1988).
2. H. Y. Wen and G. Kasper, J. Aerosol Sci., 20, 483 (1989).
3. G. Kasper, H. Y. Wen, and H. C. Wang, in "SEMICON/EAST89 Technical Proceedings", 80 (1989).
4. H. C. Hamaker, Physica 4, 1058 (1937).
5. K. L. Johnson, K. Kendall, and A. D. Roberts, Proc. Roy. Soc. Ser. A, 324, 301 (1971).
6. B. V. Derjaguin, V. M. Muller, and Y. P. Toporov, J. Colloid Interface Sci., 53, 314 (1975).
7. K. L. Mittal, editor, "Particles on Surfaces 1: Detection, Adhesion, and Removal", Plenum Press, New York (1988).
8. K. L. Mittal, editor, "Particles on Surfaces 2: Detection, Adhesion, and Removal", Plenum Press, New York (1989).
9. M. B. Ranade, Aerosol Sci. Technol., 7, 161 (1987).
10. M. C. Kordecki, J. M. Gladden, and C. Orr, Report No. B-148, Engineering Experiment Station, Georgia Inst. of Technol., Atlanta, Georgia (1959).
11. A. D. Zimon, "Adhesion of Dust and Powder", Plenum Press, New York (1969).
12. NRC Reports, "Airborne Particles", University Park Press, Baltimore, pp 11-54 (1979).
13. H. C. Wang, H. Y. Wen, and G. Kasper, in "Proceedings of Inst. of Environ. Sci. 35th Annual Technical Meeting", 406 (1989).
14. G. Kasper and H. Y. Wen, Particles in ultrapure process gases, in "Handbook of Contamination Control in Microelectronics", D. Tolliver, editor, Noyes Publications, Park Ridge, NJ (1988).
15. H. C. Wang, H. Y. Wen, and G. Kasper, Solid State Technol., 155 (May, 1988).
16. H. C. Wang and G. Kasper, Note on inertial deposition of particles on an orifice disc, J. Aerosol Sci., in press (1990).
17. H. C. Wang, Theoretical comparison of mechanisms to overcome particle-surface adhesion. presented in the European Aerosol Conf., Vienna (1989).

# FLOWRATE DEPENDENCE OF PARTICLE SHEDDING FROM A GAS DELIVERY LINE

Scott J. Alberhasky* and Richard J. McCluskey

Department of Chemical Engineering
Clarkson University
Potsdam, NY 13699-5705

Denis J. Poley

IBM Corporation
Essex Junction, VT 05452

Particle count measurements have been made on filtered $N_2$ passed through a 30 ft section of 0.18" i.d., 316L stainless steel tubing that had seen extended service as a delivery line for dry HCl. Experiments of two hours duration were run in sequence at $N_2$ flowrates of 0.14, 0.33, 0.43 and 1.0 CFM. The corresponding flow Reynolds numbers were 1200, 2900, 3800 and 8800. A PMS Model LAS-X particle counter tallied particles within a 0.3 lpm exit sample stream over 10 minute time intervals. Counts were made for fifteen equistep size classes between 0.17 and 0.62 µm, as well as for all particles >0.62 µm.

Average particle count/ft$^3$ increased steadily with increasing flowrate and ranged from 230 to 38,000 counts/ft$^3$. Within each flowrate experiment, counts generally decreased with time, but occasional 10 minute time periods gave unusually high counts. A simple exponential decay model was used to describe all results except for the 0.43 CFM data which showed great variability in particle counts between adjacent 10 minute time intervals. Least squares decay time constants ranged from 38 minutes at the highest flowrate to 83 minutes at the lowest flowrate.

Particle size distributions at the two higher flowrates showed little change with either flowrate or time. The distributions were log-normal with median sizes close to 0.35 µm. At the two lower flowrates, size distributions for time intervals showing unusually high counts were significantly different from those showing lower, and more typical, particle levels. However, if the "high" count data are disregarded, size distributions for the two lower flowrates are again very similar, with 70% of the counts for particles larger than 0.62 µm.

---

* Present address: Motorola Corporation, Mesa, AZ 85202.

*Particles in Gases and Liquids 2,* Edited by K.L. Mittal
Plenum Press, New York, 1990

## INTRODUCTION

A major cause of device failure in the manufacture of VLSI circuits is contamination by particles in the submicron size range. Future generations of semiconductor devices will have smaller feature dimensions, making them even more susceptible to contamination caused failure. Therefore, the study of factors influencing particle generation during the manufacture of semiconductor devices is important both for improving yields from current systems and for rationalizing the design of future systems.

This paper reports on experiments examining the effect of gas flowrate on particle release from the wall of a gas delivery line. Recently published papers[1-4] have described material of construction and welding considerations for minimizing pipeline contamination sources. Measurement of microcontamination in gas systems is usually made with a laser particle counter, a condensation nucleus counter, or a combination of these instruments. The precautions necessary for obtaining accurate particle counts have been described[5-8].

Nonetheless, severe problems remain when probing ultraclean gas delivery systems[9-11]. These include: difficulty in distinguishing particle counts from background noise, long times necessary to achieve a steady state count rate, and the occurrence of seemingly random "bursts" of particles. We have circumvented the first of these difficulties by examining a section tubing expected to have a high degree of particle contamination. By focusing on the short time decay in exit particle concentration, the need to establish steady state conditions was avoided. Lastly, standard statistical tests for detection of outliers were employed to filter sudden high particle counts from our data set.

## EXPERIMENTAL

A 30 ft section of 316L stainless steel tubing, 0.18 inch i.d., containing no ninety degree bends, was taken from service, carefully capped, and then installed in the experimental particle monitoring system shown in Figure 1. This 30 ft section had been employed for several years as part of a delivery line for dry HCl gas. A shorter section of similar tubing that had seen intermittent exposure to HCl in a gas analysis laboratory had been tested previously and found to be heavily contaminated with particulates[12]. Particles from the gas laboratory tubing were collected on a Nucleopore filter and found from x-ray fluorescence spectroscopy (XFS) analysis to be rich in iron and nickel chloride. Thus, it was

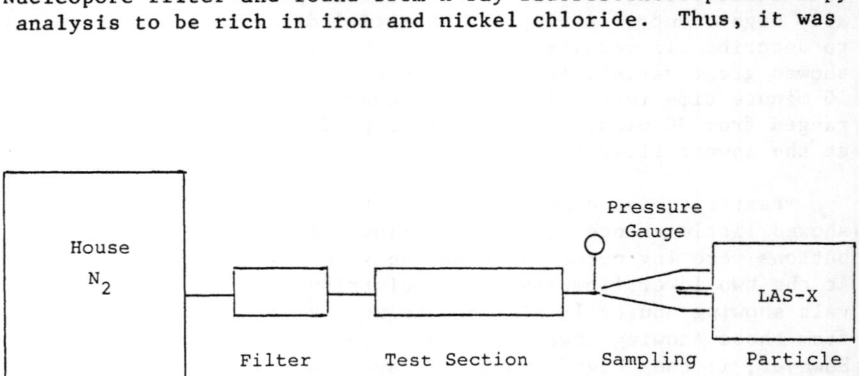


Figure 1. Experimental set-up for measuring particle release from gas line tubing.

anticipated that the 30 ft test section would also shed quantities of particulates.

House nitrogen was fed to the test section after passing through a 0.05 µm filter (Waferguard Model WGGB06WR1). After the test section, the gas flowed through a pressure reduction cone, detailed in Figure 2, from which a 0.3 lpm sample was passed into a Particle Measurement Systems (PMS) laser particle counter, Model LAS-X. The sampling cone was provided by PMS. Gas discharged from the cone at atmospheric pressure. The flowrate through the test section was found from a pressure gauge at the cone's inlet and a calibration curve, established by the use of a bubble flowmeter on the exit of a large glass cylinder.

The LAS-X counter sorted particles into fifteen equisize classes between 0.17 and 0.62 µm. The counter also gave a single tally of all particles larger than 0.62 µm. Counts in each size range were obtained over ten minute time intervals. The counter had been calibrated against polystyrene latex spheres; thus, counts are subject to error due to differences in particle shape and refractive index from those of the latex spheres[13]. There is also a small undercount due to non-isokinetic sampling[14]. An isolation transformer was always included in the power line to the LAS-X to minimize false counts due to electrical disturbances. With the isolation transformer, operation with no gas flow produced a background count of only one particle per hour. Particle count experiments were performed at four different flowrates for two hours duration at each flowrate. Experiments with the 30 ft test section were run over a two day period in sequence beginning with the lowest flowrate. Replication of the test sequences, while obviously desirable, was not possible due to other demands on space and equipment.

A comparison experiment with clean tubing was performed shortly after the flowrate experiments on the HCl service tubing. The sampling cone and LAS-X particle counter were connected to a 300 ft section of newly installed, chempolished, 0.18 inch i.d., stainless steel tubing. Immediately prior to testing, the 300 ft section was purged for two days with filtered nitrogen at 1 CFM, the single flowrate used in this particle count test.

## RESULTS

Table I lists the particle count data obtained on the 30 ft section of HCl service tubing at flowrates of 0.14, 0.33, 0.43 and 1.00 CFM, as well as data from the 300 ft section of clean tubing tested at 1.00 CFM. All counts were for particles larger than 0.17 µm, measured using the LAS-X laser particle counter. Also shown in the table are the average particle concentrations for the nitrogen leaving the tubing in particles/ft$^3$, the percentage of the particles classified by the LAX-X as less than 0.5 µm, and the flow Reynolds numbers.

The particle count data on the 30 ft section of tubing show three important phenomena: particle counts increase with increasing flowrate; counts generally decrease with time during the course of any experiment; and there are bursts or episodic periods of very high particle counts. The average exit particle concentrations increase uniformly with increasing flowrate, as does the percentage of particles less than 0.5 µm. The flow Reynolds numbers span the range from strictly laminar to fully turbulent flow.

Particle Reynolds numbers based upon a 0.5 µm diameter are a factor of 9000 less than the flow Reynolds numbers, and for the conditions of

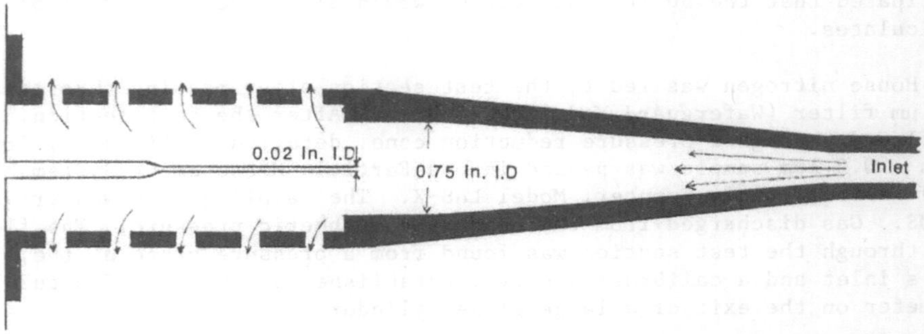


Figure 2. LAS-X Pressure Reduction Vessel.

Table I. LAS-X Particle Counts (>0.17 μm) over 10 Minute Intervals with a 0.3 lpm Sample Stream

| | 30 ft of HCl Service Tubing at Various Flowrates | | | | 300 ft New Tubing |
|---|---|---|---|---|---|
| Time (min) | 0.14 CFM | 0.33 CFM | 0.43 CFM | 1.00 CFM | 1.00 CFM |
| 10 | 15 | 54 | 448 | 5866 | 4 |
| 20 | 8 | 50 | 52 | 6664 | 0 |
| 30 | 178 | 32 | 63 | 5135 | 0 |
| 40 | 12 | 36 | 1627 | 14838 | 0 |
| 50 | 6 | 49 | 777 | 2038 | 4 |
| 60 | 46 | 787 | 45 | 5945 | 0 |
| 70 | 5 | 11 | 26 | 3648 | 0 |
| 80 | 4 | 22 | 448 | 1842 | 1 |
| 90 | 7 | 18 | 17 | 710 | 0 |
| 100 | 4 | 13 | 1536 | 803 | 2 |
| 110 | 3 | 13 | 25 | 501 | 0 |
| 120 | 4 | 14 | 37 | 758 | 0 |
| Counts/ft$^3$ | 230 | 865 | 4010 | 38300 | 8.7 |
| % <0.5 μm | 27 | 39 | 69 | 71 | 100 |
| Reynolds # | 1200 | 2900 | 3800 | 8800 | 8800 |

of our experiments are always less than one.

For the new 300 ft section of tubing, the particle counter saw no contaminants during most of the time periods examined. This illustrates the efficacy of modern techniques for chempolishing and handling gas system components. It also demonstrates that the high counts obtained with the 30 ft section of tubing were not an artifact of the sample cone or detector. The contrast between the particle counts with the new tubing versus the HCl service line is even more pronounced when it is noted that the new tubing was a factor of ten longer and, hence, had an order of magnitude greater surface area.

Before further analysis, the counts were corrected for non-isokinetic sampling effects using equation (1)[8],

$$C/C_o = 1 + (U_o/U - 1)\{1 - 1/[1 + (2 + 0.62U/U_o)Stk]\} \quad (1)$$

Where C is the observed particle count, $C_o$ is the "true" count, U is the velocity in the sample probe, $U_o$ is the velocity upstream from the sample probe, and Stk is the Stokes number characterizing each particle size range. Correction factors, $C/C_o$, found for dry nitrogen at 20°C and spherical particles of specific gravity two, are listed in Table II. The Stokes number for particles larger than 0.62 μm was arbitrarily based upon a diameter of 0.9 μm; all other size ranges were characterized using the mid-range diameter.

A simple exponential decay model describing decline in particle counts with time is given in Equation (2). This model assumes that a fixed number of particulates, removable at the test flowrate, are initially at rest on the tube's surface. The possibilities of erosion exposing fresh layers of particles, of a sequence of particle sources and sinks, or of decomposition of large particles into numerous samll ones are not considered.

$$P = P_o \exp(-kt) \quad (2)$$

where P is the number of particles larger than 0.17 μm, which can be dislodged at the test flowrate, but remain on the tube wall at time t; $P_o$ is the initial number of removable particles on the tube wall; 1/k is the mean particle lifetime within the tube; and t is time.

Since our data are of particle counts over sequential ten minute time intervals, Equation (3) is the preferred form of this model for linear least squares analysis.

$$C_{i-10} - C_i = (Q_{sample}/Q)*P_o(\exp(10k) - 1)\exp(-kt_i) \quad (3)$$

where the left hand side is the particle count for a 10 minute period ending at time $t_i$, $(Q_{sample}/Q)$ is the ratio of gas sampling rate (0.3 lpm) to gas flowrate, and k has units of $min^{-1}$.

The residuals from a linear least squares analysis of the natural logarithm of particle count versus time were subjected to the Dixon outlier test[15]. If an extreme data point failed the test, it was discarded and the least squares procedure was repeated until all data included in the analysis could be considered part of the same population to a 95% significance level. Accordingly, the two largest particle counts were removed from the 0.14 CFM data, one data point was discarded from the 0.33 CFM data, and no data points were eliminated from the higher flowrate data sets. All discarded data points would have failed the Dixon outlier test at a 99% significance level, and all retained data would have passed even at an 80% significance level. So only a few particle counts, that were clearly of an unusually large magnitude, were removed from the data set.

Table II. Non-isokinetic Correction Factors.

| | Particle Diameter in μm | | | | | | | | |
|---|---|---|---|---|---|---|---|---|---|
| | 0.20 | 0.26 | 0.32 | 0.38 | 0.44 | 0.50 | 0.56 | 0.605 | 0.9 |
| Reynolds # | | | | | | | | | |
| 1200 | 0.986 | 0.979 | 0.971 | 0.961 | 0.950 | 0.939 | 0.926 | 0.916 | 0.842 |
| 2900 | 0.979 | 0.978 | 0.970 | 0.960 | 0.949 | 0.937 | 0.925 | 0.914 | 0.839 |
| 3800 | 0.971 | 0.970 | 0.969 | 0.960 | 0.949 | 0.937 | 0.924 | 0.913 | 0.837 |
| 8800 | 0.961 | 0.960 | 0.957 | 0.957 | 0.945 | 0.933 | 0.919 | 0.908 | 0.829 |

Results of the least squares analyses are summarized in Table III. Reynolds numbers are used to identify the different flowrates. The values shown for k and its standard error were obtained directly from the analysis. Values for $P_o$ were derived from the least squares intercepts and indicate total removable particles at the onset of each run. These large numbers have been scaled by the total internal surface area, 1300 $cm^2$, to allow better insight as to their magnitude. The correlation coefficients, R, listed in the table, prove that considerable scatter remains even after discarding outliers. The results at a Reynolds number of 3800 show so much variability that Equation (3) fails as a correlation, only 9% ($R^2$) of the total variation in particle count is explained by the model.

Regression analysis using the raw data, with no correction for non-isokinetic sampling, gives the same values for k and R as appear in Table III. The results for $P_o$ are about ten percent lower.

The count of particles in different size ranges can be used to describe size distributions. Figure 3 summarizes size distribution results with a plot, on log-probability paper, of cumulative number percent less than a given size versus particle diameter. Six different distributions are shown. The two on the left are for Reynolds numbers of 1200 and 2900 with all counts associated with bursts excluded. The two middle distributions are for those 10 minute intervals at Reynolds numbers of 1200 and 2900 which were denoted as bursts by the regression analysis. The total counts in each distribution are: 78 for the Reynolds number of 1200 excluding bursts, 201 for Re = 1200 with a burst, 360 for Re = 2900 excluding bursts, and 872 for Re = 2900 with a burst. Only the last distribution has a total count that would satisfy a statistician. Nonetheless, there is evidence that the particle populations arising from bursts are richer in fine particles than the typical populations generated at that flowrate.

The two distributions shown on the right of Figure 3 are for Reynolds numbers of 3800 and 8800, and are based on total counts in excess of 5,000 and 50,000, respectively. They are close matches to each other and are quite distinct from the previous four distributions. The linearity of the data points on log-probability paper suggests that these last distributions can be modelled as log-normal. The median particle sizes are in the range 0.35 to 0.37 μm, and the slope of the data implies a geometric standard deviation of 2.0.

In summary, Figure 3 shows three different classes of size distributions: one for lower Reynolds number in the absence of bursts; another for lower Reynolds number with bursts; and a third distribution characteristic of higher Reynolds number.

Table III. Least Squares Regression Analysis of Log (Particle Count) Versus Time.

| Reynolds # | R | k ± std err ($min^{-1}$) | $P_o$ | $P_o/cm^2$ |
|---|---|---|---|---|
| 1200 | 0.85 | 0.012 ± 0.0026 | 1600 | 1.2 |
| 2900 | 0.87 | 0.014 ± 0.0027 | 14500 | 11 |
| 3800 | 0.29 | - | - | - |
| 8800 | 0.86 | 0.026 ± 0.0050 | 4600000 | 3500 |

where R is the linear correlation coefficient and $P_o$ is the initial number of removable particles on the tube wall. See Equations (2 and 3).

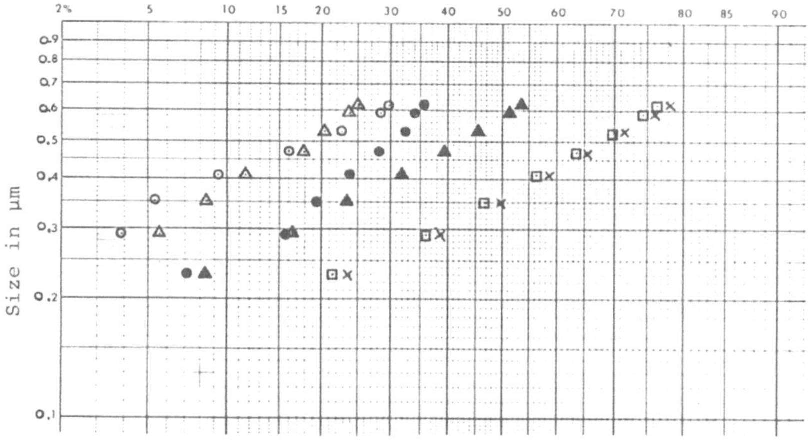


Figure 3. Log probability paper plot of LAS-X particle size versus number percentage less than the given size. ○ Reynolds number of 1200 with bursts excluded, 78 total counts. △ Reynolds number of 2900 with bursts excluded, 360 counts. ● Particle burst at Reynolds number of 1200, 201 total counts. ▲ Particle burst at Reynolds number of 2900, 872 total counts. □ Reynolds number of 3800, 5480 total counts. × Reynolds number of 8800, 52,000 total counts.

Figure 4 offers another way of viewing these size distributions; here, number fraction in any size range is plotted as a bar chart versus particle size. The distribution is only for counts in the LAS-X size range of 0.17 to 0.62 µm. The distributions are compared for Reynolds numbers of 8800, 2900 with a burst, and 2900 without bursts. For the high Reynolds number, number fraction monotonically decreases with increasing particle size. The distribution for Re = 2900 with a burst is surprisingly flat. Number fractions lie between 0.11 and 0.16 for each size range. With a Reynolds number of 2900 and no bursts, the distribution has few particles less than 0.23 µm and is rich in particles in the 0.41 to 0.47 µm size range.

At the higher Reynolds numbers, there were sufficient particle counts to allow study of variation in size distribution as a function of time. Little variation was found. Figure 5 shows number fraction versus particle size for four distributions obtained at a Reynolds number of 3800. At this Reynolds number, there were five 10 minute intervals that gave more than 400 counts and seven 10 minute intervals that each had fewer than 70 particles. The first distribution shown is for the 10 minute interval, associated with a large count, which produced the largest median particle size. Next is the distribution obtained by including all counts. Third is a distribution obtained by including only counts from the seven 10 minute intervals producing low total counts. And the final distribution is from the 10 minute interval producing the smallest median diameter. The only difference of note is that the low count distribution is unusually rich in 0.41 to 0.47 µm particles. The particles in the 0.17 to 0.62 µm size range for this distribution number only 221, so a discrepancy of 0.03 number fraction corresponds to only 7 particles. There was less variation in median particle size at a Reynolds number of 8800 than at 3800. Data on the variation of median particle size, the $d_{50}$, for all

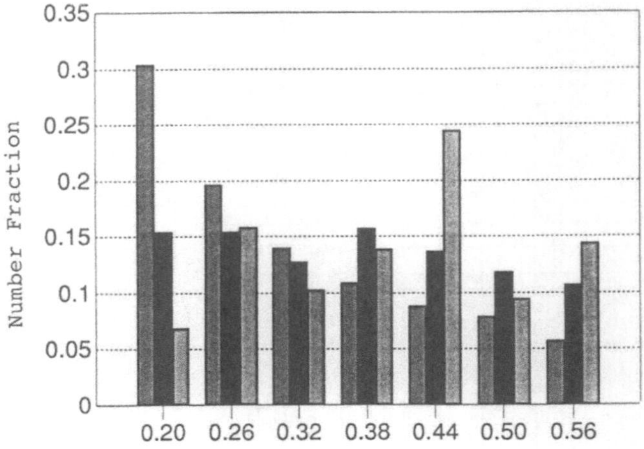


Figure 4. Size distribution of 0.17 to 0.62 µm particles. The bar on the left shows the number fraction in each size range at a Reynolds number of 8800 (40,500 total count); the center bar indicates the number fraction within a particle burst at a Reynolds number of 2900 (460 total counts); and the bar on the right gives number fraction at a Reynolds number of 2900 with bursts excluded (90 total counts).

particles greater than 0.17 µm are summarized in Table IV.

DISCUSSION

The Reynolds number characterizes the nature of flow through a tube. At Reynolds numbers less than 2000 the flow is laminar, with smooth streamlines. At Reynolds numbers above 4000, the flow is turbulent. For Reynolds numbers between 2000 and 4000, the flow is said to be in a transition regime, wherein the flow can fluctuate unpredictably between laminar and turbulent behavior. The onset of turbulence is promoted by vibration and surface roughness. In industrial pipelines, turbulence will commence at Reynolds numbers close to 2100. However, for the smooth tubing used in this experiment, turbulence might not appear until higher Reynolds numbers.

For Reynolds numbers of both 1200 and 2900, Figure 3 shows similar size distributions and Table III shows equivalent least squares slopes. This suggests an analogous, laminar flow regime, at least in the vicinity of the wall. Likewise, high particle counts and similar size distributions suggest the experiments at Reynolds numbers of 3800 and 8800 share the same turbulent flow regime. In Table III, the exponential decay constant at the highest Reynolds number is a factor of two larger than at lower Reynolds number.

A question left unresolved is why at a Reynolds number of 3800 the particle counts are so variable. It is tempting to link this with the laminar to turbulent flow transition, however, the particle count data are taken over ten minute time intervals. This is a long time frame over which to observe laminar to turbulent fluctuation at a Reynolds number as

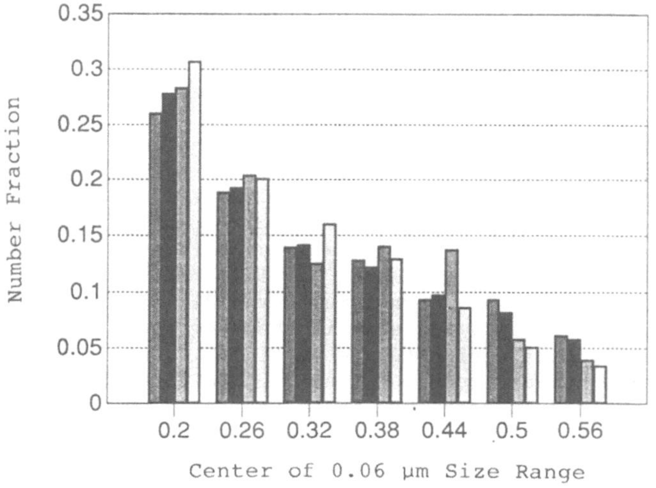


Figure 5. Size distributions of 0.17 to 0.62 µm particles at Reynolds number of 3800. First bar - size distribution with maximum median particle size, 1260 total counts. Second bar - size distribution based upon all counts, 4185 total counts. Third bar - size distribution for all time intervals with fewer than 70 particles, 221 total count. Fourth bar - size distribution with minimum median particle size, 379 total count.

Table IV. Variation in Median Particle Size for all LAS-X Counts >0.17 µm.

| Reynolds # | $d_{50}$ (µm) | Time Interval (min) | Significance |
|---|---|---|---|
| 3800 | 0.38 | 90 - 100 | Max $d_{50}$ at this Re # |
| 3800 | 0.37 | 0 - 120 | All counts included |
| 3800 | 0.355 | 7 intervals | All low count particles |
| 3800 | 0.335 | 0 - 10 | Min $d_{50}$ at this Re # |
| 8800 | 0.37 | 30 - 40 | Max $d_{50}$ at this Re # |
| 8800 | 0.35 | 0 - 120 | All counts included |
| 8800 | 0.33 | 10 - 20 | Min $d_{50}$ at this Re # |

high as 3800. Furthermore, the size distribution of particles observed during intervals of few counts is a close match to the size distributions from time intervals producing large counts.

Figure 3 suggests the released particle size distribution may be a sensitive measure of the laminar to turbulent transition. We know of no work that employed particle monitoring to study the laminar-turbulent flow transition. The point at which this transition occurs is important since hydrodynamic effects in any channel flow should scale as the ratio of Reynolds number to Reynolds number at the point of flow transition[16].

The experiments reported here are limited to particles shed from stainless steel tubing exposed to HCl gas. Our main interest was a sample of tubing which would readily shed large numbers of particles. Nonetheless, our test sample represents a type of surface corrosion of reasonably common occurrence, and is one which can be readily generated in most gas analysis laboratories. The shedding characteristics of this surface are anticipated to be similar to those of other types of steels subject to chloride corrosion. But the shedding behavior may be very different from that observed with metallic surfaces onto which foreign particles have been either accidentally or artificially deposited.

Particle shedding rate is strongly dependent on gas flowrate. This is shown quantitatively in Table I and through the values for k and $P_o$ in Table III. Even at a Reynolds number as low as 1200, our 30 ft section of test tubing shed particles at an initial rate (k x $P_o$) of 19 particles/min. Studies on removal of artificially deposited silica and polymeric particles have indicated significant particle removal occurs only under turbulent flow conditions[17]. When our Reynolds number was increased to 2900, there was a ten-fold increase in shedding rate, but little change in the size distribution of observed particles. Data from Table I show that as Reynolds number increased from 3800 to 8800 there was another increase by an order of magnitude in the particle counts, but Figure 3 indicates only a small change in their size distribution. It has long been recognized that higher flowrates will cause more particle release, but the explanation for this has been that the greater wall stress allows finer particles to overcome their adhesion forces. With increasing flowrate, our data show a significant increase in shedding but little reduction in mean particle size so long as we stay within the same laminar or turbulent flow regime.

This suggests that contaminants which are shed from our corroded test line are of two classes, a coarse class of particulates readily released even into a laminar flowing stream and a more numerous, finer class of particulate which requires the stronger shearing forces of turbulent flow to overcome wall adhesion. Within any flow regime, as flowrate increases the frequency of particle release events will increase and/or the number of particles released per event will increase, but essentially the same LAS-X size range of particles is shed.

A third type of size distribution occurs at Reynolds numbers of 1200 and 2900, for time intervals having unusually high particle counts. We refer to these as laminar bursts. Inspection of Table I shows unusually high counts in three out of the twenty four time intervals sampled. Despite this low frequency of occurrence, most of the particles observed at the lower Reynolds numbers came from bursts. A better understanding of this phenomenon is warranted because production tools within the microelectronics industry employ gas flows in the laminar regime. Unfortunately, our data are over much too limited a time frame for an adequate estimation of burst frequency, much less causality. But our results do demonstrate the existence and the importance of laminar bursts. Furthermore, it is shown in Figure 3 that laminar bursts give a smaller median particle size than the background counts.

Table III indicates that the magnitude of the mean particle residence time, 1/k, is of the order of one hour. The implication is that cleansing of lines by purging will take a reasonably long period of time. Because of the sensitivity of particle shedding to gas flowrate, purge gas velocities should always be at least as great as those used in normal delivery line service. As a point of comparison with 1/k, the largest mean gas residence time was only a few seconds.

## CONCLUSIONS

A 30 ft section of 0.18 inch i.d., 316L stainless steel tubing was removed from dry HCl service and shown to shed high levels of particulates when purged with filtered nitrogen. For Reynolds numbers of 1200 and 2900, the LAS-X median particle size was greater than 0.62 μm and most counts occurred during isolated incidents, commonly referred to as bursts. At Reynolds numbers of 3800 and 8800, median particle sizes were 0.37 and 0.35 μm, respectively, the geometric standard deviation was two, and size distribution did not change over two hours time. Three sets of size distribution were observed: laminar flow without bursts, laminar bursts, and turbulent flow. Particle count increased substantially with increasing gas flowrate. An exponential lifetime distribution was able to describe the short time particle shedding rates for both laminar and turbulent flow conditions. At a Reynolds number of 3800, considerable variability in particle counts over sequential ten minute time intervals was observed, and the exponential wall particle decay model was not suitable for modeling shedding rate.

REFERENCES


1. K. Sugiyama, F. Nakahara, and T. Ohmi, Microcontamination, 7(7), 29-32, 96-102 (1989).
2. K. Sugiyama, T. Ohmi, T. Okumura, and F. Nakahara, Microcontamination, 7(1), 37-40, 60-65 (1989).
3. A. Kumar and C.E. Dyer, Solid State Technol., 30(2), 89 (1987).
4. D. Jensen, Microcontamination, 5(5), 52 (1987).
5. U.H. Koch and J.H. Pinson, Microcontamination, 7(3), 19-21, 61-66 (1989).
6. S.D. Cheung and R.P. Roberge, Microcontamination, 5(5), 44 (1987).
7. D.W. Cooper, Solid State Technol., 29(1), 73 (1986).
8. W.C. Hinds, "Aerosol Technology: Properties, Behavior and Measurement of Airborne Particles," Ch. 10, p. 191, J. Wiley and & Sons, New York, 1982.
9. M.L. Malczewski, J.D. Borkman, and G.T. Vardian, Solid State Technol., 29(4), 151 (1986).
10. R. Sitze and W.J. Thorn, "Sampling System Contribution to Particle Concentrations Measured in Nitrogen from 2200 psig Cylinders," Report from the Particle Measurement Laboratory, Keasbey, New Jersey, Linde Division, Union Carbide Corporation, April 25, 1986.
11. J.M. Davidson, Solid State Technol., 30(7), 63 (1987).
12. S.J. Alberhasky, "Particle Contamination in Compressed Gas Systems," M.S. Thesis, Clarkson University, Potsdam, NY, August, 1987.
13. W.C. Hinds and G. Kraske, J. Aerosol Sci., 17(1), 67-72 (1986).
14. M.D. Durham and D.H. Lundgren, J. Aerosol Sci., 11(2), 179-188 (1980).
15. N.L. Johnson and F.C. Leone, "Statistics and Experimental Design in Engineering and the Physical Sciences," Vol. 1, 2nd Ed., pp. 252-254, John Wiley & Sons, New York, 1977.
16. N.T. Obot, "Determination of Incompressible Flow Friction in Smooth Circular and Noncircular Passages," ASME Paper 88-WA/FE-1, Winter Annual Meeting, Chicago, 1988.
17. J.W. Cleaver and B. Yates, J. Colloid Interface Sci., 44(3), 464-474 (1973).

# APPARATUS FOR MEASURING ULTRAFINE PARTICLE EMISSIONS FROM AIR IONIZATION EQUIPMENT

Michael Yost, Arnold Steinman
and Al Lieberman*

Ion Systems Inc.
2546 Tenth St.
Berkeley, CA 94710

*Particle Measuring Systems Inc.
46729 Fremont Blvd.
Fremont, CA 94538

Recent work has established that electrostatically charged surfaces can increase particle deposition rates. This increase in particle deposition can become especially critical in cleanrooms. Air ionization is one technique that can effectively neutralize surface charges on products in cleanrooms. However, a few papers have suggested that some corona discharge air ionization equipment may release particles. A need exists for a test method to measure and characterize the emission of particles from air ionization equipment.

This paper describes a unique test apparatus designed to provide measurements for either the flow-through or static chamber test methods. Details of the hardware, software, and mechanical aspects of the test apparatus are presented, along with test results that demonstrate the performance of the chamber. The calculated decay constant for the sealed box is compared to empirically measured values. The results of some preliminary tests of ionization equipment are presented. The discussion addresses the relevance of sealed box and flow-through test results to actual operating conditions for ionizers in cleanroom situations.

## INTRODUCTION

Semiconductor manufacturers must go to great lengths to maintain a clean processing environment and to prevent particle contamination of critical wafer surfaces. A number of studies have identified electrostatic forces as having a major influence on the rate of particle deposition onto charged surfaces. In many instances, electrostatic forces can be more significant than gravitational, aerodynamic, or thermal forces[1]. Bipolar air ionization can effectively reduce surface charges, and as a result decrease the rate of particle deposition on surfaces[2,3,4]. Several tests have shown a significant decrease in surface particle counts on test wafers when the surface potential was reduced from 5 kV to <100 volts[5,6,7].

However, other authors have reported that some high voltage corona discharge ionizers generate large numbers of submicron particles in the 3 to 30 nanometer size range[8,9]. The small size of these particles would explain the apparent contradiction with reports of lower surface contamination in ionized areas, since these small particles would remain undetectable by most surface scanning equipment. Little published research has been done since these papers appeared.

These papers describing particle generation have drawn criticism in several respects. Some results were based on accelerated testing at corona currents up to 50 times the normal operating levels. Other tests failed to report such factors as the placement and grounding of the particle sampling probe, the air flow past the ionizer and probe, and the background particle counts. Some of the tests used emitter materials that ionizer manufacturers do not use because these materials erode rapidly. Overall, the lack of standardization in the test methods prevented any comparison of the particle readings from different materials and ionization devices. Finally disagreement exists as to the source of the particles; an analysis of the deposits formed on ion emitters in cleanrooms shows that the nonmetallic residue probably comes from the environment.[10]

A number of methods have been used in the past to obtain a measurement of particle emission by semiconductor processing equipment. One approach is to take local particle samples from near the equipment while it is operating in a cleanroom. A sampling probe attached to an aerosol particle counter is used to draw an air sample from near the device being measured. The total particles being emitted are calculated based on the sample counts. This method fails to define the probe location and volume of dilution air surrounding the source. As other authors have noted,[11] it produces highly variable results.

A second approach is to conduct the sampling in a flow bench. In this method the device is placed within a tunnel with a known flow rate of HEPA filtered air continuously moving past it. A sampling tube is placed downstream from the source to measure the concentration of particles. Device performance can be evaluated based on a clean air standard such as Federal Standard 209D. With adequate air mixing this method can provide reproducible results, but it lacks the sensitivity to detect low source emission rates.

A third approach may be called static box sampling. In this method the device is enclosed in a sealed chamber where the particle concentration is measured over time. The particle concentration is allowed to reach steady state levels to determine the particle generation rate of the device. This technique provides reproducible results and has the highest sensitivity for detecting low levels of particle emissions.

Ultrafine particle generation concerns both equipment manufacturers and their customers. Clearly a standardized test method is needed for measuring particle emissions from cleanroom equipment. We have developed a test method that defines the dilution volume that the ionizer operates into and provides reproducible background counts that are nearly zero. The ionizer can be operated at normal voltage and current levels, with surrounding HEPA filtered air, or in "dead air" conditions. This test method has applications for testing a variety of process equipment used in cleanrooms.

## EXPERIMENTAL DETAILS

### Apparatus

The test chamber (Figure 1) is an open ended rectangular box, 1.8m x 0.61m x 1.22m (6ft x 2ft x 4ft) constructed of 1cm (3/8 inch) clear plexiglas with a large hinged door attached to one side. Threaded studs seal the door against a rubber gasket. Two matching VLSI-grade HEPA filters (Flanders #9z1952su) attach to the open ends of the chamber, with an exterior fan unit and associated plenum on the inlet side of one filter. The entire chamber is housed in a separate room which also has HEPA filtered air.

A key feature of the chamber is the inclusion of 6 compressed air jets, arranged on opposite walls along the top and bottom face of the chamber. These jets supply about 2 air changes per hour of make-up air and insure thorough mixing inside the box. The compressed air supply passes through a 2-stage pressure regulator, a rotometer, and 6 successive stages of filtration, with the final 0.02 $\mu m$ membrane filters exhausting directly into the box. The make-up air maintains a slight net flow across the HEPA filters which helps prevent outside particles from leaking into the box.

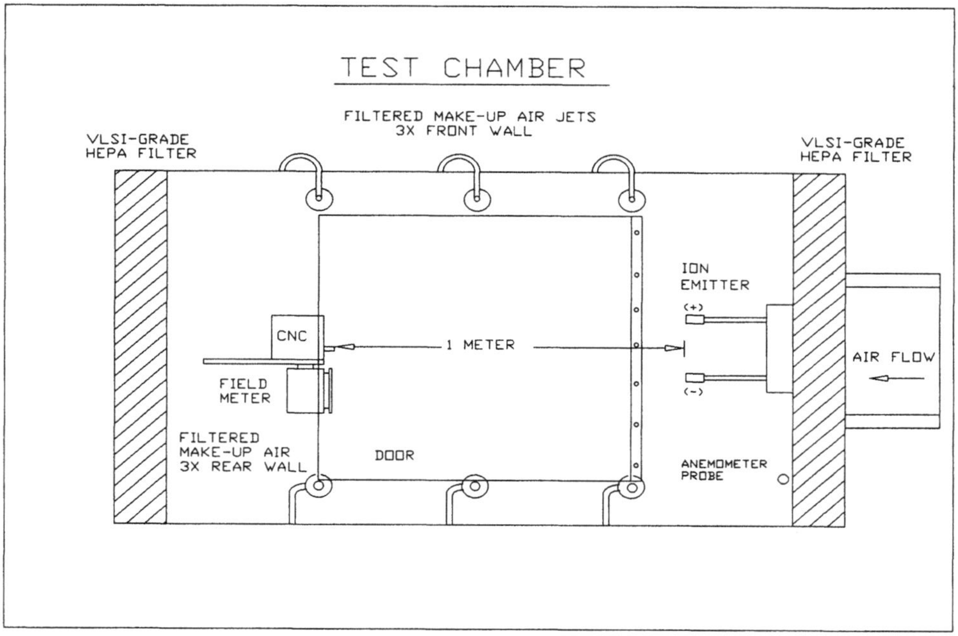


Figure 1. Diagram of the experimental test chamber, showing the location of the HEPA filters, fan, air jets, emitter unit, and particle monitor.

Particle counting is accomplished with a TSI Inc. model 3760 Condensation Nucleus Counter (CNC). This counter detects particles larger than 0.014 $\mu m$ at a sample rate of 1.42 L/min (0.05 $ft^3$/min). The CNC sits inside the chamber and air is drawn directly into the counter to minimize sampling losses. The inlet of the CNC is centered 1m downstream from the ion emitter points (one negative, one positive point). A polyethylene tube connects the butyl alcohol inlet of the CNC to an exterior valve, so the CNC reservoir can be refilled from outside the chamber. A KNF Neuberger (#N022) vacuum pump located outside the chamber draws the air flow for the CNC.

The CNC signal output connects to a TSI Inc. model 3704 single channel processor and a personal computer for data logging. The data logging software was developed specifically for this project. The software gathers data at selected intervals (usually 2 minutes) and saves the data to a disk. The files are written in an ASCII format that is compatible with commercial spreadsheet programs. The software provides a real time graphic plot of particle concentration and a moving time average. Data files can be recalled from disk for plotting, and specific time segments of a test can be selected in a "window" for detailed statistical analysis and plotting.

Additional particle count and size distribution information was obtained from a Particle Measuring Systems micro LPC-HS laser optical counter. This device measures particles larger than 0.05 $\mu m$ and classifies them into 8 size categories. The optical counter was placed outside the test chamber and a 2.83 L/min (0.1 $ft^3$/min) air sample was drawn through a tube attached to a sample probe located adjacent to the CNC inlet. This counter makes it possible to obtain high resolution, real time size distribution information.

A variety of other instruments monitor the test environment. A DASIBI model 1003-AH meter measures ozone concentrations to ppb levels. A Kurtz thermoanemometer measures the face velocity of the HEPA filter. A Trek field meter can be installed below the particle counter to read the charge on a 20 pF isolated plate to test the ionizer operation. Temperature and humidity are monitored with a digital thermohygrometer. An X-10 powerhouse (model CP-290) outlet controller switches selected AC outlets and serves as a computer programmable remote controller for the test chamber, ionizer, and instruments.

The ionization power source is a Ion Systems NilStat model 5000 controller and model 5100 emitter unit. The emitter points are located 30 cm from the HEPA filter face, and 1m from the CNC inlet. This is a pulsed DC self-regulating ionization system which allows the user to preset a constant output current at each emitter point. The emitter can be adjusted to maintain a constant current anywhere between 0.1 and 3 $\mu A$ at each point. Prior to testing, the output current for different settings of the emitter was measured in "free space" conditions (i.e. at least 2m from the nearest wall) and then with the unit placed in the chamber. This test showed essentially no difference (<1%) in emitter current between the two measurements. Subsequently tests were conducted with the output set to maintain a 2 $\mu A$ output current at each point.

## Test Methods

The chamber design allows for emitter testing in either a flow-through box or in a static box. In both cases particle levels are monitored continuously during the test sequence. The flow-through arrangement surrounds the emitter with HEPA filtered air, thus simulating the actual operating conditions with the emitter mounted on a cleanroom ceiling. For this test, the HEPA fan unit is switched on, and after a 30 minute purge period, the ionizer is switched on to begin the test.

The static box test sequence is slightly more complicated than the flow-through test. This test also begins with a 30 minute purge period with only the HEPA fan unit operating. The HEPA fan unit is then shut off and the background count is checked for 60 minutes with only the make-up air on. Next, both the ionizer and HEPA filter unit are turned on for 30 minutes to purge the chamber with bipolar ionized air. This step neutralizes any static charge on the chamber walls. Then the HEPA fan unit is switched off so the ionizer operates with only the make-up air on. This portion of the test typically lasts 1 to 3 days, and measures the particle production rate. During the final 120 minutes of the test, the ionizer is switched off and the particle concentration is allowed to decay. The decay period provides a direct measure of the particle residence time in the chamber after the ionizer has equilibrated the charge in the chamber.

## Chamber Dynamics

Previous work has shown that the static test chamber can be modeled as a well mixed box.[12,13] An advantage of this technique is that the test results can be used to determine the particle generation rate of the test device. The basic differential rate equation for the concentration of a contaminant in a well-mixed box is given by:

$$V\frac{dC}{dt} = P - kC - QC \qquad (1)$$

$V$ = chamber volume ($L^3$)

$P$ = particle production ($t^{-1}$)

$C$ = particle concentration ($L^{-3}$)

$k$ = wall loss rate ($L^3t^{-1}$)

$Q$ = make-up air flow rate ($L^3t^{-1}$)

Applying appropriate initial conditions, equation(1)can be integrated to obtain some familiar equations describing the filling and emptying of the box. Assuming a constant particle production begins at t=0 when C=0, the concentration rise toward steady state is given by:

$$C(t) = \frac{P}{Q^*}\left[1 - \exp\left(\frac{-tQ^*}{V}\right)\right] \qquad (2)$$

where $Q^* = k + Q =$ the equivalent air flow rate.

As t becomes large, the exponential term becomes small and equation (2) simplifies to the steady state equation:

$$P = C_\infty Q^* \qquad (3)$$

If the chamber is filled with particles and the source is suddenly switched off, the concentration will decay toward zero according to:

$$C(t) = C_0 \exp\left(\frac{-tQ^*}{V}\right) \qquad (4)$$

Equations (2-4) show that the chamber has a transient behavior similar to a first order low-pass filter. When a device with variable particle emission is placed in the box, the concentration fluctuations will be damped with a time constant:

$$\tau = \frac{V}{Q^*} \qquad (5)$$

These equations imply that we need to wait about 3 time constants to approach (within 5%) steady state conditions. Equation (4) indicates that slope of the decay curve is the inverse of the box time constant. Agglomeration losses have been ignored since the particle concentrations are never allowed to reach levels where Smoluchowski coagulation becomes significant[14]. The above equations assume perfect mixing in the box and constant wall losses. How closely the real box approaches this ideal behavior depends on how well these conditions are met.

**RESULTS**

The test chamber must have very low background counts that remain stable during the entire test period. After initial assembly, the chamber walls were wiped clean periodically with methanol while running the HEPA fan unit and the make-up air system. This process was repeated for several days. Finally, the HEPA fan unit was turned off leaving only the 60 L/min (2.12 $ft^3$/min) of make-up air operating.

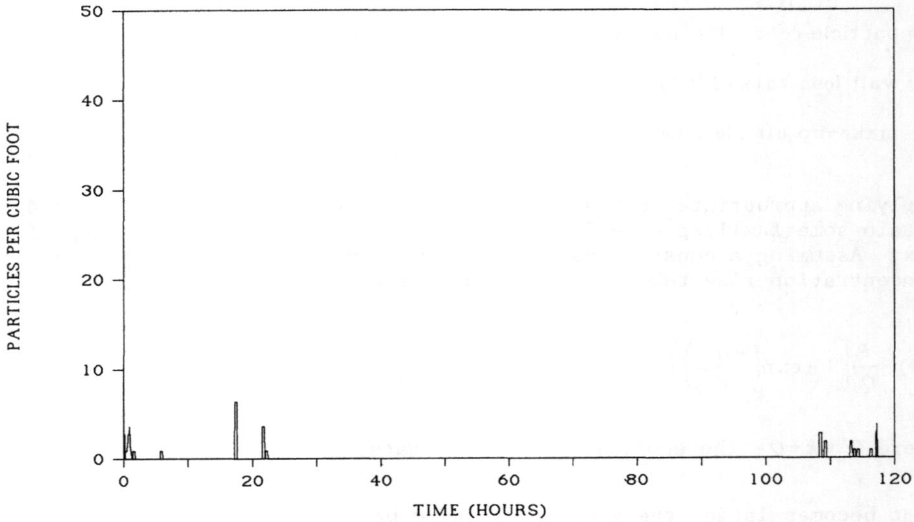


Figure 2. Typical background particle counts measured in the test chamber over several days of operation. The sample volume for the counts was 1 cubic foot.

After the initial treatment, the chamber easily maintained low background counts. Typically, the background is less than one count per hour on the CNC with long periods of zero counts. Figure 2 shows a typical background count recorded over a 5 day period. The average concentration recorded by the CNC during this test was less than 0.03 particles/L (<1 particle/$ft^3$). The chamber door can be opened while keeping the HEPA fan unit on, and after a brief 30 minute purge period, the long-term background returns to previous levels.

Operating the chamber in the flow-through mode greatly increases the dilution air flow. We measured the air velocity at 16 points in a vertical plane at 30 cm from the HEPA filter and 130 cm from the filter. These locations correspond to a cross section near the emitter points and the particle counter inlet. The readings were conducted with the normal compliment of working instrumentation, that is with the emitter and particle counter in place but without the trek meter. The air velocity through the chamber cross section at 130 cm from the HEPA filter averaged 28 cm/s (55 ft/min) $\pm$ 19%. The average volume flow rate through the chamber, calculated from these readings, was 12,460 L/min (440 $ft^3$/min). Thus the flow-through mode provides about 210 times more dilution air than the static box.

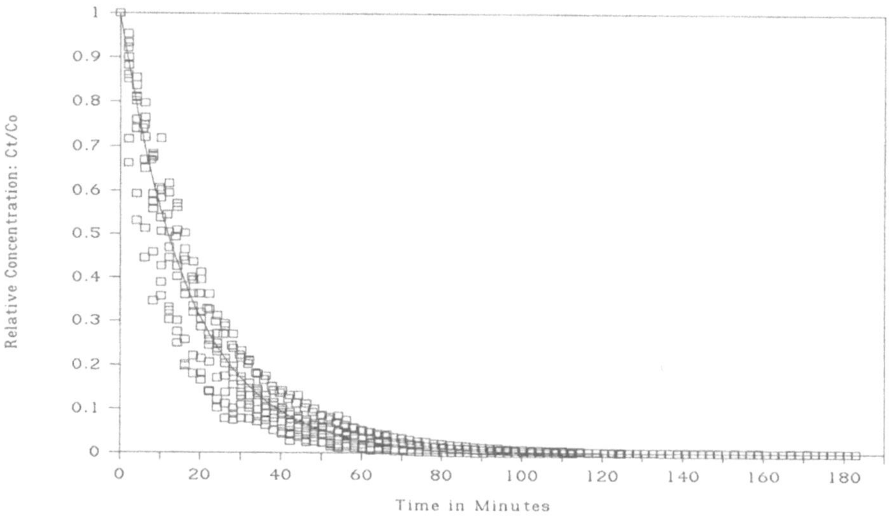


Figure 3. Results of 9 particle decay tests, normalized for initial concentration. The squares represent individual counts gathered by the CNC. The line was fitted to the data using a least squares model based on equation (4). The sample volume for all counts was 0.1 cubic foot.

Decay tests were conducted to determine the wall losses and time constant of the chamber when used as a static box. Partially filtered room air was introduced into the box to create initial particle concentrations between $3x10^6$ per L ($10^5/ft^3$) and $3x10^4$ per L ($10^3$ /$ft^3$). The make-up air remained constant in all cases at 60 L/min. A total of nine decay curves were normalized for initial concentration and a least squares semi-log regression line was fitted to the data. The fitted line slope was 0.057 ($R^2$ = 0.86), giving a box time constant of 17.5 minutes and a $Q^*$ value of 77.5 L/min. Figure 3 shows a plot of the particle decay data and the fitted line.

Two plots are provided to illustrate the operation of the chamber in each test mode. Figure 4 shows the results of an emitter tested in the flow-through mode. Particles were measured using both the CNC and the laser counter sampling a volume 1 cubic foot. The plot of the CNC data shows a fairly "noisy" appearance, indicating that the particle production was quite erratic. The average concentration recorded by the CNC during the test was about 2 particles/L ($50/ft^3$), while the laser counter was essentially at background levels. This indicates that virtually all the particles are smaller than the 50 nm detection limit of the laser counter.

Figure 5 presents a plot of the results of the same emitter tested in a static box mode. The particle sampling conditions and other factors remain the same as in the previous plot (Figure 4). The much higher particle concentrations clearly show that the sealed box provides superior sensitivity. The average particle concentration during this test was 42 particles/L ($1200/ft^3$). As before, the data suggest that almost all the particles are too small to detect by the laser counter. This finding agrees with previous researchers who found a mean particle size of about 30 nm and a narrow size distribution (geometric S.D. = 1.6)[9].

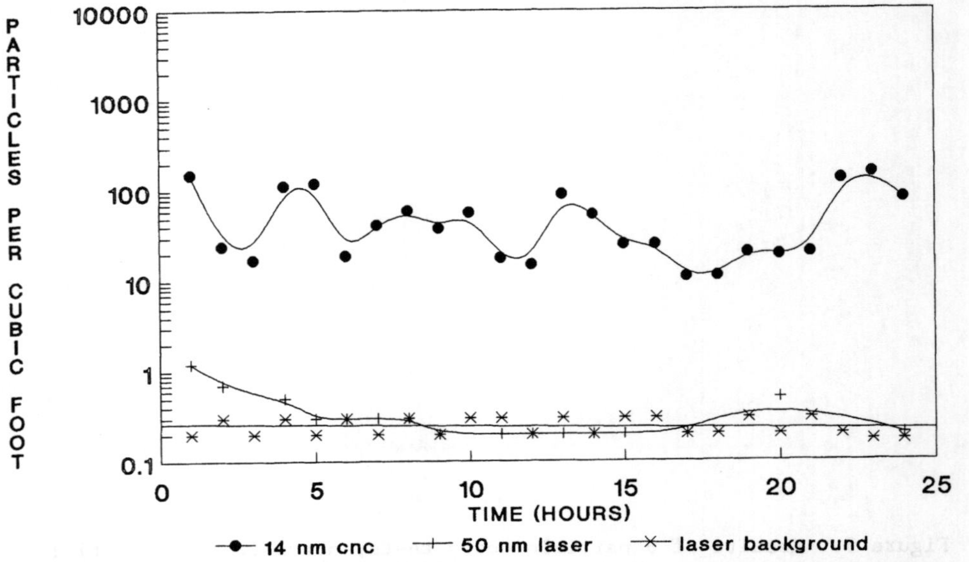


Figure 4. Results of an emitter tested in a flow through box. The upper line (dots) shows data for particles >14 nm obtained from the CNC. The lower curve (+'s) shows data for particles >50 nm obtained from the laser counter during the same time. The horizontal line (*'s) shows the typical background counts obtained from the laser counter. The sample volume for all counts was 1 cubic foot.

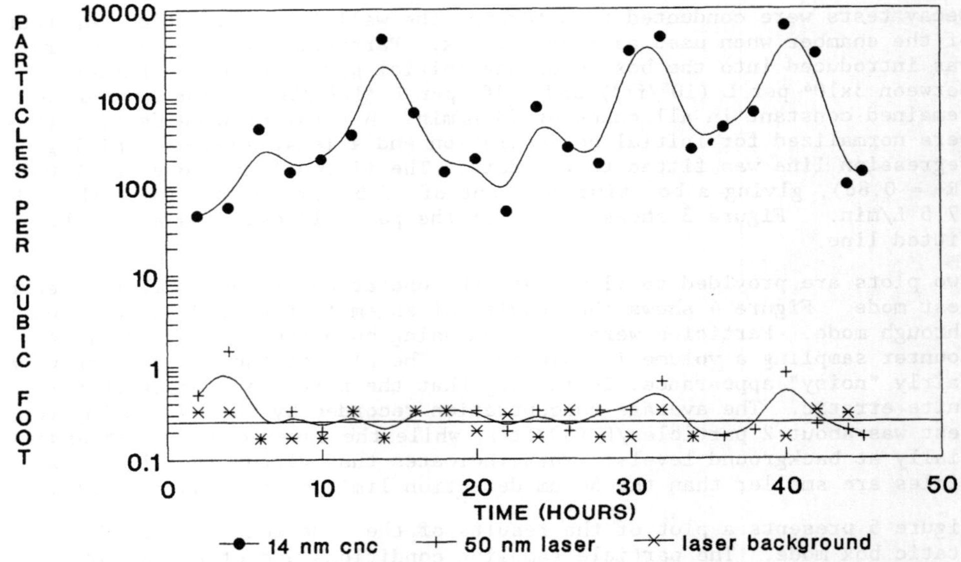


Figure 5. Results of an emitter tested in a static box. The upper curve (dots) shows data for particles >14 nm obtained from the CNC while the lower curve (+'s) shows particles >50 nm measured with the laser counter. The horizontal line (*'s) shows the background counts obtained from the laser counter. The sample volume for all counts was 1 cubic foot.

## DISCUSSION

The results demonstrate that the chamber is a sensitive tool for measuring particle emissions. This design has advantages over previous methods, since it can operate as a static box or flow-through chamber. Using filtered air jets to stir the box avoids the additional contamination or particle losses contributed by mixing fans. While the static box is a more sensitive particle detector, it should be remembered that the flow-through chamber is more representative of the actual operating conditions for corona discharge ionizers.

The particle concentrations measured in the flow-through mode are only about 24 times less than in the static box. This is much less than the 210 fold difference in the volume flow rate of dilution air . This discrepancy suggests that the particle emissions do not fully mix with the air passing through the box when operating in flow-through mode, but rather stay in a narrow plume. We conducted horizontal and vertical sampling traverses and found that the particles stay within a plume about 1 foot across at the CNC location. If we take into account the 10:1 ratio of the plume area to the chamber cross section area, the expected dilution factor for the flow-through mode is about 21 times more than the static box. This agrees fairly well with the observed dilution factor of 24.

The relatively large volume of make-up air flushed into the static box is crucial to the background stability of the system. The measured value of $Q^*$ was 77.5 L/min, of which 60 L/min was due to the make-up air. The increase in $Q^*$ over the make-up air flow is (presumably) due to wall losses. This implies that wall losses account for less than 30% of the decay constant. This is important since wall losses generally depend on poorly controlled or unknown factors such as the particle size distribution. Increasing the make-up flow air would further reduce the influence of wall losses, but at the expense of sensitivity.

Equation (3) for the steady state concentration shows that the sensitivity of the measuring system is dictated by the detection limit of the particle sampler and $Q^*$. If the lower detection limit for the CNC is 1 particle over 20 minutes (the box averaging time), then the detection limit is 0.04 particles/L (i.e. 1 particle in a 1 $ft^3$ sample). This implies that the chamber system can detect particle emission rates on the order of 3 particles/minute. Changing the make-up air flow has the dual effect of altering the averaging time and the detection limit of the system. Increasing the make-up air lowers the box time constant (equation (5)), but the additional dilution raises the minimum detection limit.

The decay curve plotted in Figure 3 is an average over many tests. Consequently, the fitted value of $Q^*$comes from a heterogeneous data set representing a variety of starting concentrations and particle size distributions. In any particular test the fit of the exponential decay curve is typically tight ($R^2 > 0.9$) and smooth over several decades. Such close adherence to ideal behavior is indicative of a well mixed chamber.

The observed increase in $Q^*$ over the known make-up flow rate is probably a result of particle deposition in the chamber. Since the particles from ion emitters in these tests are predominantly $<0.05\ \mu m$ in size, deposition is primarily due to diffusion which is a function of particle size. Thus experimental values of $Q^*$ may provide some indication of the particle size distribution when most of the particles are very small.

The fact that this chamber is designed to test ionization equipment raises some rather unique issues. Foremost is the need to consider the electrostatic environment created in the chamber. The ions make the air conductive so that air with $>10^4$ ions/cc has a Debye length of <1 cm. Debye shielding insures that boundary effects will diminish rapidly over much larger distances as typically found in the chamber. However, one must remain scrupulously aware of grounded and charged objects that can alter the electric field when placed close to the emitters. A plexiglas box was chosen to avoid having grounded walls near the emitter which might alter the equipotentials. The box was made sufficiently large so that the distance to the walls was at least as much as the separation between the two emitter points.

Plexiglas surfaces can become highly charged, and this could alter the particle losses due to wall deposition. However, in practice the walls have a nearly zero net charge because of the presence of bipolar ions. Likewise, particles emitted from a corona source could be highly charged, but because the particles are in contact with a bipolar ion cloud for several minutes they should quickly reach an equilibrium charge distribution with a net zero charge. The rate at which the aerosol is neutralized depends on **Nt**, where **N** is the ion density and t is the time. A value of **Nt** of $6x10^6$ is required to neutralize a highly charged aerosol[14]. Since the ion density in the chamber is about $2x10^5$ ions/cc, neutral conditions should be reached in about 30 seconds, which is far shorter than the typical particle residence time.

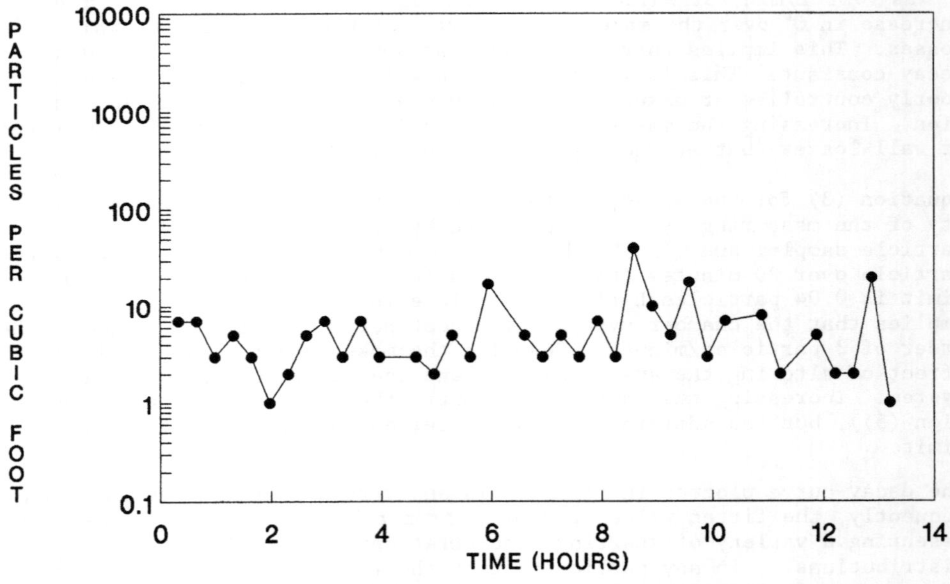


Figure 6. Particle counts from a low particle emission, non-tungsten emitter material tested in a static box. The particle counts average about 6 per cubic foot during the period, which is more than 10 times lower than the emitter test results presented in figure 5. The sample volume for the counts was 1 cubic foot.

Figure 6 provides an example of an extremely low level of particle emission that a pulsed DC corona ionizer achieved when using a different emitter material. Essentially, the particle counts are approaching the background for the static box. The average concentration in this test was 0.3 particles/L ($8/ft^3$), more than 100 times lower than the material tested in Figure 5 using the same emitter current. These data strongly demonstrate the possibility of producing air ions by corona discharge with a mimimum of particle emissions in the CNC size range. However, the data presented in this paper should only serve as an indication of the performance of the test method, and is not generally indicative of all corona ionizers. So far, the evidence points to vast differences in the rate of particle production using different emitter materials, geometry, and environmental conditions.

**SUMMARY AND CONCLUSIONS**

This paper describes the construction and testing of a chamber suitable for use in measuring particle emissions from cleanroom equipment. The chamber was developed to test air ionizers, but it could also be used for other devices. The chamber can be operated as either a flow-through box to simulate cleanroom conditions, or as a static box to provide greater sensitivity.

We present test data that demonstrate the utility of the chamber for particle emission testing. The chamber can maintain low background counts for long periods of time in either flow-through or static box mode. Decay test data show that the static box mode behaves with a predictable time constant primarily set by the make-up air flow. Tests with ion emitters show that the chamber provides a useful tool for studying the ultrafine particles from ionizers, and that the test method has sufficient sensitivity to detect very low levels of particle emission.

**REFERENCES**


1. D.W. Cooper, R.J. Miller and J.J. Wu, Deposition of submicron particles during integrated circuit manufacturing: Theory, Proc. 9th ICCCS meeting held in Los Angeles, 19, (Sept. 1988).

2. R.A. Welker, Equivalence between surface contamination rates and class 100 conditions, Proc. 1988 Annual Meeting of the Inst. of Environmental Sciences, p. 449.

3. R.P. Donovan, A.C. Clayton and D.S. Ensor, The dependence of particle deposition velocity on surface potential, Proc. 1988 Annual Meeting of the Inst. of Environmental Sciences, p. 473.

4. M. Inoue, S. Sakata and S. Chirifu, Aerosol deposition on wafers, Proc. 1988 Annual Meeting of the Inst. of Environmental Sciences, p. 423.

5. J.J. Wu, R.J. Miller and D.W. Cooper, Deposition of submicron aerosol particles during integrated circuit manufacturing: Experiments," Proc. 9th ICCCS meeting held in Los Angeles, 27 (Sept. 1988).

6. H.J. Fissan and J.R. Turner, Control of particle flux to prevent surface contamination, Proc. 9th ICCCS meeting held in Los Angeles, 33 (Sept. 1988).

7. R.P. Donovan, T. Yamamoto and A.C. Clayton, Experimental study of particle deposition on silicon wafers under the combined effects of electric fields and thermal gradients, Proc. 9th ICCCS meeting held in Los Angeles, 37 (Sept. 1988).

8. B.Y.H. Liu, W.O. Kinstley, and W.G. Fisher, Aerosol charging and neutralization and electrostatic discharge in clean rooms, J. Environmental Sci., 42, (March/April 1987).

9. R.P. Donovan, P.A. Lawless and D.D. Smith, Polarity dependence of electrode erosion under DC corona discharge, Microcontamination, 38 (May 1986).

10. M. Blitshteyn and S. Shelton, Contamination and erosion of cleanroom air ionizer emitters, Microcontamination, 28 (August 1985)

11. R.P. Donovan, B.R. Locke and D.S. Ensor, Measuring particle emissions from cleanroom equipment, Microcontamination, 36 (October 1987).

12. G.R. Rodgers and L.G. Bailey, A closed chamber method of measuring particle emissions from process equipment, Microcontamination, 43 (February 1987).

13. K.T. Whitby, G.R. Anderson and K.L. Rubow, A dynamic method for evaluating room-size air purifiers, ASHARE Trans., 89, p. 172, (1983).

14. W.C. Hinds, "Aerosol Technology," Wiley Interscience, New York (1982).

# MEASUREMENT OF SUBMICROMETER PARTICLE DEPOSITION ON SILICON WAFERS IN CLEAN-ROOM ENVIRONMENT

Ahmed A. Busnaina and Camilla M. Saviz

Microcontamination Research Laboratory
Department of Mechanical Engineering
Clarkson University
Potsdam, NY 13676

The rate of deposition of submicrometer particles from a particle source onto a silicon wafer in a VLF cleanroom is investigated. Particle diameters ranging from 0.32 to 0.99 $\mu m$ were studied. The effect of wafer surface voltage, time of exposure and particle concentration are also considered. High surface voltage resulted in greatly increased particle deposition rates. The minimum particle deposition rate was observed to be at 0.8 $\mu m$. For particles with $d < 0.5\mu m$, specifically $0.32\mu m$ particles, Brownian motion and electrostatic attraction forces were found to have a substantial effect on particle deposition. The measured deposition rate from a particle source above a wafer presented in this paper resemble realistic situations in many cleanrooms.

## INTRODUCTION

Surface contamination on semiconductor wafers is of great concern in the electronic component manufacturing industry[1]. Particle deposition on semiconductor wafers is currently one of the major problems facing the industry. Particle contamination on the wafer surface obstructs the proper functioning of microcircuits by interrupting or shorting conduction paths or by altering the capacitance of a circuit component. This causes losses of billions of dollars every year for the semiconductor industry.

Since circuit feature sizes on semiconductor wafers are approaching 0.7 $\mu m$ and lower, contaminant particles within the size range of 0.1 to 1.0 $\mu m$ are of primary concern. Knowledge of the mechanisms causing deposition helps in devising effective measures to prevent deposition of submicrometer particles on the wafer surface. The forces acting on each particle in a cleanroom can vary greatly depending on the particle size, mass, concentration and the airflow velocity.

Many studies have been performed[2–5] to predict particle behavior under the action of various external forces. However, there have been very few experimental studies to

*Particles in Gases and Liquids 2,* Edited by K.L. Mittal
Plenum Press, New York, 1990

verify the theoretical results. There is a need for experiments particularly in the realistic flow situations found in a cleanroom and in the submicrometer particle range. The present study considers deposition **from a particle source** above the wafer. This resembles realistic situations in many cleanrooms.

## BACKGROUND

The rate of deposition of particles from the flow near the wafer onto its surface is influenced by the thickness of the boundary layer. The velocity near the wafer is needed to describe the flowfield. The two most common wafer/flow configurations are the stagnation point flow and parallel flow. The flowfield near a horizontal wafer under a vertical laminar flow is an example of an axisymmetric stagnation point flow. The flowfield near the wafer (boundary layer) is described by the similarity solution of the Navier-Stokes equations[6].

The flow geometry for vertical wafers in a vertical laminar flow can be modeled as flow through a series of circular parallel plates (or channels). Fuchs[7] calculated the fraction of particles deposited on the surface of the parallel plates from the air flow. He obtained a solution using the fractional particle penetration through the channel in terms of a dimensionless parameter. The parameter relates the diffusion coefficient to the mean air velocity between the plates and the dimensions of the channel.

Calculations performed by Liu and Ahn[3] showed that the deposition rate of particles is dependent on the particle size. The deposition velocity for uncharged particles ranging in size from $0.01\mu m$ to $10\mu m$ was determined for horizontal and vertical wafers in vertical laminar flow. The first two flow configurations showed a decrease in the deposition velocity with increasing particle size within the diffusion regime, $d < 0.1\mu m$. The minimum velocity occurred between $d = 0.15\mu m$ and $d = 0.24\mu m$. For vertical wafers in vertical flow, the deposition velocity decreased with increasing particle size.

Busnaina *et al.*[8] predicted three-dimensional flowfield inside a typical cleanroom configuration. The trajectory of contaminant particles in the range of 1 to 20 $\mu m$ was predicted to show the possibility of microcontamination from various particles sources. A Lagrangian approach was used to model the particle transport in the room accounting for the drag and gravitational forces only.

Liu *et al.*[9] extended their previous calculations to include the effects of electrostatics on particle deposition. The deposition velocities for uncharged and charged particles of various charge densities were calculated. The velocity was shown to increase greatly as the charge on the particle increased. Figure 1 compares their results to experimental results of Hayakawa *et al.*[10], but there appears to be some disagreement between the two. This was attributed to the possible presence of a charge on the particles during the experiments resulting in higher deposition rates overall.

Work performed by Cooper *et al.*[2,11] to predict the deposition rate of submicrometer particles used both analytical and computational (Brownian dynamics simulation) techniques. The effects of electrostatic forces, inertial forces and diffusion were considered. The electrostatic forces included both the particle image force and the

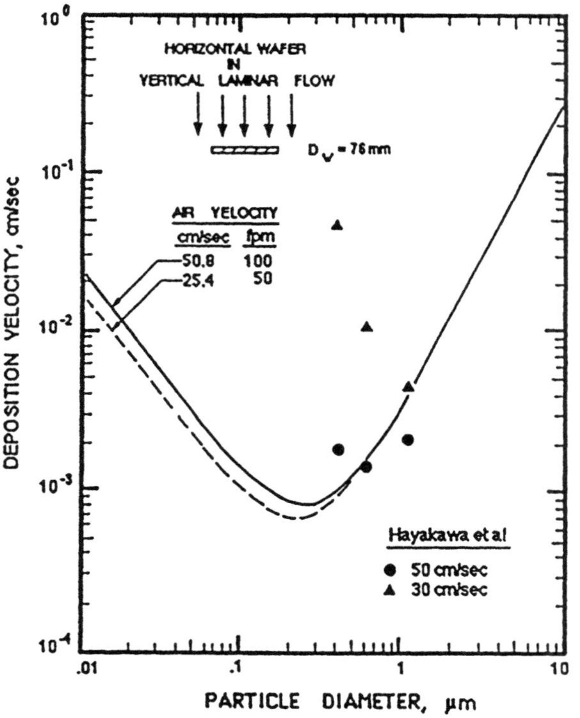


Figure 1. Comparison of deposition velocity with experimental data of Hayakawa *et al.*[3]

Coulombic electrostatic attraction of the charged particle toward the surface. Their results show that the presence of a uniform electric field or an electrostatic force has a dramatic effect on particle deposition. For a given diameter, these forces can increase the deposition rate by an order of magnitude over the deposition rate of uncharged particles (Levich-Smoluchowski solution).

In two related papers, Chari and Rajagopalan[12,13] presented the results of a theoretical study of colloidal particles in plane stagnation point flow. The presence of attractive electrical double-layer, van der Waals forces and the effect of fluid convection were considered. Although the interaction layer (or spatial range) in which these electrostatic forces act is generally small, the influence on the particle motion can be substantial. The same applies to the effect of convection within the range of influence of these electrostatic forces (due to the motion of the fluid close to the surface). Previous studies had neglected the influence of convection in the interaction layer. As a result, they were only valid for low fluid convection or a small interaction layer thickness. Their solution was accurate under conditions of high fluid convection and for increasing attractive double-layer forces. The results showed considerable increases in the deposition rate in the presence of large fluid convection, *i.e.* large Peclet numbers $Pe$, and thick interaction layers.

This study was based on the assumption of particle deposition onto a clean surface. However, one study[14] has shown a time dependency where the interactions with particles previously deposited on the surface affect the rate of particle deposi-

Inoue *et al.*[16] measured the deposition velocity on semiconductor wafers. They considered the effect of wafer electrostatic charge, particle size and relative direction of air flow to the wafer surface. However, only three particle sizes were used 0.3, 1.0 and 3 $\mu m$. They showed that the deposition velocity due to electrostatics was three orders of magnitude more than that due to diffusion only for 0.3$\mu m$ and one to two orders of magnitude more than velocity due to inertial and gravitational forces for 1.0 – 3.0 $\mu m$. Their measurements exhibited disagreement with analytical calculations and similar measurements made by Donovan *et al.*[17] for 0.6 and 1.0 $\mu m$, although the latter measurements considered a wafer in an enclosure.

This paper offers a comprehensive experimental study that evaluates the effect of electrostatic charge and other external forces on submicrometer particle deposition. This is particularly crucial since the semiconductor industry is now using circuit path lines on the order of 0.7$\mu m$ and smaller. Much smaller path lines are planned for future memory and logic chips.

## EXPERIMENTAL APPROACH

The experimental procedure involved three stages, all of which had to be performed within a cleanroom environment. The first and third stages required scanning the wafers before and after particle deposition to determine the difference in the particle concentration on the surface. A Tencor 4000 Surfscan (laser surface scanner) was used to determine the particle concentration on the wafer. Scanning was performed in a class 100 cleanroom under a clean hood with an airflow velocity of 100 ± 10% feet/min . The second stage involved deposition of submicrometer particles onto bare silicon wafers. This was performed in a class 10 cleanroom, under a clean hood with an airflow velocity of 100 ± 3% feet/min . Room conditions for each stage are given in Table I.

Table I. Deposition and Scanning Locations and Conditions

| | Particle Deposition | Wafer Scanning |
|---|---|---|
| Location | Contamination Control Laboratory | Manufacturing Line |
| Classification | Class 10 cleanroom | Class 100 cleanroom & Class 10 clean hood |
| Aisle ceiling coverage | 100% ULPA filter | 50% HEPA filter |
| Hood ceiling Coverage | NA | 100% HEPA filter |
| Airflow pattern | VLF | Mixed Flow |
| Avg. air velocity under clean hood | 100 ± 3% fpm | 100 fpm ± 10% |
| Workstation | Solid tabletop | Perforated tabletop |
| Air return | Perforated floor | Sidewall return |
| Temperature | 22°C ± 3°C | 24°C ± 3°C |
| Relative Humidity | 40% ± 5% RH | 60% ± 5% RH |

The wafers used were unetched, polished, 125 mm (5 inch) silicon wafers that were stored vertically in a wafer carrier inside a carrier box. The wafers were cleaned (using tion. Because of this effect, Chari and Rajagopalan also presented an expression to estimate the time required for the surface to be saturated with particles, assuming a constant initial rate of deposition. The authors assumed that the particles form a monolayer rather than forming particle chains (aggregates) on the surface since the long-range interactive forces repel particles from one another. The estimates showed that the time required for the surface to reach saturation was quite large for dilute suspensions, *e.g.* $\tau$ = 266.2 min. for Pe = 1.0 and bulk concentration, $c_o \approx 10^7 particles/cm^3 \approx 2.8 \times 10^{11}$ particles/c.f. . This concentration is extremely large compared to a cleanroom environment. Thus, the particle interaction effect on the surface can be neglected for a long time for low particle concentrations.

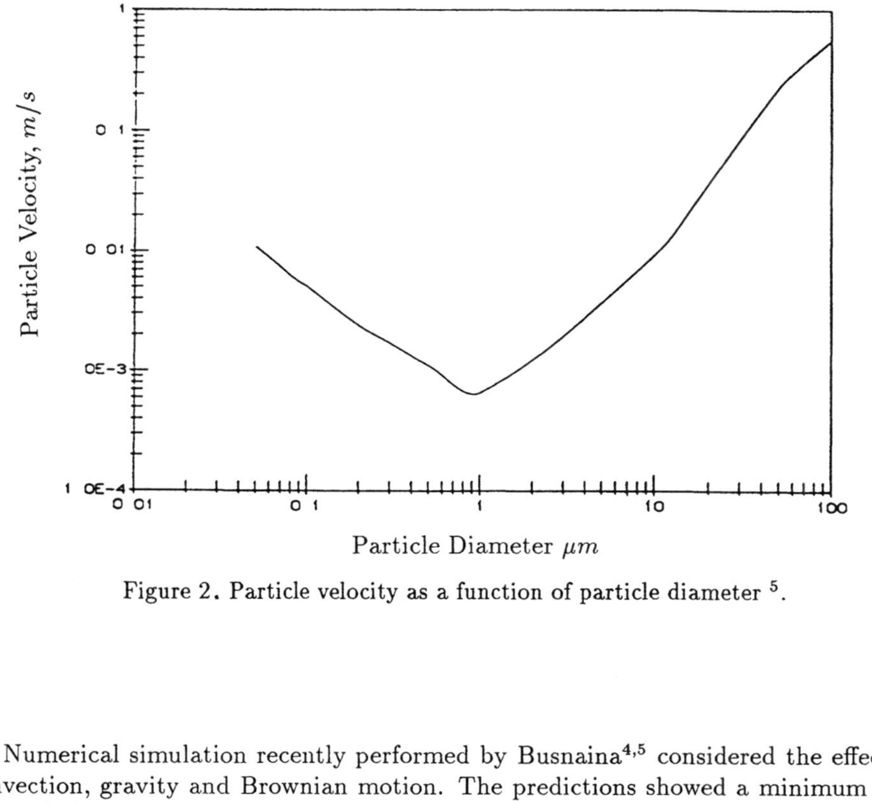


Figure 2. Particle velocity as a function of particle diameter [5].

Numerical simulation recently performed by Busnaina[4,5] considered the effect of convection, gravity and Brownian motion. The predictions showed a minimum particle velocity at approximately $0.8\mu m$ < d < $1.0\mu m$ for fluid velocities of 25, 50, and 100 fpm as shown in figure 2. The shown velocity is the actual airborne particle velocity in air (at 100 ft/min) before deposition and not the deposition velocity as calculated based on the deposition flux. These results disagree with the numerical work performed by Liu and Ahn[3] where the minimum was predicted (for the deposition velocity based on deposition flux) to occur at approximately $0.2\mu m$. Liu used a reduced form of the convective diffusion equation and assumed linear superposition with the sedimentation velocity. This may cause about 30 % error in the prediction of where the minimum deposition velocity occurs (with respect to the particle diameter). It should be noted that the calculated minimum for the deposition velocity depends on the freestream velocity[15].

a Convac brush cleaner) prior to each run to remove most of the excess particles on the surface and then scanned using the Tencor 4000 Surfscan. This determined the original number of particles on the surface before the deposition phase. The location of each wafer in the carrier was recorded.

To measure the rate of submicrometer particle deposition onto silicon wafers, monodispersed polystyrene latex (PSL) spheres were used. Fastek AccuBead PSL particles with mean diameter 0.32 $\mu m$, 0.48 $\mu m$, 0.78 $\mu m$, and 0.99 $\mu m$, and Particle Measuring Systems (PMS) PSL particles with mean diameter 0.804 $\mu m$ were used. The particle suspension was atomized using a DeVilbiss model 40 Nebulizer. The particles were emitted continuously downward from the source located 35 cm above the wafer. The wafer is positioned on a flat Static Control Services Charge Plate Monitor (CPM). The charge plate is positioned on a solid cleanroom work bench. There were no equipment nearby and no personnel in the cleanroom except for the person conducting the experiment.

The concentration and size of the particles generated were measured continuously using a Particle Measuring Systems $\mu m$LPC–555 laser particle counter, with the probe placed 25 cm above the wafer. The probe did not affect the particle deposition. Deposition measurements made with and without the sampling probe did not reveal any significant differences. The probe was sampling the aerosol flow at the edge of the enveloping aerosol cone above the wafer. Data from the particle counter were stored using the PMS Facilities Monitoring System software that logs and plots particle counts per cubic foot of sample air for 1 minute intervals. The concentration of particles generated was maintained on the order of 100–300 particles per cubic foot of air for all tests to simulate typical cleanroom conditions. However, one set of results was also obtained using 0.32 $\mu m$ particles and a concentration on the order of 1000 particles/ c.f. air to study the effect of the concentration of airborne particles on the deposition rate.

To measure the effect of electrostatics on the deposition rate, the wafer was placed on a Static Control Services Charge Plate Monitor (CPM). The CPM has two metal plates: the bottom is the charge plate and the top is the isolated plate. The voltage applied to the plate was measured using a SCS Model 970 non-contact electrostatic fieldmeter. The voltage applied can be varied resulting in a field intensity within the range 0-4.5 kV/cm. Within the manufacturing environment, wafer surface potentials up to 5.0 kV may be encountered[18]. The effect of six different field intensities: 0, 0.65, 1.0, 2.0, 3.0, and 4.0 kV/cm was investigated for each particle size. Positive field intensities were tested for all particle sizes, and the effect of a negative value was also tested for 0.99$\mu m$ particles.

The plate was initially zeroed, then the required voltage was applied. The particle source was located directly above the center of the charge plate. Once the voltage had stabilized, the pre-scanned silicon wafer was transferred from the carrier box to the center of the charge plate using a vacuum wand. The wafer remained exposed to the particles for the following time durations: 1, 3, 5, 10, and 15 minutes. These represent typical time increments during which the wafer is exposed during the manufacturing phase. The charged plate and wafer were removed from beneath the source and left to neutralize. After deposition, the wafers were scanned a second time where a particle map, histogram, and summary data of particles on the surface were obtained for each wafer.

To determine the deposition rate of particles onto vertical wafers in a vertical flow, wafers were placed in a carrier on a table in the cleanroom. Two different class cleanrooms were used. Electrostatic effects were not considered for this portion of the study; however, time effects were included. The wafers remained exposed to the cleanroom air for the following times: 10, 15, 20, 25, and 30 minutes. The wafer positions are shown in Figure 3. Three wafers were placed far apart (wafer numbers 22, 13, and 4) and three were placed close together (wafer numbers 3, 2, and 1) to determine the effects, if any, of wafer location on shielding. The wafers were scanned before and after tests and all particle sizes greater than $0.3\mu m$ were considered.

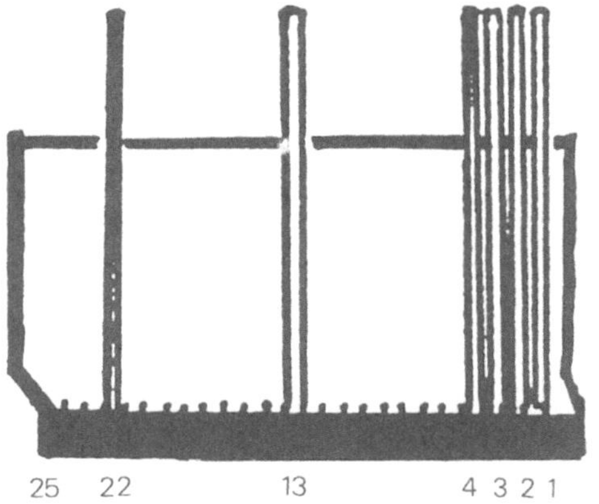


Figure 3. Wafer positions in the carrier for vertical wafer tests.

## RESULTS AND DISCUSSION

It should be noted that the following results are expressed in terms of deposition flux. The deposition flux is quantitatively comparable to the deposition velocity if the particle concentration for each run and particle size were the same. The concentration for all the following results was maintained at an average of 100 particles per cubic foot with a standard deviation of 25 over a period of three hours.

Figures 4 shows the steady state results for the deposition rate as a function of particle diameter. This graph shows that the deposition rate at an electrostatic field intensity of 4.0 kV/cm is approximately 10 times larger than at 0 V/cm for 0.32, 0.48, and 0.99 $\mu m$ particles. Most notable in these curves is that the deposition rate of particles close to $0.8\mu m$ in size is consistently much lower than all other sizes studied, regardless of the wafer surface voltage. This shows that the minimum deposition velocity occurs at $0.8\mu m$. The average charge on the generated particles in this study was between -20 and -40 V/cm. Fissan and Turner[19] showed that for a charged particle (between -10 and -100 V/cm), the minimum deposition velocity occurs between 0.4 and 0.9 $\mu m$. This presents a reasonable agreement within the range for the minimum velocity between the present experimental results and their theoretical calculations. However, it should also be noted that Fishkin and Baker[20] reported that PSL spheres of size 0.8$\mu m$ on polished silicon wafers had low scattering cross section compared to 0.6 and 0.9$\mu m$ on a Tencor 5000. This may explain why the minimum remains at 0.8$\mu m$ for different wafer surface voltages.

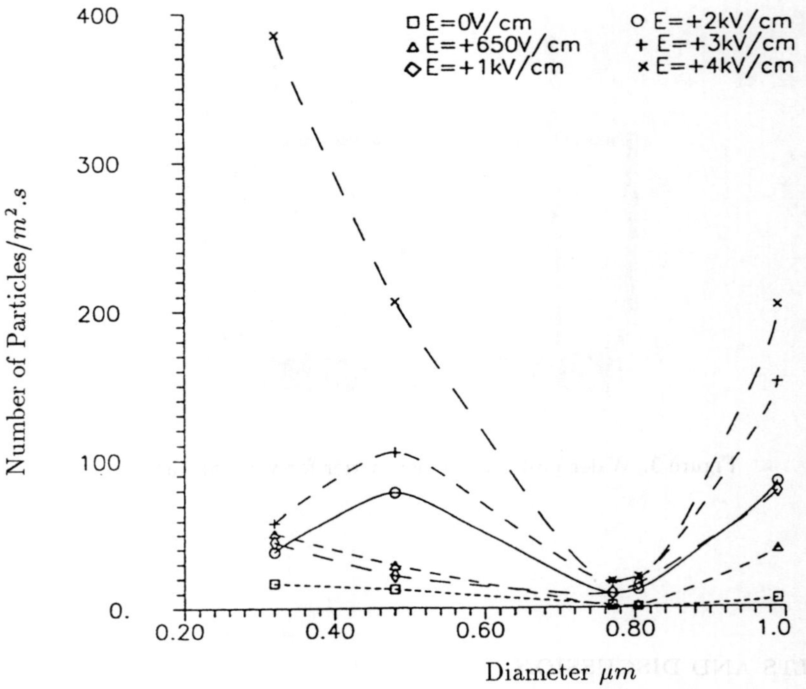


Figure 4. Particle deposition rate as a function of diameter after $t$ = 15 min.

As the wafers are placed on the plate, the flow field in the immediate vicinity of the wafer is disturbed and becomes transient. This explains the sudden increase in the first minute which indicates a higher deposition rate initially as shown in figure 5. The figure shows the rate of 0.32$\mu m$ particles deposited as a function of time. Each curve represents a different electrostatic field intensity. The rates for E $\leq$ +3.0 kV/cm do not vary greatly as the voltage is increased. The flat portion of each curve at these field intensities shows that the rate is almost constant over time. However, the rate of particles deposited sharply increases with time when E = 4.0 kV/cm. The authors can not explain this behavior and more experiments are planned to further investigate particles sizes less than 0.3 $\mu m$ when E = 4.0 kV/cm.

The results indicate that the transition occurs for a field intensity, 3.0 kV/cm < E ≤ 4.0 kV/cm. This is exhibited by the sudden increase in the deposition rate when the field intensity is raised to 4.0 kV/cm. Figures 6 and 7 clearly show this behavior where the curves do not continue to increase as expected when d < 0.5$\mu m$. The pattern does not extend to Figure 8 where the deposition is plotted as a function of size for E = 4.0 kV/cm.

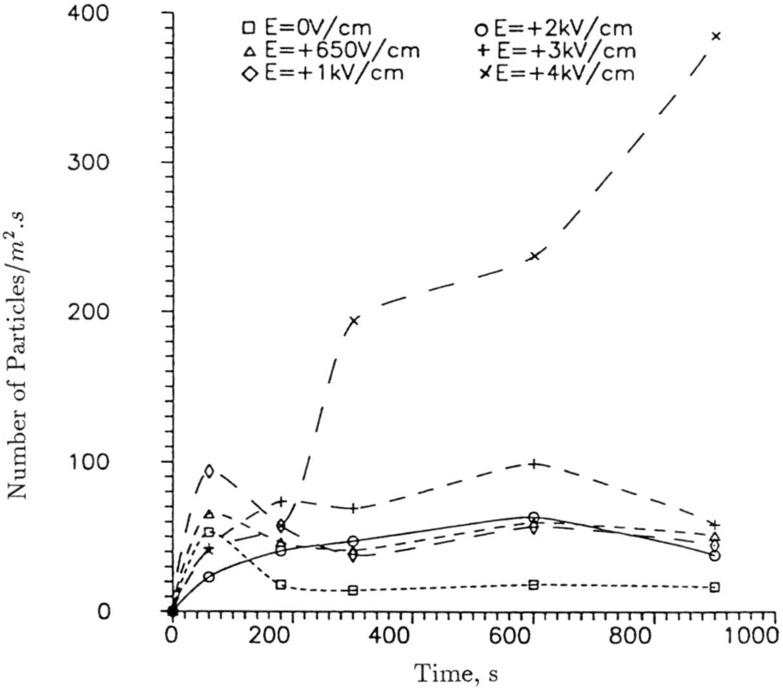


Figure 5. Particle deposition rate as a function of exposure time; particle diameter, $d$ = 0.32$\mu m$.

Figure 8 illustrates the difference in the deposition rate for the 0.78 and 0.804 $\mu m$ particles in comparison to all other particle sizes tested. It shows the particle deposition rate as a function of the field strength at $t$ = 5 min. The rate for particles in the 0.8$\mu m$ range does not significantly increase with field strength, remaining consistently low at all applied voltages (see above comment). For particles larger than 1.0$\mu m$, the gravitational force and the fluid drag are the dominant forces of deposition. For particles diameters less than 0.5$\mu m$, Brownian diffusion is the dominant deposition mechanism. Therefore, within the range 0.5$\mu m$ < d < 1.0$\mu m$, particles are too light to deposit onto the surface, but too heavy to be affected by collisions with gas molecules (Brownian motion). For these particles, the combined effects of the drag force and fluid convection are large enough such that the particles maintain a path along the flow streamlines.

As a result, the deposition rate reaches a minimum within this range as the particles are carried away by the fluid. These results disagree with the numerical work performed by Liu and Ahn[3] where the minimum was predicted to occur at approximately $0.2\mu m$. Liu's calculations assume linear superposition between the convection and diffusion velocity and the sedimentation velocity. This may cause about 30% error in the prediction of the deposition velocity near the minimum deposition velocity.

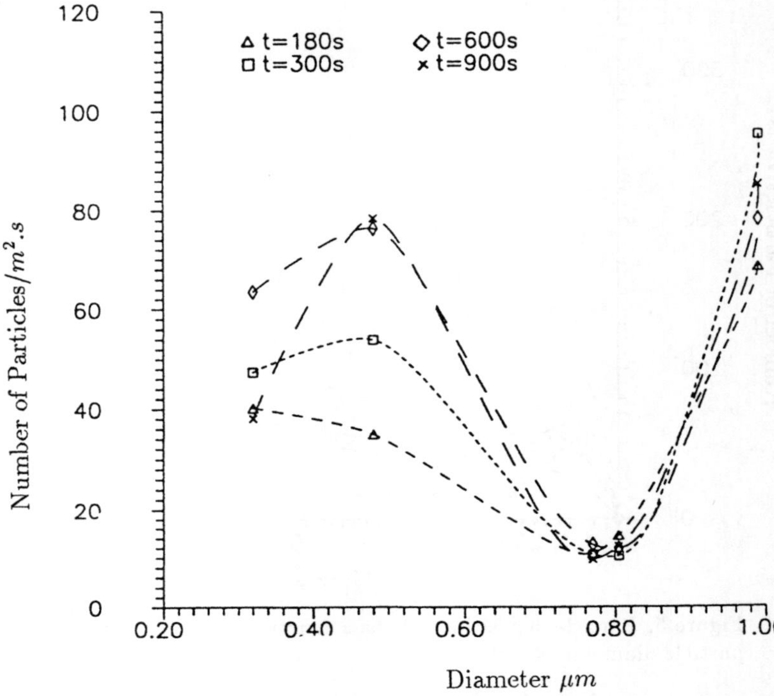


Figure 6. Particle deposition rate as a function of diameter; wafer surface potential E = 2 kV/cm.

However, the experimental results compare favorably with computational analysis recently performed by Busnaina and Abuzeid[5]. Their results considered all forces, including drag, convection, and Brownian motion. Their model predicts a minimum particle velocity at approximately $0.8\mu m$ < d < $1.0\mu m$ for fluid velocities of 25, 50, and 100 fpm. It should be noted that the predicted velocity is the actual airborne particle velocity in air before deposition and not the deposition velocity as calculated based on the deposition flux. Behavior consistent with the experimental observations was predicted for particles above and below this size range.

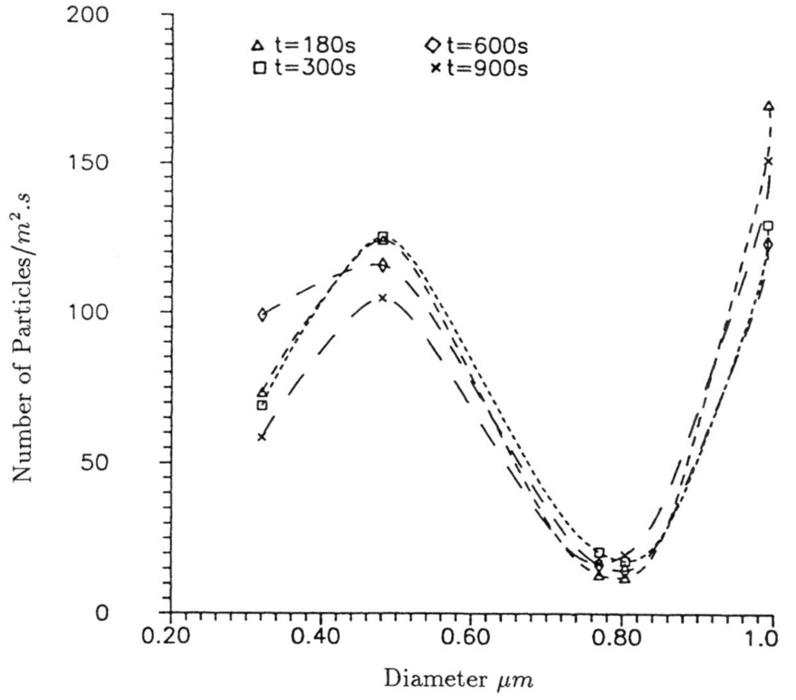


Figure 7. Particle deposition rate as a function of diameter; wafer surface potential E = 3 kV/cm.

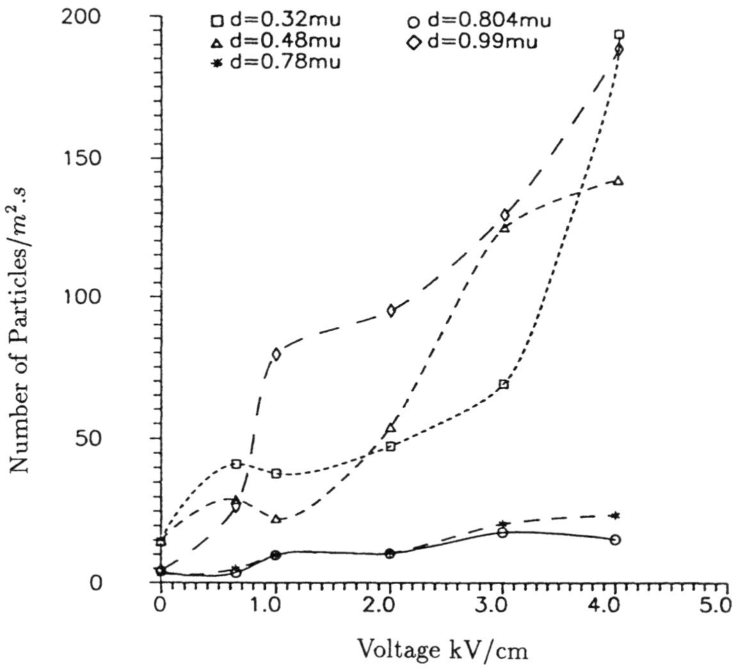


Figure 8. Particle deposition rate as a function of surface potential; after exposure time, t = 5 min.

Figure 9 compares the effect of the polarity of the surface potential on the deposition rate. Results for ± 650 V/cm and ± 2.0 kV/cm are similar. Since the average charge on the particles was low (- 0.03 kV/cm), the polarity of the voltage applied did not have a substantial effect on the deposition rate. The particles were not neutralized or passed through a classifier for the measurements. This indicates that the particles may not necessarily feature a Boltzmann distribution.

Figures 10 compares the deposition rates of $0.32\mu m$ particles as a function of the field intensity for different airborne particle concentrations. The dashed lines represent low concentration (~ 100 particles/c.f.) and the solid lines represent high particle concentrations (~ 1000-2000 particles/c.f.). In each case, the rate is almost constant up to E = 1.0 kV/cm. At E = 4.0 kV/cm, the concentration of particles deposited on the surface was so large that it exceeded the detection capabilities of the Surfscan. These portions have been estimated (shown as a dotted line). When the concentration of airborne particles increased by an order of magnitude, the number of particles deposited on the wafer increased by an order of magnitude for E ≥ 3.0 kV/cm. Also, when such a high concentration of particles is introduced into the 'clean' air, diffusion forces become much larger.

## VERTICAL WAFERS IN A CARRIER

Particle deposition on vertical wafers was measured in two different class cleanrooms. Particles were not generated for these tests. Results have been plotted in Figures 11 and 12. Each figure shows the total number of particles deposited on the wafers plotted as a function of the wafer position. Wafers 1–3 were placed in consecutive slots, while wafers 4, 13, and 22 were placed far apart. The solid and dashed bars represent results obtained in the Class 10 and Class 100 cleanrooms, respectively. The class 10 cleanroom is the so called Vertical Laminar Flow cleanroom (VLF). The class 100 cleanroom is a mixed flow cleanroom.

Figure 11 shows the number of particles deposited on the vertical wafers in 10 minutes. In both cases, approximately 5–10 particles were added to wafers 1–3 in the Class 10 room, but this number increased to 10–20 for the Class 100 room tests. For the wafers placed further apart, over 20 particles were added to the surface in the Class 100 room. These numbers are quite large considering the fact that 10 minutes is a relatively short time period during a manufacturing step. Figure 12 shows results for a 30 minutes time period. In this case, the number of particles deposited in the Class 100 environment is approximately twice that of Class 10. This varies from the results observed for a horizontal wafer where an increase in the airborne particle concentration by a factor of ten resulted in a corresponding ten fold increase in the number of particles deposited. This major difference between the two configurations occurs because sedimentation and impaction are no longer an important factor when the wafers are placed vertically. Brownian diffusion becomes important for this configuration.

Both figures show the effects of 'shielding'. The wafers placed far apart experienced an increase in the number of particles twice that of those placed in consecutive

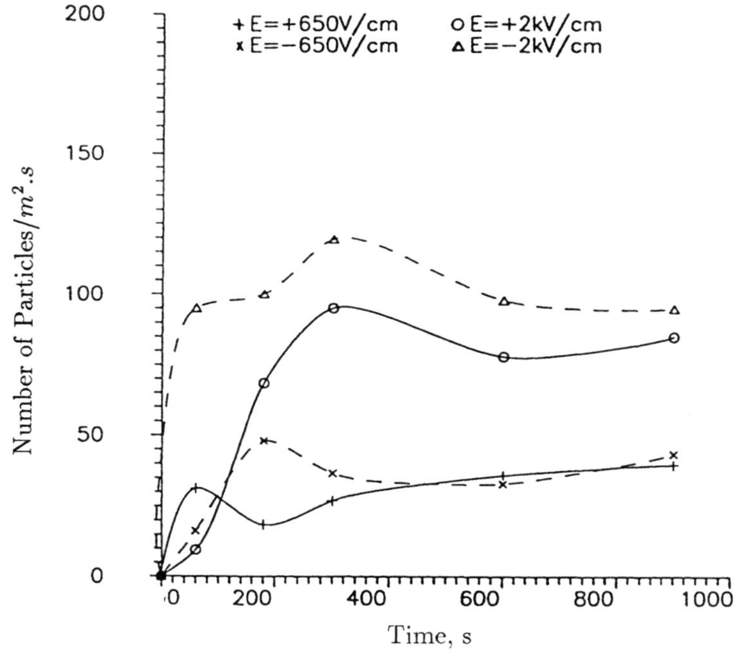


Figure 9. Particle deposition rate as a function of time for positive and negative surface potential; particle diameter, $d = 0.99\mu m$.

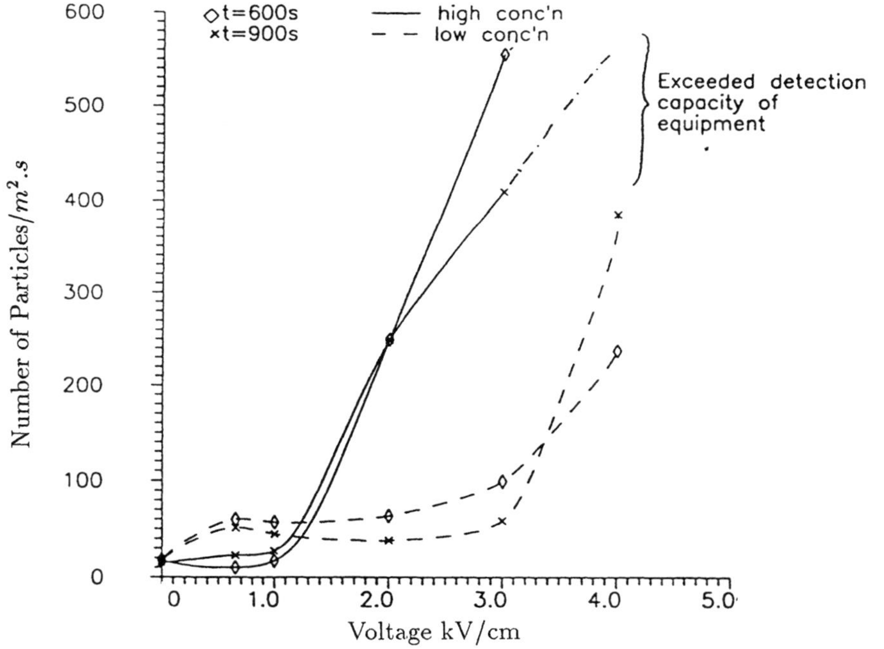


Figure 10. Particle deposition rate as a function of voltage with varied concentration of airborne particles above the wafer, Particle diameter, $d = 0.32\mu m$, $t = 10, 15$ min

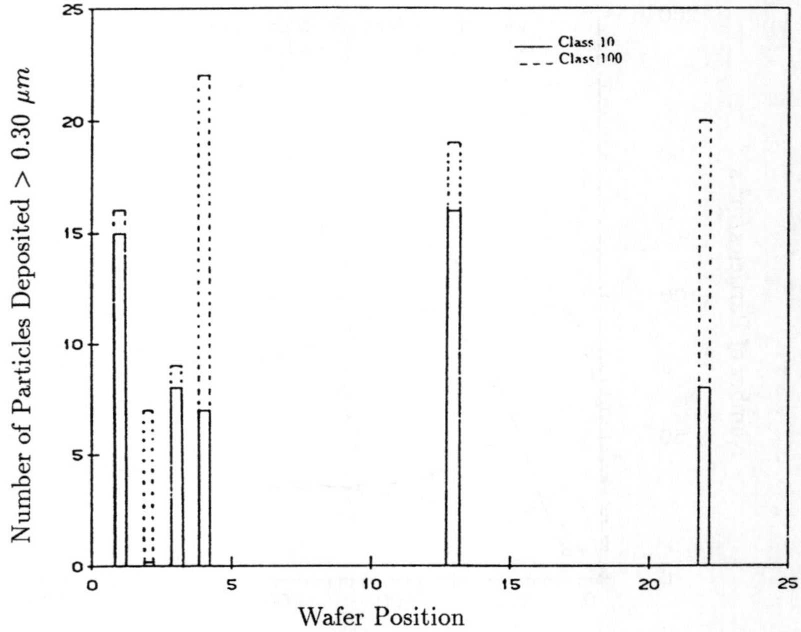


Figure 11. Number of particles deposited on vertical wafers as a function of wafer position in the carrier; exposure time, $t$ = 10 min.

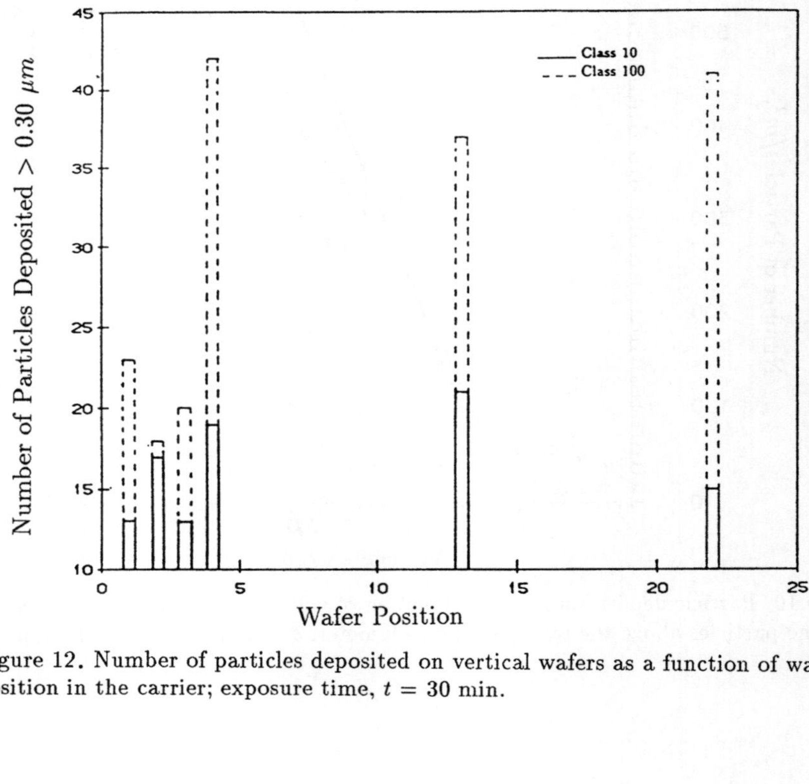


Figure 12. Number of particles deposited on vertical wafers as a function of wafer position in the carrier; exposure time, $t$ = 30 min.

slots. This was particularly true in the Class 100 cleanroom where the overall rates were much higher. This behavior follows the predicted pattern of increased fractional deposition as the distance between the two parallel plates, $h$, is increased. From these observations, use of a 'shield' wafer when possible, or placing wafers more closely together in the carrier, would be highly recommended for manufacturing purposes. This would aid in reducing deposition of airborne contaminants on the product surface.

## CONCLUSIONS

The deposition behavior of submicrometer particles in the range $0.1\mu m \leq d \leq 1.0\mu m$ was studied experimentally. The study considers deposition from **a particle source** above the wafer. This resembles realistic situations in many cleanrooms. The presence of electrostatic forces between 0 and 4.0 kV/cm was shown to increase the particle deposition rate substantially, particularly for E > 1.0 kV/cm. Brownian forces were found to have a significant effect on particle motion up to approximately $d = 0.5\mu m$. Both were found to substantially increase the rate of particle deposition onto silicon wafer surfaces.

For particles with $d < 0.5\mu m$, specifically $0.32\mu m$ particles, Brownian motion was found to have a substantial effect on the particle behavior. Consequently, the overall deposition rate of these particles is greater than that of larger particles, up to the sizes where sedimentation becomes important.

## REFERENCES


1. A. A. Busnaina, J. Edler, G.W. Gale and F.W. Kern, A fluid dynamics study of microcontaminant particle removal from silicon wafers, Proceedings, Institute of Environmental Sciences 33rd Annual Technical Meeting, San Jose, CA, May 4-8, 1987.
2. D.W. Cooper, M.H. Peters and R.J. Miller, Predicted deposition of submicron particles due to diffusion and electrostatics in viscous axisymmetric stagnation point flow, J. Aerosol Sci., 20, 123 (1989).
3. B.Y.H. Liu and K.H. Ahn, Particle deposition on semiconductor wafers". Aerosol Sci. & Technology., 6 , 1987, pp. 215-224.
4. A. A. Busnaina, Three-dimensional modeling of fluid flow and particle transport in cleanrooms, Presented at the Fine Particle Society Annual Technical Conference, Santa Clara, CA, July, 1988.
5. A. A. Busnaina and S. Abuzeid, Accuracy of numerical modeling of fluid flow in cleanrooms, Proceedings, Institute of Environmental Sciences 35th Annual Technical Meeting, Anaheim, California, May 1-5, 1989, pp. 245-250.
6. H. Schlichting, "Boundary Layer Theory", 6th ed. McGraw-Hill, New York (1968).
7. N.A. Fuchs, "The Mechanics of Aerosols", Macmillan, London (1964).

8. A. A. Busnaina, S. Abuzeid, and M.A.R. Sharif, Three-dimensional numerical simulation of fluid flow and particle transport in a cleanroom, Proceedings, Institute of Environmental Sciences 34rd Annual Technical Meeting, King of Prussia, Pennsylvania, May 3-5, 1988, pp. 375-381.

9. B.Y.H. Liu, B. Fardi and K.H. Ahn, Deposition of charged and uncharged aerosol particles on semiconductor wafers, Proceedings, Institute of Environmental Sciences 33rd Annual Technical Meeting, San Jose, CA, May 4-8, 1987.

10. I. Hayakawa, S. Fujii and K.Y. Kim, Studies on particulate behavior and adhesion in laminar airflow cleanrooms, Proceedings, Institute of Environmental Sciences 33rd Annual Technical Meeting, San Jose, CA, May 4-8, 1987.

11. M.H. Peters, D.W. Cooper and R. J. Miller, The effects of electrostatic and inertial forces on the diffusive deposition of small particles onto large disks: viscous axisymmetric stagnation point flow approximations, IBM T.J. Watson Research Center, Yorktown Heights, New York, August, 1987.

12. K. Chari and R. Rajagopalan, Deposition of colloidal particles in stagnation point flow, J. Chemical Soc. Faraday Trans. 2, 81 , 1345-1366 (1985).

13. K. Chari and R. Rajagopalan, Transport of colloidal particles over energy barriers, J. Colloid Interface Sci. 107(1) , 278-282 (1985).

14. D. Gupta and M.H. Peters, A Brownian dynamics simulation of aerosol deposition onto spherical collectors, J. Colloid Interface Sci. 104(2) , 375-389 (1985).

15. S. Abuzeid,A. A. Busnaina, and G. Ahmadi, Lagrangian simulation of particle deposition in a turbulent channel flow, Proceedings, ASME International Computers in Engineering Conference, Anaheim, California, July 30-August 2, 1989, pp. 127-134.

16. M. Inoue, S. Sakata, S. Chirifu, T. Yoshida and T. Okada, Aerosol deposition on wafers, Proceedings of the 9th ICCCS 1988 International Symposium on Contamination Control, Los Angeles, CA, September, 1988.

17. R. P. Donovan, A.C. Clayton and D.S. Ensor, The dependence of particle deposition velocity on the surface potential, Proceedings, Institute of Environmental Sciences 33rd Annual Technical Meeting, San Jose, CA, May 4-8, 1987.

18. R. Wilson, Nuclear air ionization for contamination control, Proceedings, Institute of Environmental Sciences 33rd Annual Technical Meeting, San Jose, CA; May 4-8, 1987.

19. H. J. Fissan and J. R. Turner, Electrostatic effects in particle deposition onto product surfaces, Proceedings, Institute of Environmental Sciences 34rd Annual Technical Meeting, King of Prussia, Pennsylvania, May 3-5, 1988, pp. 400-404.

20. B. Fishkin and E.J. Baker, Particle performance evaluation of CVD and epitaxial processes and equipment, Proceedings, Institute of Environmental Sciences 34rd Annual Technical Meeting, King of Prussia, Pennsylvania, May 3-5, 1988, pp. 517-523.

# PART III. PARTICLES IN GASES: DETECTION, CHARACTERIZATION AND CONTROL

# CHARACTERIZATION OF INDIVIDUAL PARTICLES IN GASEOUS MEDIA BY MASS SPECTROMETRY

M. P. Sinha

Jet Propulsion Laboratory
California Institute of Technology
Pasadena, California 91109

A method known as Particle Analysis by Mass Spectrometry (PAMS) for the chemical characterization of particles suspended in gases is described. In this method, the mass spectral measurements of individual particles are made on a continuous, real-time basis. PAMS involves a combination of the particle beam technique with mass spectrometry. A beam of particles is produced by the expansion of the gas laden with particles through a nozzle in a vacuum and is directed into the ion source of a focal plane mass spectrograph. Each particle of the beam is individually volatilized and ionized in the ion source. The resulting ions are spatially separated, according to their masses, along the focal plane and are measured with a linear array detector. The array detector enables the simultaneous measurement of all ions along the focal plane. Complete mass spectra of individual particles are obtained. PAMS has been applied for the chemical characterization of organic, inorganic, and biological particles. Individual particle composition, in addition to providing the toxic properties of particles, contain chemical signatures of their sources. PAMS can be used to detect and characterize particles in semiconductor fabrication facilities in real time and the results can be used to determine the emitting sources of these particles for their subsequent control and elimination.

## I. INTRODUCTION

Particles suspended in gases/air vary in size and chemical composition. A number of methods exist for determining the size and concentration of these particles.[1] Individual particles larger than 0.1 μm can be sized on a continuous, real-time basis by measuring the light scattered by them. However, no such method for chemical analysis of particles in a gaseous medium is commercially available at the present time. The chemical composition of aerosol particles is generally determined by collecting them on a suitable substrate (e.g., filter paper, impactor stages). The accumulated material is then extracted, prepared, and analyzed. The results obtained are the chemical compositions averaged over the mass of the particulate material and over the collection time. The various steps involved in the sample acquisition and the analytical procedure are also time consuming, and different artifacts might be introduced during this time.[2,3]

The chemical composition of the individual particle[4] contains detailed information about the particles in an aerosol mass, and may differ significantly from the average composition. The characteristic compositions of individual particles are distributed over a large number of particles in the analysis of the bulk material. Particles rich in toxic elements or compounds have different effects on health or the environment from particles of average composition. The presence of biological particles, and of particles containing asbestos fibers, for example, in different protected and field environments, illustrates the importance of determining individual particle composition.

Individual particles also carry chemical signatures of their sources.[5] Knowing the chemical composition of individual particles in a mixture of aerosols from different sources can, therefore, provide their respective source information, and thus be valuable in controlling their emissions. Particulate contamination is becoming increasingly important in semiconductor industries.[6] The yield in the device fabrication is significantly affected by the particle contamination.[7,8] The sources of particles may be within or without the process equipment. Similar conditions exist in the operating environments of a high density, high volume data storage device. A particle finding its way in between the read/write head and the rotating disk may prove to be catastrophic. Here also, particles may be generated either within the disk environment (e.g., moving motor, bearings, lubricants, etc.) or may enter from outside. In such processing and operating equipment, it is important to detect these particles and to determine the sources of particles for their subsequent elimination. Single particle chemical composition can provide such information.

Various physical methods[9] have been used for the chemical analysis of individual particles. These methods, in general, require the collection of particles on a substrate. Some techniques (such as FTIR, and Raman scattering from single particles) presently do not possess the sensitivity to analyze small particles (<1.0 μm). Other methods such as energy dispersive x-ray analysis provide only elemental composition and lack information about the organic content of the particles. The method of laser microprobe mass analyzer (LAMMA) has been used to analyze individual particles.[10] It also requires collection of particles on a substrate and may involve matrix interferences in the analysis. Consequently, a method for the chemical characterization of individual particles without collection on a substrate is needed. The method should be rapid (real-time), and sensitive (single particle) to analyze particles of less than 1.0 μm diameter, and also be capable of analyzing both organic and inorganic particles.

Recently, a novel method known as Particle Analysis by Mass Spectrometry (PAMS) has been developed in our laboratory for the analysis of individual aerosol particles on a continuous, real-time basis. Particles are directly introduced into the mass spectrograph in the form of a beam where they are individually volatilized, ionized, and their complete mass spectra measured. PAMS methodology and some of the results of particle measurements are described in this paper. PAMS application to the measurement of particles present in low concentration as found in semiconductor industries is also discussed.

## II. PAMS METHODOLOGY

Figure 1 shows the schematic of the PAMS System. It uses the technology of a miniaturized focal plane mass spectrograph (MS). The particles are introduced directly in the form of a beam into the ion source. The MS is equipped with an array detector for the simultaneous measurement of ions of different masses.

### A. Introduction of Particles into the MS

The particle beam generator used to introduce particles in the mass spectrograph consists of a capillary nozzle (100 $\mu$m diameter, 5 mm length) and two skimmers.[11,12] The regions between the nozzle and the first skimmer, and between the skimmers are differentially pumped. A beam of particles is produced by an aerosol expanding into a vacuum through the nozzle. The flow rate through the nozzle is about 60 atm-cm$^3$ min.$^{-1}$. The particles, being much heavier than the gas molecules, remain concentrated along the central axis. The skimmers strip off most of the accompanying gas, and allow for an efficient transmission[13] of the particles into the MS region. A particle in the ion source is converted into vapor and ions, either by thermal volatilization followed by electron impact ionization, or by a laser-induced volatilization and ionization (V/I) process.

### B. Particle Volatilization and Ionization

In the method of thermal volatilization, the particle beam is directed into a resistively-heated (250-1500 C) V-shaped rhenium filament in the ion source. Here the particles are volatilized, one at a time, and the resulting vapor is ionized by electron impaction. Complete vaporization of organic as well as of relatively volatile inorganic particles (e.g., ammonium sulfate, nitrate) takes place in the V-type filament[11]. However, particles made up of high melting and boiling point materials are not volatilized efficiently or completely. In such cases (e.g., for sodium chloride particles) a miniaturized Knudsen cell

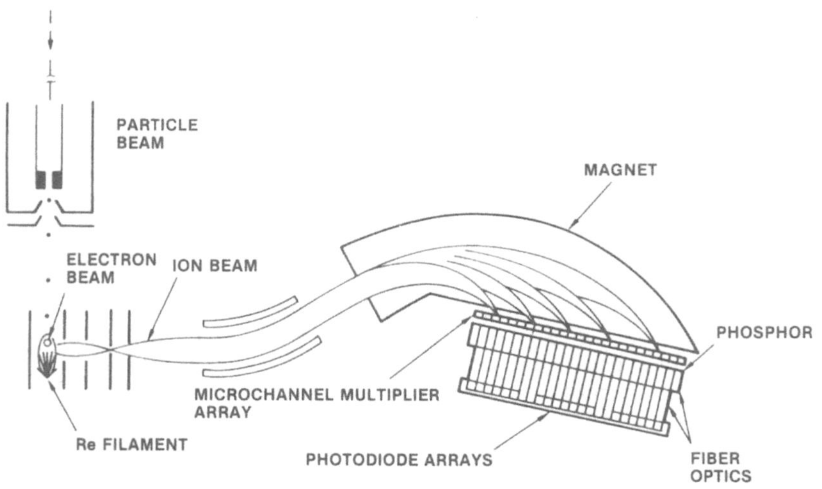


Figure 1. Schematic of Particle Analysis by Mass Spectrometry System. The main components of the system include the particle beam generator, the miniaturized mass spectrograph, and the electro-optical ion detector.

fabricated from rhenium metal sheets was found to be more suitable[12]. A Knudsen cell traps the particle for its complete vaporization. Each particle produces an ion pulse (2-5 ms wide). The ions are then mass separated by the focal plane mass spectrograph.

Lasers have been found to provide a softer method of ionization with high efficiency for substances collected on a substrate[14]. Involatile and fragile molecules can be evaporated by a laser pulse due to its extremely fast heating rate.[15] Ionization efficiencies in the range of $10^{-1}$ - $10^{-2}$ have been reported at power densities in the range of $10^{9}$ - $10^{10}$ W $cm^{-2}$. This high efficiency of ionization by laser, when used in the PAMS system, will increase its sensitivity for analyzing smaller size particles and for species present in low concentrations in a particle.

Preliminary work on the volatilization and ionization of particles in the beam by a pulsed YAG laser has been performed in our laboratory using earlier versions of PAMS[16]. In these measurements a quadrupole mass spectrometer was used. The arrival of a particle, and of the laser pulse in the ion source, were synchronized by sensing the particle upstream from the ion source using light scattering and from the knowledge of the particle velocity. The YAG laser was operated in a free-running mode that produced pulses of approximately 100 $\mu$s width. The long pulse width facilitated its synchronization with the particle. Ions at a few masses (m/e = 23, 28, and 40) were observed from potassium acid phthalate particles (1.9 $\mu$m diameter) by the laser interaction. Their intensities were weak and are believed to have been due to the low power density ($10^{5}$ - $10^{6}$ W $cm^{-2}$) of the focused laser pulses used in these experiments. For high efficiency of ionization, the laser power density needs to be increased. This can be achieved by Q-switching the laser pulse. Proper synchronization needs to be made of the Q-switched laser pulse (15 - 20 ns) with the particles in the ion source of the MS.

C. Mass Spectrograph

A miniaturized double-focusing focal plane mass spectrograph (Mattauch-Herzog type) is used for the mass analysis of the ion pulse generated from a particle. The different mass ions are spatially separated along the focal plane of the (permanent) magnet after passage through the electric and magnetic sectors. The spectrograph has a 12.7 cm long focal plane and covers a mass range of 25-500 amu.

The sensitivity of a focal plane mass spectrograph (nonscanning type MS) is inherently higher than that of a scanning mass spectrometer because the latter discards signals from all other masses while measuring the signal at any one mass. In the past, one of the important reasons for not using the former type of mass spectrograph for measurements that require high sensitivity has been the lack of a suitable ion detector at the focal plane. A photographic plate was conventionally used to measure these ions. A photographic plate, although it measures all the ions simultaneously along the focal plane, lacks the sensitivity of an electron multiplier detector. Over the last several years, an array detector known as an Electro-Optical Ion Detector (EOID) has also been developed at JPL for a focal plane mass spectrograph.[17] This combines the simultaneity of measurement of a photographic plate and the sensitivity of an electron multiplier.

The EOID comprises a microchannel plate (MCP), phosphor-coated fiber optic window, and fiber optic couplers to transmit the ion images to a photodiode array (PDA). Figure 2 shows the functional diagram for the detector. An ion exiting from the magnet impinges on the MCP at the focal plane, initiating an electron cascading process along the channel. The electrons coming out at the other end of the channel are accelerated to a phosphor-coated fiber-optic window. Photon images of different mass ions are thus generated (Fig. 3). The images are coupled to photodiode arrays by fiber optic couplers. These couplers, together with the PDA's, (each with a 2.54 cm active length in a 1 x 1024 array) are interdigitated to cover the entire focal plane with a minimum of discontinuity at the junction of two couplers.

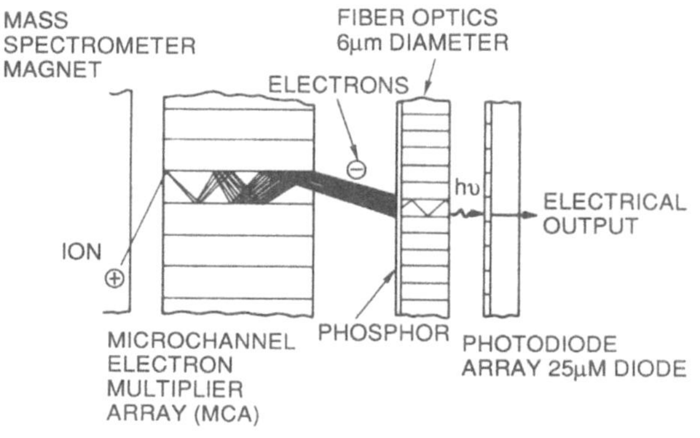


Figure 2. Functional diagram of the electro-optical ion detector. The ions exiting the magnet strike the microchannels of the electron multiplier array. The conversion of the ion-electrons-photons takes place. The photon signal is then integrated by the photodiodes for the predetermined length of time.

Figure 3. Photograph of ion images on the fiber-optic window. The photodiode arrays and their electronics are removed.

D. Data Acquisition

The photodiodes are integrating detectors, and accumulate the photon signal (proportional to the ion signal) for the desired period of integration. The position of the photodiode along the focal plane determines the mass of the ions producing the ion image at the location. The signals stored in the photodiodes are read serially, (at a rate of 220 kHz), by a Masscomp computer (which uses a Motorola 68010 microprocessor) after a predetermined integration time (Figure 4). Each readout, called a scan, provides a mass spectrum of all the ions accumulated during the integration period. The MS with the EOID possesses "100% duty cycle" for signal measurements. Consequently, the brevity of the ion signal generation from a transient sample does not put any limitation on its mass spectral measurement capability. Complete mass spectra from individual aerosol particles are thus measured in the PAMS method.

Two modes of data acquisition have been implemented on the PAMS System:

1. Continuous Mode

In this mode, the PDA's are read repeatedly by the computer after a predetermined signal integration time (typically 50 - 200 ms). A photodiode starts signal integration again just after its readout, and the stored signals in the photodiodes are read at the lapse of the integration time.

In the continuous mode of data acquisition, a large volume of data in the form of scans is acquired and written on the computer disk. The mass spectra from particles are contained in only some of these scans, (see Section III), and in some cases the particle mass spectrum is divided into two scans. The latter condition arises because of the fact that there is no synchronization in time for the ion generation from the particle with the initiation of the integration time.

2. Trigger Mode

A preferred way to acquire mass spectra from particles would be to synchronize the arrival of a particle in the ion source with the start of the signal integration and acquire data only when a particle is present in the ion source.

This method would be particularly useful in the analysis of particles from low-particle-number-density aerosols, where the frequency of the arrival of particles in the ion source will be low (approx. 2 $min.^{-1}$). Only particle mass spectral data will be written on the computer disk in this mode of data acquisition. Such a scheme of data acquisition needs the detection of the arrival of the particle in the ion source. The particle detection signal then triggers the initiation of the signal integration in the PDA's.

In the laser V/I scheme, the He-Ne laser pulse scattered by the particle, and used for the synchronization of the particle with the YAG laser pulse, can also be used for the initiation of the signal integration. For thermal V/I, a He-Ne laser beam is incorporated in the source chamber for particle detection. The photodiodes are integrating detectors, and are reset by their readout. These diodes can get saturated by integrating their own dark current while waiting for the

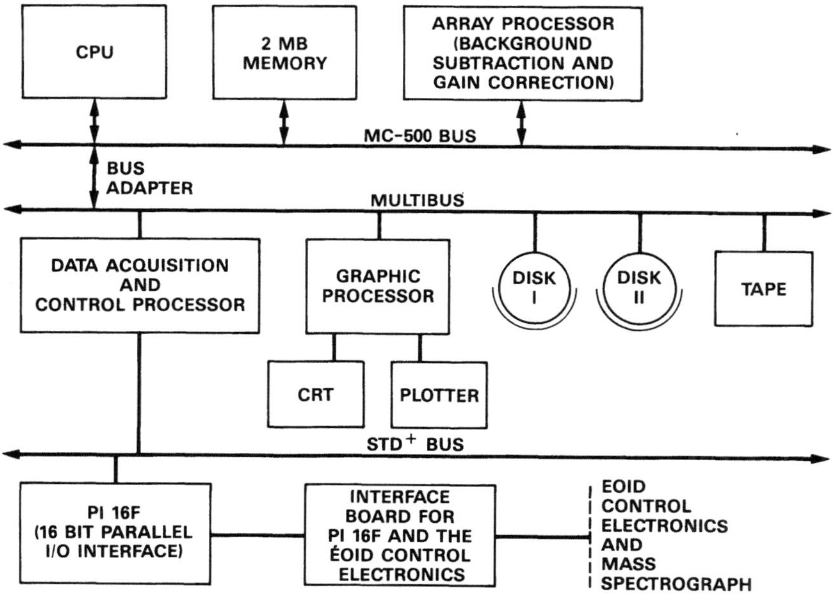


Figure 4. PAMS Data System. The Data Acquisition and Control Processor (DA/CP) collects data from the photodiodes in the direct memory access mode.

arrival of a particle. Consequently, they need to be reset just prior to the particle detection. This is accomplished by reading the PDA's at a faster rate of 1.1 MHz continuously. In this mode, the dark current integration is for <4.5 ms, and is small compared to the particle signal. At the occurrence of the trigger pulse, the PDA readout is stopped, the signal is integrated for the predetermined length of time, and is read by the computer at a slow rate (220 kHz). After the acquisition of the particle mass spectral data, the reading of the photodiodes reverts back to the faster rate while waiting for the next trigger pulse. The data read at the 1.1 MHz rate are not acquired. Its main purpose is to reset the photodiodes and keep them ready for the particle signal acquisition. Such a trigger mode of data acquisition has been developed for particle spectrum measurement.

## III. RESULTS AND DISCUSSION

Monodisperse particles of 4,4'-dibromo-biphenyl (DBBP) were used for determining the performance of the PAMS System. The aerosol was generated using a vibrating-orifice aerosol generator from a solution of DBBP in isopropanol. A 0.01% (V/V) solution of DBBP was prepared in AR grade isopropanol and forced through a vibrating orifice (20 $\mu$m). This produced liquid droplets of uniform size by the breakup of the liquid jet. These droplets were carried along in a stream of dry nitrogen gas. Monodisperse particles (1.8 $\mu$m diameter) of DBBP in the nitrogen stream were obtained by the evaporation of the solvent from the droplets. These particles were passed through a neutralizer containing $Kr^{85}$ for the neutralization of any electrical charges on them, before introducing them into the particle beam generator. The beam of particles produced by the expansion of the aerosol through the capillary nozzle was directed into the ion source of the mass spectrograph where they were thermally evaporated and the resulting vapor was ionized by electron impaction. The mass spectra of ion pulses from single particles were measured. The signal integration period for the diodes was set at 100 ms and the accumulated signals in the photodiodes were read every 100 ms. In this mode of data acquisition there is no synchronization of the arrival of the particle into the ion source with the start of the integration. The frequency of the arrival of the particles in the source was maintained low. Consequently, only one particle was present in the ion source at one time, and only some of the scans contained signals from the particles. Other scans contained the mass spectra of the background gases and of the solvent isopropanol that are always present. Figure 5 shows a mass chromatogram. It is a plot of scan versus the intensity of mass peaks at m/e 73, 151, and 312. These are some of the characteristic peaks in the mass spectrum of DBBP molecules. The absence of these peaks from a scan shows that no particle was present during the integration for such a scan. The purpose of the mass chromatograph is to demonstrate the presence or absence of particles in a given scan, showing the discrete nature of the particle mass spectral measurements. The background signal is present in all the scans, and therefore needs to be subtracted from the scan containing the particle spectrum in order to get the true particle mass spectrum.

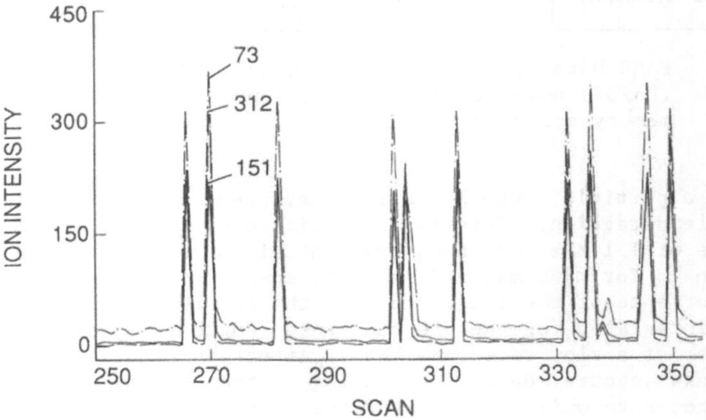


Figure 5. Mass chromatogram of different mass peaks at m/e of 73, 151 and 312, measured from 1.8 $\mu$m diameter dibromo-biphenyl (DBBP) particles. The intensities of the above mass peaks are plotted against the scan number. Each scan is the readout of the signal accummulated in the photodiodes for 100 ms. These peaks in a scan show the presence of a DBBP particle during the signal integration period of the scan. Complete mass spectrum for each san is acquired during the measurement.

A complete mass spectrum from one particle (1.8 $\mu$m diameter) of DBBP present in scan 558 is shown in Figure 6. The mass of the particle is about $4 \times 10^{-12}$ g. The mass spectrum contained in scan 557 is also shown in this figure. It shows the spectrum from the background gases only. The sensitivity of the PAMS system at present, based on DBBP particle measurements, is found to be approximately $5 \times 10^{-13}$ g. In order to determine the reproducibility of the particle mass spectrum, intensity ratios of various mass peaks in a number of single particle spectra were calculated. The relative standard deviation of intensity ratios of different mass peaks varies between 10 and 19%. Table I lists the performance characteristics of the PAMS System.

Table I. Performance Characteristics of the PAMS System

| | |
|---|---|
| Mass Range: | 25-500 amu |
| Minimum Particle Mass for Analysis: | $5.0 \times 10^{-13}$ g |
| Spectral Readout Time: | 23 ms |
| Maximum Frequency of Particle Analysis: | 44 $sec.^{-1}$ |

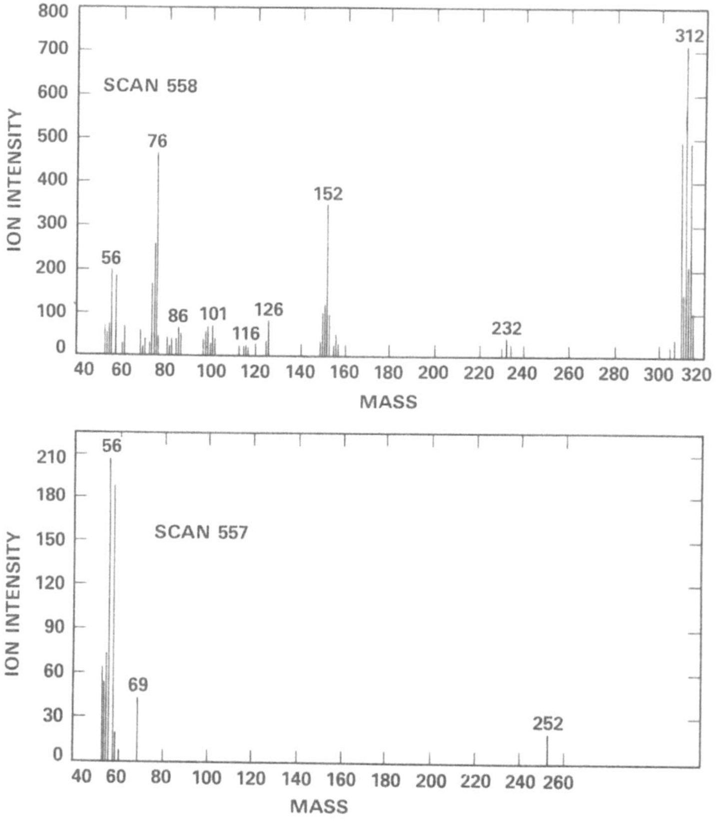


Figure 6. Scan 558 shows the complete mass spectrum of a 1.8 μm diameter DBBP particle. Spectrum from scan 557 contains mass peaks of the background gases only.

PAMS has been applied to the measurement of a wide variety of aerosol particles, both organic and inorganic in nature. These include particles generated in the laboratory[11,12,18] as well as atmospheric particles outside the laboratory[16]. One of the important applications of PAMS lies in the real-time detection/identification of airborne microorganisms[20,21]. Mass spectra from bacterial particles have been measured. Bacterial aerosols were generated in the laboratory by the nebulization of their suspensions in ethanol. A micronebulizer was used

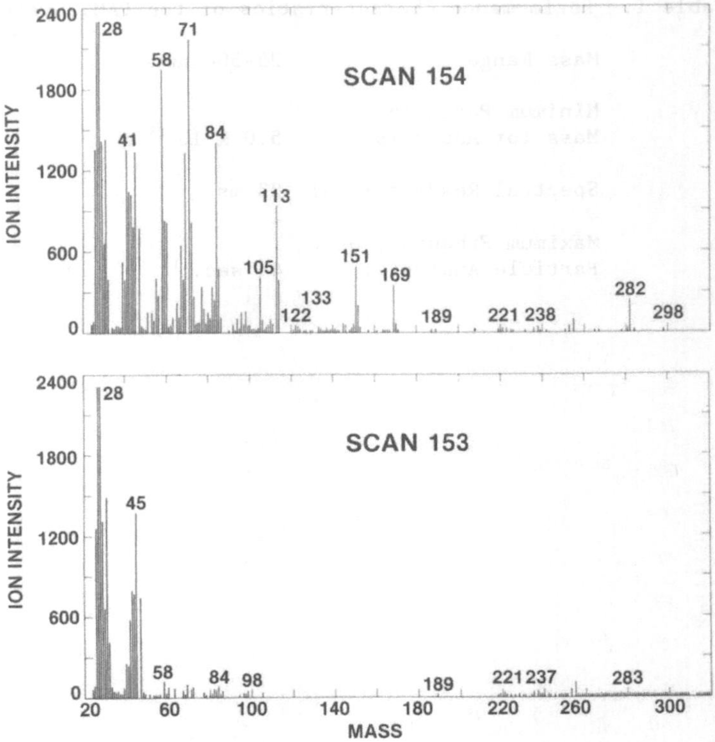


Figure 7. Mass spectrum measured from a single Bacillus cereus particle from its aerosol is shown in scan 154. No particle was present in scan 153 and therefore the spectrum contained in this scan arises from background gases only.

for this purpose. These particles were introduced into an ion source of the MS in the form of a beam. Figure 7 shows a mass spectrum measured from a single Bacillus cereus particle. The spectra from biological particles contain mass peaks up to 300 amu, which enable the differentiation of biological from nonbiological particles usually present in the atmosphere.

Several different bacterial samples were studied. Similarities in their spectra are observed. This is to be expected because of the nature of the chemical building blocks of all these bacterial cells. However, some distinguishing features were also observed in the mass spectra of different bacterial particles (B. cereus, B. subtilis, B. Anthracis sterne and Pseudomonas putida). These features may be exploited to differentiate between the different biological particles[20].

IV. CURRENT WORK

At the present time, a significant effort is being devoted to applications of the PAMS System for the measurement of particles in aerosols of low particle concentrations (1 - 10 $ft^{-3}$). These include clean rooms, and semiconductor fabrication and processing environments. Chemical compositions of particles will help identify their sources.

As described previously, the volume flow rate of the aerosol sample through the nozzle in the PAMS System is 60 atm.-cm$^3$ min.$^{-1}$. At this flow rate, it will take a long time (approx. 47 min.) for a particle to reach the ion source from an aerosol sample containing 10 particles per cubic foot (assuming 100% efficiency of the particle introduction system). Thus the analysis of aerosols containing a low number density of particles by PAMS requires suitable sample preparation and introduction. These include concentration of particles, and efficient introduction of particles in the MS without any loss in the process.

The particles in the aerosol should be concentrated without sacrificing their aerosol nature; i.e., the particles should remain suspended in gases. A virtual impactor is being developed for this purpose in our laboratory. An impactor is an inertial separator. The aerosol is passed through a nozzle and from the emerging flow, the central (minor) flow is sampled in a collection probe. Particles with sufficient inertia (size) pass into the minor flow (5-20% of the total flow), and thus the impactor concentrates the large particles by a factor of 5-20.

For achieving a high concentration factor, a new design of a two-stage virtual impactor will be used. Here the aerosol from the minor flow of the first stage is further concentrated. This can be accomplished by redirecting the minor flow from the first stage to the second stage of the impactor. A concentration of >100 fold can be achieved after the two stages. The flow parameters at both stages will be adjusted, such that a size cut-point of 0.3 $\mu$m is achieved; i.e., particles above this size enter the minor flows. The particles from the concentrated aerosol will be introduced into the ion source as a highly focused particle beam. In some semiconductor fabrication and disk operating environments, refractory particles composed of silicates and metals have been found. Laser vaporization and ionization will be a method of choice for the analysis of such particles by PAMS. The method, because of its high ionization efficiency, would also enhance the sensitivity of PAMS for the analysis of submicron size particles.

## V. SUMMARY

The technique of particle analysis by mass spectrometry (PAMS) can measure the mass spectra from individual particles suspended in a gas/air on a continuous, real-time basis. PAMS has been applied to the analysis of particles of different origins. Particles composed of both inorganic and organic compounds, including bacterial particles, have been analyzed. These measurements provide the chemical characteristics of particles present in a mixture of particles from various origins. The results can form an important basis for their source apportionment, and for planning their control strategy. The real-time response of the PAMS system can be used for process control, and for taking countermeasures in the event of a threat from toxic or biological particles in both protected and field environments. PAMS also provides a direct method for studying heterogeneous/catalytic reactions taking place on particles in the atmosphere or in the laboratory. For keeping the real-time attributes of PAMS for the measurement of particles in the low number density environments, a two-stage virtual impactor for the concentration of particles along with a particle beam generator producing highly focused particle beams is in the process of development in our laboratory. Work on the application of laser-induced volatilization and ionization of particles is also in progress.

ACKNOWLEDGMENTS

The author wishes to acknowledge the late Mr. Charles Giffin for his help in the design of the mass spectrograph and the detector, and Mr. James Hill for his role in the construction of the PAMS system. The work was supported in part by grants from the National Science Foundation, U. S. Army Chemical Research and Development Center, and by the NASA-JPL Equipment Grant.

REFERENCES

1. H. G. Barth, S. T. Sun, and R. M. Nickol, Anal. Chem., 59, 142R (1987) and references therein.

2. B. R. Appel, Studies in atmospheric particulate characterization techniques, in "Air/Particulate Instrumentation and Analysis," P. N. Cheremisinoff, editor, pp. 25-87, Ann Arbor, MI. (1981).

3. R. E. Clement, and F. W. Karasek, Int. J. Environ. Anal. Chem. 7, 109 (1979).

4. S. K. Friedlander, "Smoke, Dust and Haze: Fundamentals of Aerosol Behavior," Wiley-Interscience, New York (1977).

5. M. S. Miller, S. K. Friedlander, and G. M. Hidy, J. Colloid Interface Sci., 39, 165 (1972).

6. K. L. Mittal, editor, "Particles in Gases and Liquids 1: Detection, Characterization and Control," Plenum Press, New York, 1989.

7. S. Batchelder, Applications of particulate inspection of wafers in semiconductor manufacturing, in "Proceedings of the 34th Annual Technical Meeting," pp 359-362, Institute of Environmental Sciences, King of Prussia, PA, May 3-5 (1988).

8. J. R. Monkowski, Particulate surface contamination and device failures, in "Treatise on Clean Surface Technology," K. L. Mittal, editor, pp 123-148, Plenum Press, New York, 1987.

9. T. Y. Toribara, J. R. Coleman, B. E. Dahneke, and I. Feldman, editors, "Environmental Pollutants - Detection and Measurements," Plenum Press, New York (1978).

10. P. Wieser, R. Wurster, and V. Haas, Z. Anal. Chem., 308, 260 (1981).

11. M. P. Sinha, C. E. Giffin, D. D. Norris, T. J. Estes, V. L. Vilker, and S. K. Friedlander, J. Colloid Interface Sci., 87, 140 (1982).

12. M. P. Sinha, and S. K. Friedlander, Anal. Chem., 57, 1880 (1985).

13. T. J. Estes, V. L. Vilker, and S. K. Friedlander, J. Colloid Interface Sci., 93, 84 (1983).

14. I. D. Kovalev, G. A. Marksimov, A. I. Suchkov and N. V. Larkin, Int. J. Mass Spectrom. Ion Phys., 27, 101 (1978).

15. R. N. Zare, and R. D. Levine, Chem. Phys. Lett., 136, 593 (1987).

16. M. P. Sinha, Rev. Sci. Instrum., 55, 886 (1984).

17. H. G. Boettger, C. E. Giffin and D. D. Norris, Electro-optical ion detector in mass spectrometry: Simultaneous monitoring of all ions over wide mass ranges, in "Multichannel Image Detector," Y. Talmi, editor, pp 292-317, ACS Symposium Series No. 102, American Chemical Society, Washington, DC, (1976).

18. M. P. Sinha and S. K. Friedlander, J. Colloid Interface Sci., 112, 573 (1986).

19. M. P. Sinha "Particle Analysis by Mass Spectrometry," NASA-JPL Document No. D-6319, March 30, 1989.

20. M. P. Sinha, R. M. Platz, S. K. Friedlander, and V. L. Vilker, Appl. Environ. Microbiol., 49, 1366 (1985).

21. M. P. Sinha, Mass spectral measurement of bacteria particles suspended in air, in "Proceedings of the 1986 Scientific Conference on Chemical Defense Research," Nov. 18-21, 1986, U.S. Army Armament, Munitions, and Chemical Command, Aberdeen Proving Ground, MD.

## CHARACTERIZATION OF PARTICLES IN HIGH-PURITY GASES

W. R. Gerristead Jr., E. F. Ezell, and R. Sherman

The BOC Group Inc. Technical Center
Murray Hill, N.J. 07974

Sub-micron particle analysis methods for VLSI gas distribution systems include laser particle counting and SEM/microprobe evaluation of particles collected on membrane filters. In this paper, the performance of automated SEM/microprobe methods is compared to the performance of laser particle counters in counting and sizing submicron particles. This comparison of particle counting methods involves analyses of membrane filters exposed to gas streams that were simultaneously sampled by laser particle counters. Experimental results demonstrate the ability of the SEM/microprobe technique to count, size, and compositionally identify submicron particles that are typically found in high purity gas streams. The advantages and disadvantages of SEM/microprobe analysis and laser particle counter analysis are also discussed.

### INTRODUCTION

The elimination of submicron particles which reduce chip yields is of concern to integrated circuit manufacturers. Gases used in fabrication processes, e.g. nitrogen, argon, oxygen, hydrogen, and specialty gases, as well as their associated distribution systems may contribute to the overall particle load. Bulk and specialty gas suppliers and equipment manufacturers are continually improving methods of producing and delivering process gases with the highest chemical purity and the lowest particle content. At the Technical Center of The BOC Group, both laser particle counters and an automated SEM/microprobe are used to detect and count particles present in high purity gas streams. In addition, the SEM/microprobe compositionally identifies each particle.

Two steps are involved in particle analysis of membrane filters by an automated SEM/microprobe. The first step in automated SEM/microprobe particle characterization is image analysis for particle detection. The electron beam is rastered over each frame using digital beam control. Image analysis is relatively simple and fast. If a particle is present, the output of the backscattered detector increases and exceeds a threshold. This initiates the sizing and shaping of the detected particle.

*Particles in Gases and Liquids 2,* Edited by K.L. Mittal
Plenum Press, New York, 1990

The second step involves x-ray characterization of the particles for compositional analysis. This is done in the automated SEM/microprobe by either energy dispersive spectroscopy (EDX) or wavelength dispersive spectroscopy (WDX). EDX offers the advantage of speed and ease of automation for detection and peak identification, while WDX offers lower detection limits and better resolution for peak separation. Two EDX detectors are used for automated analysis. One detector allows for routine detection of elements heavier than neon, the other detector is capable of light-element detection. Various adaptations of this technique have been discussed elsewhere.[1,2,3,4,5]

Automated SEM/microprobe filter analysis is a relatively new technique for submicron particle analysis in the industrial gas field. For its acceptance as a new technique, it is necessary to demonstrate its relative efficiency for particle detection as compared to laser particle counters. In this paper, the efficiency of SEM/microprobe filter analysis for detecting submicron particles of different sizes is compared with that of laser particle counters.

## REVIEW OF APPLICABLE INSTRUMENTAL TECHNIQUES

Laser aerosol particle counters (LPCs) have become commonplace for monitoring process gases and liquids. Most commercial laser particle counters operate in the same fashion. Laser light is focused into a narrow beam and the light path intercepts the flowing gas. Scattering occurs as a particle passes through the light beam and measurement of the intensity of this scattered light provides particle size data. Commercially available LPCs can detect particles 0.05 µm and larger. This sensitivity requires low gas flow rates, about 0.6 scfh to 6.0 scfh. Generally, the smaller particle size detection limits correlate with reduced flow rates. LPC analysis of high purity gas lines having low particle populations requires long sampling times at low flow rates in order to obtain reasonable statistics. Laser particle counters count and size particles as they flow past the light source, making the data representative of real-time events. However, LPCs provide no information about particle composition or origin. In order to determine particle origins by LPC, multiple-sample-point studies, using several particle counters, are commonly employed. Particle counters cannot operate at gas flows or pressures normally encountered in gas delivery systems.

Another approach to particle counting incorporates particle collection on membrane filters. Membrane filters with nominal pore sizes of 0.2 µm are regarded as highly efficient for particles larger than 0.02 µm when used for gas filtration. LPC measurements performed before and after a number of sampling filters confirmed their efficiency. Filter sampling can be done at much higher gas flow rates and pressures than for LPC analysis and such sampling is therefore more representative of gas flows found in practice. The filters are then analyzed by an automated SEM/microprobe which provides information on particle populations, sizes, and additionally, compositions. The resulting data help identify the source of particles.

All filter handling procedures are performed under better than Class 100 conditions so as to minimize additional contamination. A Nuclepore type membrane is placed within a Millipore in-line filter housing. Generally, 47 mm diameter filters with nominal pore sizes of 0.2 µm are used. The filter housing is then placed into the gas line. Particles are collected on the filter from a measured volume of flowing gas. The filter is removed from the housing, immediately attached to a 47 mm Lexan substrate, and coated with a 0.02 μm conductive thin film, usually carbon. The sample is then transferred to a JEOL 733i Electron Probe Micro-Analyzer (EPMA or SEM/microprobe) for particle evaluation.

The automated SEM/microprobe from JEOL (Peabody, MA), shown in Figure 1, allows particle analysis through a combination of computer control of the electron beam, sample stage, backscatter detector, and x-ray detector. This capability is part of the Tracor/Northern (Middleton, WI) software known as PRC--Particle Recognition and Characterization. The critical sequence of events for particle recognition occurs when the electron beam is scanned across an image frame of the filter surface at 4000X magnification. Imaging is done using a backscattered detector so that the image intensity is primarily related to average atomic number. A threshold is set above the backscattered intensity of the filter surface. If the backscattered intensity during analysis exceeds this threshold, the program considers a particle to be detected. When this happens, the particle is sized and shaped and an EDX spectrum is acquired from the particle. Measurement of up to 25 elements is provided by the PRC software.[6] A separate peak-finding module, based on the Tracor/Northern program IDENT, evaluates peaks not explicitly measured by the PRC program. This sequence of detection and identification continues within each frame over several hundred frames for each filter.

The counting errors of the EDX peaks are calculated so that peaks which are statistically insignificant can be excluded. X-rays are produced randomly with time, and thus observed intensities follow Poisson statistics.[7] The error treatment of the EDX spectrum is based upon the counting statistics of independent events, i.e. peak intensity and background intensity measurements. For each element, the relative composition is determined from the net counts of the EDX peak after background subtraction. We have found that the PRC treatment of peak

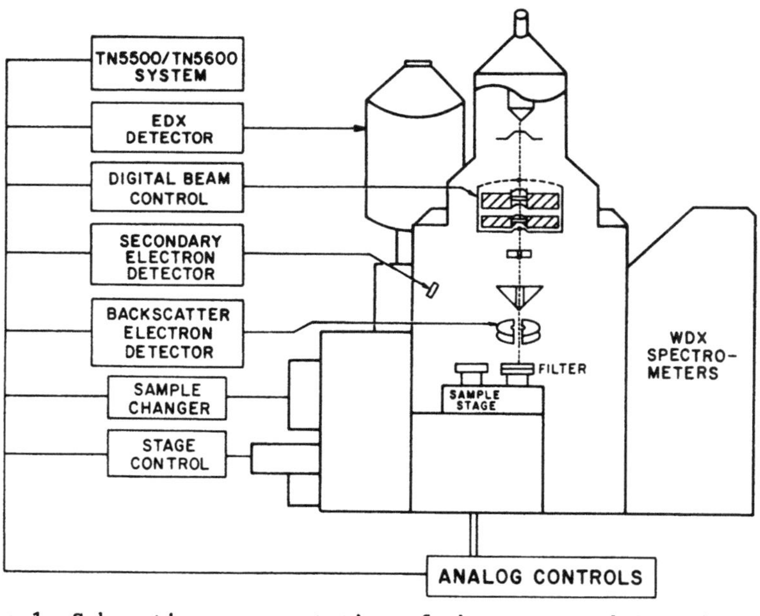


Figure 1. Schematic representation of the automated SEM/microprobe

count data does not adequately describe the error for each measurement and have modified the software to provide improved error treatment. The method used is described below.

Each of the elements is assigned distinct energy windows, or regions-of-interest (ROI's). The channels at the ends of the ROI's are used by the PRC program to establish the background correction. Only the gross counts, $I_g$, and the net counts, $I_n$, for the windows are available for statistical treatment. The number of channels, $N_c$, in each of the windows is also known. We have used this information to estimate the statistical counting errors for the net count measurements.

The counts accumulated in the two end channels, $I_e$, are computed from

$$I_e = 2(I_g - I_n)/N_c.$$

The total counts in all channels excluding the two end channels, $I_m$, are computed from

$$I_m = I_g - I_e.$$

These two counts are independent, and have independent estimated errors

$$\sigma(I_e) = (I_e)^{1/2} \quad \text{and} \quad \sigma(I_m) = (I_m)^{1/2}.$$

Since the net counts, $I_n$, evolve from the independent counts $I_m$ and $I_e$ as

$$I_n = I_m - (N_c-2)/2\ I_e,$$

the estimated statistical counting error for $I_n$ is computed to be

$$\sigma(I_n) = [I_m + (N_c-2)^2 I_e/4]^{1/2}.$$

This standard deviation is translated into a percentage of the element intensity, and this percentage is carried along through subsequent calculations for each element. Errors computed for each elemental measurement are used to determine the accuracy of each composition determination. For this method, low concentrations generally have high relative errors while the major peaks are statistically relevant even for counting times of 15 seconds per particle. Generally 2-sigma is used as the confidence level for statistical significance. This error treatment enhances the ability to confidently group large numbers of particles with subtle elemental differences into a small number of discrete compositional classes.

In this paper, experiments are discussed which verify the use of the automated SEM/microprobe for submicron particle analysis as applied to high purity gas lines. The experiments performed include 1) determination of minimum particle size in order for the backscatter electron signal to exceed threshold for recognition; 2) minimum particle size necessary to achieve satisfactory elemental analysis by EDX; and 3) comparison of submicron particle counting statistics of the automated SEM/microprobe with LPCs.

## EXPERIMENTAL

The determination of minimum particle size necessary to achieve identification was performed by modeling the process with thin films of known composition and thickness. An aluminum alloy was deposited in known thicknesses of 0.05, 0.10, 0.55, and 1.0 µm on a polymer substrate. The output signal of the solid-state backscattered electron detector

(BSD) was monitored for each film and substrate in order to determine if the signal was above the threshold value set relative to the substrate. From a plot of the BSD output of the films and the settings for the threshold and filter background, the minimum film thickness that can exceed the recognition threshold is determined.

Minimum particle size x-ray detection limits were determined by depositing particles of known composition on membrane filters. Aluminum oxide, iron oxide, copper oxide, and titanium oxide polydispersed powders were deposited onto separate membrane filters by dusting in a clean-room plastic glove-bag. The powders were blown into the bag and were allowed to deposit on the exposed filters. The filters were then prepared and analyzed by the SEM/microprobe using standard procedures.

The comparisons between the counting efficiency of the automated SEM/microprobe and the LPCs were achieved by placing membrane filters destined for SEM/microprobe analysis on one run of a Tee fitting and an LPC on the other run of the fitting. LPC measurements made on the outlet of the sampling filter showed that the filter was capturing virtually all of the particles present on its inlet side. Flow rates for the LPC were at 6 scfh while the flow rates through the membrane filters were between 340 and 400 scfh. Total gas volumes through the membrane filters for these tests ranged from 7500 to 29,400 scf. An isokinetic sampler for the LPC was used for one comparison. The instruments used in this study included an HPGP-101 and 101-HP manufactured by Particle Measurement Systems of Boulder, Colorado.

In order to compare the particle populations found by each technique, a relative efficiency is defined. The relative efficiency is the number of particles detected by the SEM/microprobe (adjusted for the ratio of the total filter area to the examined filter area) per cubic foot of gas flow through the filter divided by the number of particles detected by the LPC per cubic foot of gas flow through the LPC. This definition normalizes the large difference in gas volumes sampled by both, with the membrane filter usually intercepting 100 times the gas flow of the LPC.

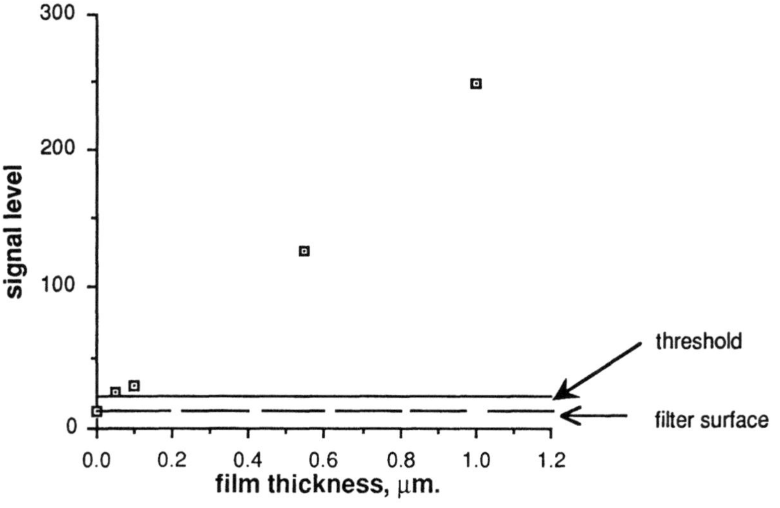


Figure 2. Plot of signal output of backscattered detector with (Al 70%, Si 30%) thin films of known thicknesses, showing level of filter surface and threshold for recognition.

## RESULTS AND DISCUSSION

### Particle Size for Recognition

A plot of the backscattered detector output signal versus the film thickness, in Figure 2, indicates that for the thresholds employed, films as thin as 0.05 µm can be detected. The threshold value is routinely set at 10% of the full scale of the digitized backscattered electron detector output. This is done with the contrast set at its maximum value. With the contrast and threshold set, the brightness of the filter (or in this case, the polymer substrate used for these aluminum based films) is adjusted until the BSD output for the polymer is one half of the threshold value.

Experimental results based upon these settings indicate that the BSD output for the 0.05 µm and 0.10 µm thin films exceeded the threshold values -- implying detection. A thin film of known thickness is a reasonable analog to a particle of similar dimensions only if few multiple scattering events occur in the thin film beyond the boundaries of a particle of the same diameter. Monte Carlo thin film calculations indicate that few scattering events occur outside such boundaries.

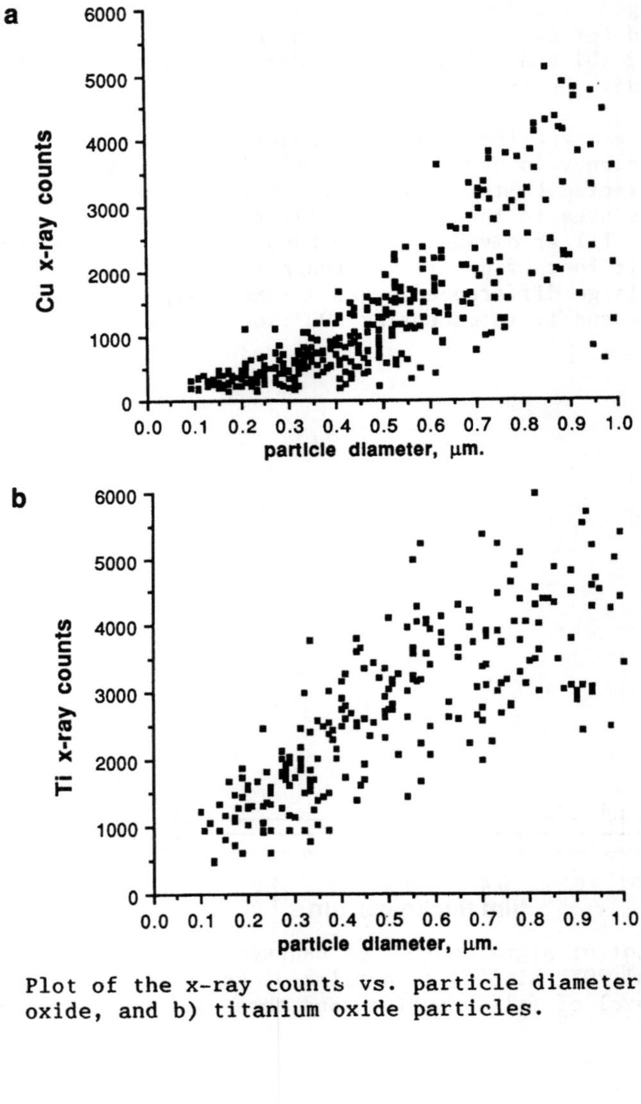


Figure 3. Plot of the x-ray counts vs. particle diameter for a) copper oxide, and b) titanium oxide particles.

Detection limit estimates based on thin film results are therefore reasonable. As the average atomic number increases for other materials, the backscatter cross section increases, implying reduced minimum particle size detection limits. With the potential minimum thin film detection limit on the order of 0.05 μm for a thin film with average atomic number of 13.3, we feel confident that particles of the same atomic number will be detected at the 0.1 to 0.2 μm sizes.

## Particle Size for Elemental Classification

Experiments were done to determine the minimum particle size that yields sufficient x-ray counts for elemental classification. X-ray yield was correlated with particle size for powders of known materials. In these experiments, particles of aluminum oxide, titanium oxide, iron oxide, and copper oxide were deposited onto membrane filters and then analyzed in the SEM/microprobe. X-ray counts were acquired from particles that were detected using the above techniques, and the number of x-ray counts for the major constituent versus particle size for the four powders was plotted.

Figure 3 displays the above data for the copper oxide and titanium oxide powders. Reasonable x-ray counts are found for all particle sizes, however, the scatter is much larger than expected from counting statistics alone. There are several physical reasons for this. Generally, it is assumed that the reported diameters represent spherical particles. The calculated average diameters are derived from eight diameters taken 22.5° apart. Actual area and volume estimates are likely to be correct for spherical particles. However, for other particle shapes -- needles, plates, or agglomerates -- the reported average diameter may not be representative of the particle.

The orientation of the particle on the membrane filter surface may also influence x-ray data. Shape factor information will only explain effects which result from particle orientation on the plane of the filter surface. A needle shaped particle lying on its side will yield a calculated volume much larger than actual. The EDX spectrum will give counts that appear low for the calculated spherical shape since the true scattering volume reflects only the thickness of the needle, not an average of its length and width. A needle shaped particle embedded into the filter will yield a smaller calculated diameter and the resulting x-ray counts will be much larger than expected. A similar situation will develop from platelets or agglomerates lying flat on or embedded in the filter.

The scatter observed in Figure 3 primarily reflects the limitations of particle size and shape descriptions; limitations shared by other particle analysis techniques where all three dimensions (or volume) are not measured. A set of sample filters (or polycarbonate substrates) prepared with known mono-dispersed particles of graduated sizes would allow for a critical study of the effect of particle size, shape, and orientation on x-ray yield. Overall, the scatter does not prevent estimation of a practical size limit for EDX characterization of particles. The data in Figure 3 clearly indicate that qualitative x-ray analysis of particles as small as 0.1 μm is practical.

## Relative Efficiency

The comparison of relative efficiencies of the automated SEM/microprobe to the LPC was accomplished by setting up two gas flows from the same gas supply. One flow went to the LPC and the other to a membrane filter for subsequent SEM/microprobe analysis. After

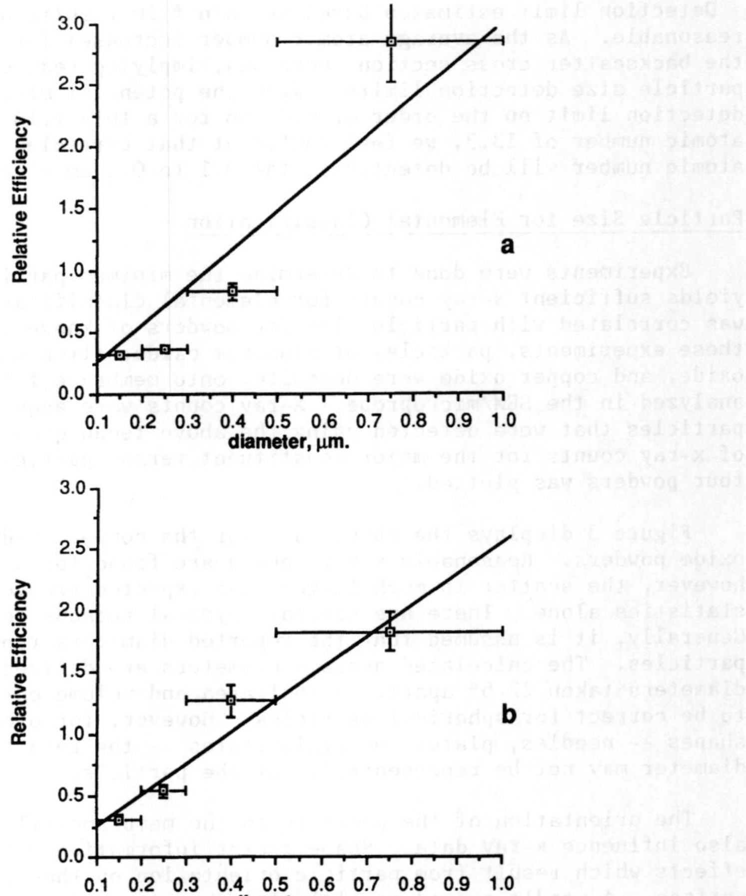


Figure 4. Efficiency of SEM/microprobe submicron particle detection relative to laser particle counter detection vs. size under the following experimental conditions: a) house nitrogen, Tee probe, 29,400 scf total volume through filter; b) house nitrogen, isokinetic probe, 21,600 scf total volume through filter; c) house nitrogen, Tee probe, with oily valve upstream of both filter and LPC, 7500 scf total volume through filter.

SEM/microprobe analysis, the number of particles per unit volume of gas was calculated for the same size ranges as the LPC reported.

The number of particles found in each size range per unit volume of gas for the two techniques is used to calculate the relative efficiency versus particle size. Relative efficiency is defined as the quotient in each size range of particles per cubic foot detected by the SEM/microprobe divided by particles per cubic foot detected by the LPC. The plot (Figure 4a) indicates that the relative efficiency is dependent on particle size. At particle sizes of 0.1 to 0.3 µm, the automated SEM/microprobe has only counted 0.3 to 0.4 times the particles counted by the LPC. At higher particle sizes of 0.3 to 0.5 µm, the relative efficiency increases to about 0.75--0.85 and for particle sizes of 0.5 µm and greater, the relative efficiency is greater than 1.0. This test was repeated with other gas flows and sampling configurations with similar results.

For the isokinetic probe, the same trend was observed but the relative slope of the line was somewhat flatter, as shown in Figure 4b. At the smallest particle size, the relative efficiency was about 0.3 from 0.1 to 0.2 µm. For particles in the size range of 0.2 to 0.3 µm, the efficiency of the SEM/microprobe relative to the LPC was found to be about 0.5, greater than the 0.35 efficiency obtained with the Tee probe for the same particle sizes. For the next particle size range, from 0.3 to 0.5 µm, the relative efficiency of the probe with the isokinetic sampler exceeded 1.0, as compared to only about 0.8 with the Tee sampler. Above 0.5 µm, the SEM/microprobe had an efficiency greater than 1.0, but the relative efficiency was not as high as in the Tee test.

Another test of the relative efficiency of the SEM/microprobe to the LPC involved placing an oily valve upstream of both the LPC and membrane filter and collecting particles with the Tee assembly. The results of the comparison in this case, in Figure 4c, indicate that the SEM/microprobe has poor efficiency for all particle sizes relative to the LPC ranging from about 0.1 to 0.5. The low efficiency for this case is related to poor detection of organic particles by the SEM/microprobe because of the low average atomic numbers of the organic particles.

These results suggest that the LPC is more sensitive than the SEM/microprobe in the smaller size ranges and less sensitive in the larger size ranges. LPC resolution and calibration may result in a shifting of a particle of a given size to either the adjacent larger or smaller size range. An analyzer with more size channels would reduce the uncertainty associated with size classification. Differences in light-scattering properties between actual particles and latex calibration spheres may result in a downward shifting of the LPC measured particle size, thus contributing to a reduced relative efficiency of the SEM/microprobe in the smaller size ranges. An alternate approach to LPC calibration using monodispersed metal and metal-oxide particles would address the uncertainty due to calibration in these experiments.

SEM/microprobe limitations may also result in the variation of relative efficiency with particle size. The light-element particle content and size distribution of this gas system is not known, and may be a source of decreased SEM/microprobe efficiency, as demonstrated in the oily valve experiment. Establishing a threshold for SEM/microprobe analysis that excludes detection of filter surface features also limits small particle sensitivity. In addition, particles smaller than the filter pore size which are trapped in the filter pores are not detected. This problem is related to filter sampling methods, and is probably sensitive to gas velocity. Enhanced counting of larger particles may be a result of the tendency of the SEM/microprobe software to break up large, multi-lobed particles into two or more smaller particles. Alternately, large agglomerates may fragment upon impacting the membrane surface.

The different results obtained with the isokinetic probe and the simple Tee test arrangement are not fully understood, but undoubtedly reflect the differences in sampling methods. This implies that sampling methods influence the particle counting behavior of both instruments. The isokinetic probe for the LPC reduces bias against large particles, although such bias should be rather small in the submicron size range. Our experience indicates that gas flow characteristics may affect observed submicron particle size distributions through gas lines.

## CONCLUSIONS

These studies show that the SEM/microprobe can detect and characterize submicron particles above 0.1 μm. The efficiency for detection relative to an LPC depends on particle size. This relative efficiency is a consequence of several factors, one of which is the fraction of particles which have low atomic number densities, (e.g., organic particles). SEM/microprobe detection is not very efficient for small organic particles because their backscattered intensities are similar to that of the filter background. The LPC does not have such limitations in counting particles of different compositions, although sizing may be influenced by the light-scattering properties of individual particles. The above factors lead to a reduced relative efficiency for the smallest particles during SEM/microprobe analysis.

At the same time that the SEM/microprobe is missing a percentage of the smaller particles, the LPC may be omitting some particles larger than 0.5 μm. The source of this phenomenon is not fully understood, but may, in part, be related to sampling technique, as suggested by improved detection during isokinetic sampling.

## REFERENCES


1. J.F. Kelly, R. J. Lee, and S. Lentz, Automated characterization of fine particulates, in "Scanning Electron Microscopy 1980 (I)", pp. 311-322, SEM Inc., AMF O'Hare, IL.

2. J.F. Kelly, and R. J. Lee, Overview of SEM-based automated image analysis, in "Scanning Electron Microscopy 1980 (I)", pp. 303-310, SEM Inc., AMF O'Hare, IL.

3. R.M. Edwards, J. Lebiedzik, and G. Stone, Fully automated SEM image analysis, Scanning, 8, 221-231 (1986).

4. G. J. Stone, Automated SEM/EDS image analysis of particles on filter blanks, in "Particles on Surfaces 1: Detection, Adhesion, and Removal" K.L. Mittal, editor, pp. 293-306, Plenum Press, New York, 1988.

5. W.R. Gerristead Jr., R. Sherman, and J. Davidson, Automated SEM/microprobe finds gas contaminants, determines sources, Research and Development, 29, 93-100 (1987).

6. G.S. Fritz, J.J. McCarthy, and R.J. Lee, Interactive software for automated particulate analysis, Proceedings of the 16th Annual Conference of the Microbeam Analysis Society, Vail, CO, July 1981, pp. 57-60.

7. M. Cox, Experimental determination of x-ray intensities, in "Quantitative Electron-Probe Microanalysis", V.D. Scott and G. Love, editors, pp. 137-140, John Wiley & Sons, New York, 1983.

# PARTICLE MEASUREMENT IN GAS SYSTEM COMPONENTS: DEFINING A PRACTICAL TEST METHOD

Ulrich H. Koch and Michael C. Kovach

Nupro Company
4800 East 345th Street
Willoughby, Ohio 44094

The objectives of this test program were to define a valid, practical test method to measure the cleanliness of process gas piping components and to illustrate the use of gas velocity as the basis for a cleanliness safety factor. Tests were conducted at three velocities using two particle counting methods, two valve designs, two cleaning methods, in both static and dynamic modes. A sufficient number of samples were tested to provide statistically valid results. The data show that high velocity flow tests are necessary to detect differences in high-purity cleaning methods and dynamic tests are required to evaluate valve designs. An industry standard test method for evaluation of systems and components is presented. Additional research is suggested to establish the critical velocity to remove particles of known size from a surface.

## INTRODUCTION

The semiconductor industry has long recognized that valves, fittings, and other mechanical components in process gas systems are potential sources of particle contamination and wafer defects. The process gas contacts the wafers directly during fabrication. Any particles in the gas stream are potential "killer" particles. The rapid shift in the industry toward high-density circuits on larger wafers has created a critical need for ultraclean gas delivery systems.

Steps taken to eliminate particle sources in gas systems are usually elaborate (and expensive). Redesigning valves and regulators, using 316L stainless steel, electropolishing, and similar measures are common. Unfortunately, there is no way to measure the cost-effectiveness of such changes short of actually producing chips.

---

This paper includes material from an article which appeared in <u>Microcontamination</u>, March 1989. Used with permission.

Therefore, there is a longstanding need for a practical way to measure the cleanliness of a component or system before putting it in service. Furthermore, there is a need to design the system with a safety factor for cleanliness. All other branches of engineering use safety factors to protect against failures from unknown stresses. Gas systems are likewise subject to many unknown stresses such as shock, vibration and flow surges. It is well known that such stresses generate particles but are not readily quantified. A safety factor for cleanliness can protect against such unknowns.

In this program, two different valves that are widely used in high purity gas systems were tested for particle contamination. Each valve was cleaned by two different methods. All the valves were tested by two particle counting methods and at three flow rates. Static and dynamic tests were run on each valve. The tests demonstrated the importance of using high velocity flow when testing very clean parts for sub-micron particles. They also showed that valve design is more important than cleaning method when a valve is operated.

From these results one can develop true performance specifications based on a test that measures the final cleanliness of a component. Now users can specify results, not methods. Both manufacturers and users can measure the cost-effectiveness of design changes and cleaning improvements. Most importantly, the user's system performance becomes independent of the manufacturer's production and cleaning processes.

## BACKGROUND

The earliest and simplest particle counting method used in evaluating such system components consisted of flushing a system and trapping particles on filter paper for later manual counting under a microscope. Today's more sophisticated instruments use laser light sources and computerized data collection to automate the counting process.

Figure 1 illustrates the detection range of different counting methods superimposed on the particle-size distribution expected in machined stainless-steel systems [1]. As the figure clearly shows, each method observes only a small portion of the total spectrum. In addition, none of these methods counts the particles on a component's surface directly -- they only count the particles that are removed by flushing a gas or liquid stream through the system.

The particles that are the most difficult to remove during testing are the "killer" particles that will damage wafers in production. Therefore testing must be done in a way that removes more particles from system surfaces than can come loose during operation. Only then will the user have an accurate indication of how clean the valve is and know the margin of safety for operational cleanliness.

Figure 2 illustrates the force (relative to particle size) required to remove particles from a polished optical surface [2]. Although it is easy to remove large particles, small particles require a very large force to dislodge them. Worse yet, the smaller the particle, the more difficult it becomes to apply the needed forces [3].

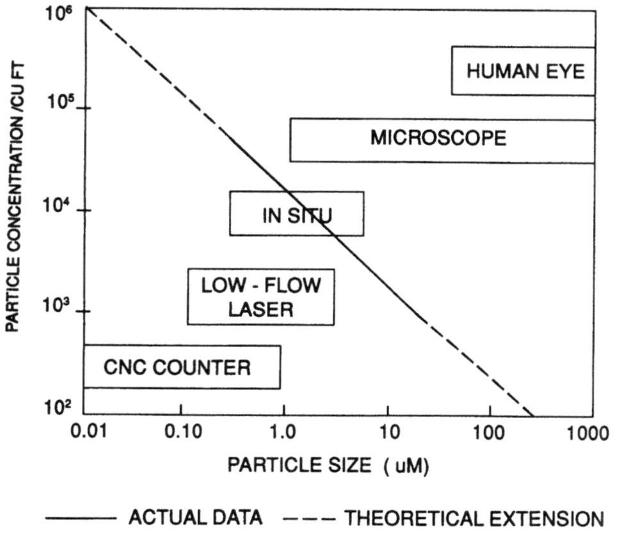


Figure 1. Detection range of various particle-counting methods compared with expected particle-size distribution of machined stainless steel (the straight line).

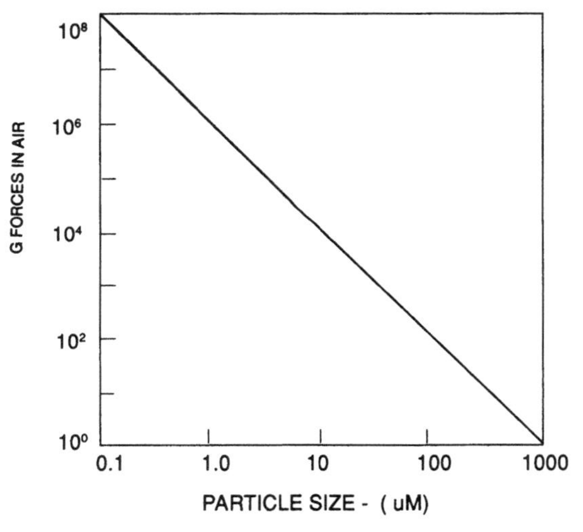


Figure 2. Shear force required to remove particles of various sizes from an optical surface.

## Drag Force

In all particle-counting methods based on system flushing, the energy to remove particles from component surfaces is the kinetic energy of the fluid. The fluid applies a drag force ($F_d$) to the particle as given by the following equation, illustrated in Figure 3:

$$F_d = C \cdot \rho \cdot D^2 \cdot V^2$$

Where:
- C = a shape constant for the particle
- $\rho$ = fluid density
- D = effective particle diameter
- V = fluid velocity at the particle

The drag force available to remove a particle decreases with the square of the particle diameter. The particle shape determines the constant C. The fluid density, $\rho$, is limited by the practical fluid choices. Thus, the only controllable factor is velocity.

## Average Velocity

An average system velocity can be calculated using the following equation:

$$V_{avg} = \frac{Q}{A_{avg}}$$

Where:
- Q = Flow rate, volumetric
- $A_{avg}$ = Cross-sectional area of the flow path

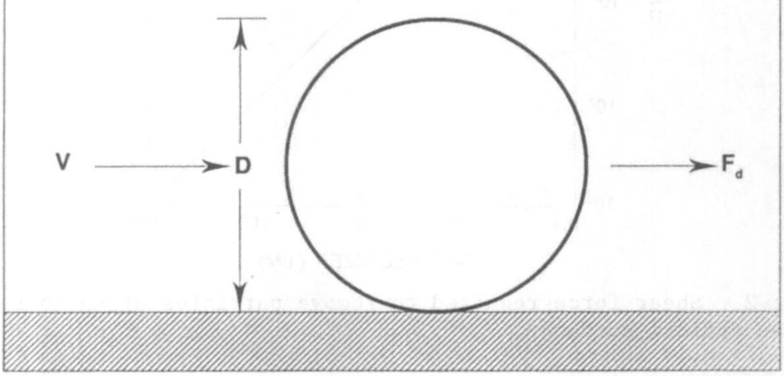


Figure 3. The drag force applied to a particle by a fluid as a function of velocity and particle size.

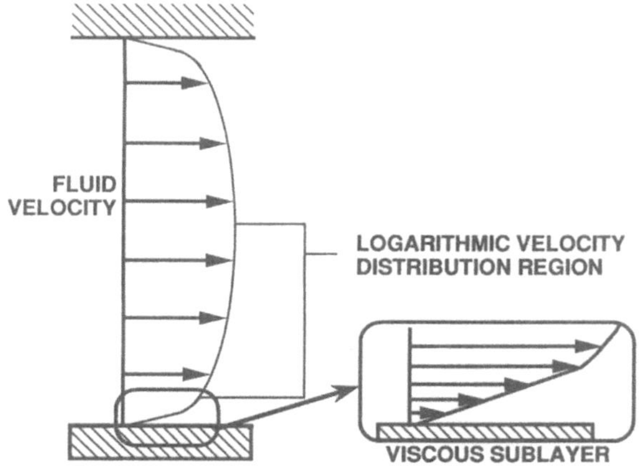


Figure 4. The boundary layer effect at the pipe wall in a turbulent flow stream.

Boundary Layer Velocity

The presence of a boundary layer in a flow stream magnifies the significance of velocity. As shown in Figure 4, the fluid velocity decreases logarithmically from the center line of the piping to the boundary layer. Even in turbulent flow, a viscous sublayer of laminar flow exists where the velocity rapidly decreases to zero at the pipe wall [4-7].

The local velocity that affects the particle depends on the average system velocity and the distance from the pipe wall, Figure 5. This viscous sublayer can be several microns thick, enabling micron-size particles to "hide".

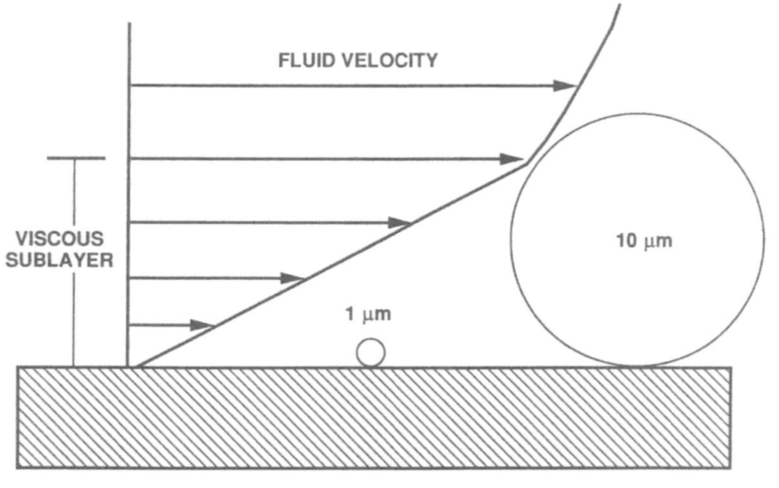


Figure 5. The viscous sublayer at the pipe wall in a turbulent flow stream.

This local velocity at sub-micron distances from the wall within the viscous sublayer is:

$$V = \frac{\tau y}{\rho \nu}$$

Where:

$\tau$ = Fluid shear stress
$\rho$ = Density of the fluid
y = Distance from the pipe wall
$\nu$ = Kinematic viscosity of the fluid

Fluid shear stress ($\tau$) at the pipe wall is:

$$\tau = \frac{f\rho(V_{avg})^2}{8}$$

Where:

f = Friction Factor
$\rho$ = Density of system fluid
$V_{avg}$ = Average system Velocity

The friction factor is:

$$f = \frac{0.316}{R_e^{¼}}$$

For $3,000 < R_e < 1000,000$

Where:

$R_e$ = Reynolds Number

and:

$$R_e = \frac{(V_{avg})\ (d_{avg})}{\nu}$$

$V_{avg}$ = Average velocity
$d_{avg}$ = Average diameter
$\nu$ = Kinematic viscosity of system fluid

To summarize, in order for a surface particle measurement to be of practical value the gas velocity used during testing must be significantly higher than the normal operating velocity. The high velocity will ensure that all particles that might come loose in service are removed and counted during the test. Thus, velocity can be used as the basis for a "Safety Factor" when specifying the cleanliness requirements for a process gas system.

## TEST APPARATUS

Two particle-measurement systems were evaluated in our study. A diffuser was added to an aerosol counter to measure particles in a piping system gas stream. A laser-based counter with a viewing module was used for in situ particle counting.

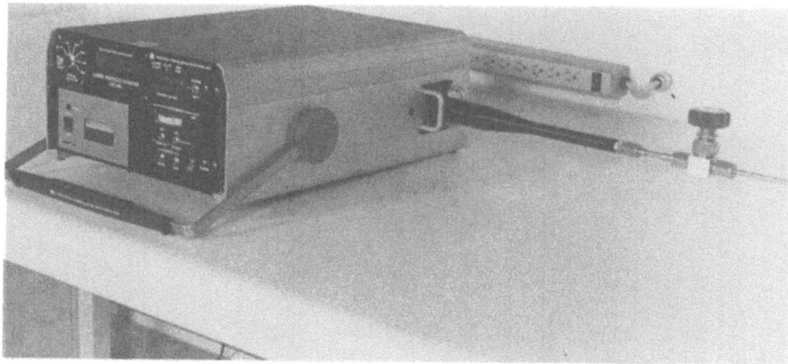

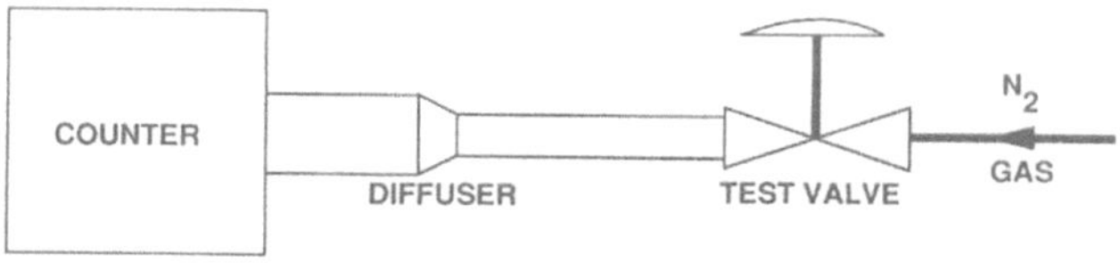


Figure 6. The standard PMS aerosol counter and diffuser used for particle counting at low flow rates.

## Aerosol Counter

Figure 6 illustrates the aerosol probe system, consisting of a PMS Model LPC-101 counter with HPD-101 diffuser (Particle Measuring System Inc., Boulder, CO). An orifice tap in the diffuser extracts a gas sample from the flow stream and reduces the pressure. The diffuser nozzle expands the sample stream to atmospheric pressure and exhausts excess gas. The particle counter samples the gas at a flow rate of 0.28 to 28 slpm (0.01 - 1.0 scfm).

Industry standards for bulk gas specifications use this method because it is appropriate for checking gas streams of installed systems [8]. It has the advantage of detecting particles in the 0.1 to 3.0 micron range. However, it has several limitations for testing valves and other piping components:

- The diffuser must take an isokinetic sample to yield a valid count [9].
- Isokinetic flow is not maintained over a wide range of flow rates.
- Gas velocity in the component is limited.
- Dynamic (cyclic) testing of valves is difficult.

## In Situ Counting Method

The in situ counting method evaluated in this study, Figure 7, measures the particles carried in a gas stream without interrupting or otherwise changing the flow pattern [10]. A viewing module in the piping system enables the counter to observe particles carried by the gas stream. Particles are sized and counted by light-scattering techniques using a laser beam projected through the window. Reducers accommodate

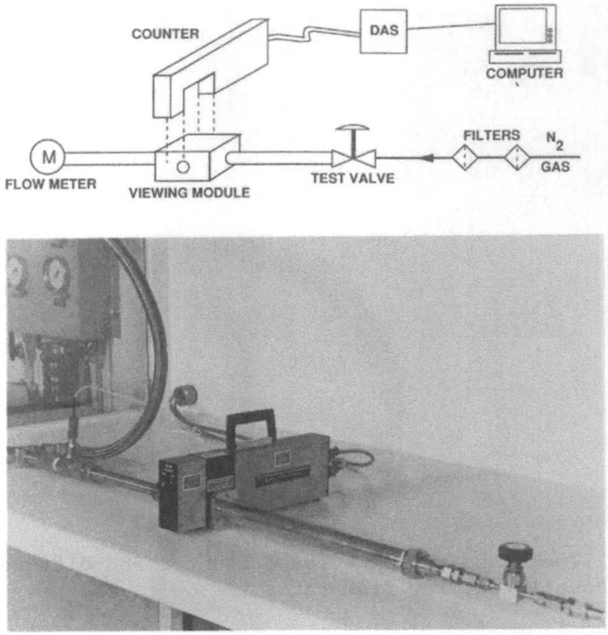


Figure 7. The standard PMS in situ particle counter used for high velocity static and dynamic tests.

many sizes of test samples (e.g. valves), and face-seal fittings prevent contamination during sample installation. A flowmeter downstream of the window measures the total gas-flow rate.

The equipment is all standard, commercially available hardware consisting of:

- PMS HPLAP (high-pressure laser aerosol probe) counter.
- PMS Model VMG viewing module 38 mm dia. (1.5 in.), 50-530 slpm flow range, 0.24-2.40 slpm (0.045%) sampling volume.
- Two 0.2 and 0.05 μm filters (Millipore Corp., Bedford, MA).

The principal limitation of this test system is the counter's inability to detect particles down to 0.1 μm. However, we found that it offered several important advantages over the aerosol sampling method:

- It can count particles above 0.3 μm to 5.0 μm.
- The counter does not disturb the flow stream, which eliminates the error of a non-isokinetic sample.
- The counter samples the flow optically rather than mechanically.
- High velocity gas flow can be used to test valves.
- A single window permits measurements over a 10:1 range of flow rates.
- The sample fraction is constant over the entire range of the window.
- Counting is done in a closed system to eliminate external contamination of the sample.
- Dynamic tests (valve cycling) are practical and yield valid, repeatable results.

Both test systems are permanently installed at Nupro Co. (Willoughby, OH) in a Class 1000 cleanroom under a Class 100 laminar-flow workbench to prevent contamination of the system or test specimens. All system tubing lines and components were cleaned by the company's ultraclean procedure before assembly [11]. Liquid-source nitrogen, pumped to high pressure and reduced at the point of use, is the clean gas used. All tubing lines from the gas source forward are stainless steel.

## TEST PLAN

A series of tests compared the aerosol probe and in situ test methods by testing both valve designs at various flow rates. Static tests with continuous flow through an open valve were run by both methods. High velocity dynamic tests, with the valve cycling open-shut, were run by the in situ method.

The two valve designs tested are designated as Types BK and BN. The BK valve, Figure 8, is a conventional bellows valve with pressure inside the bellows and internal guiding (Nupro Model SS-4BKV51-1C). The BN valve, Figure 9, is an inverted bellows valve with pressure on the outside of the bellows and external guiding (Nupro Model SS-BNV51-C). Both valves have normally closed actuators and a 125 psi rating. The two cleaning processes were Method 11, the standard method for high-purity cleaning, and Method 55, an ultra-pure cleaning method that yields certified test results [11, 12].

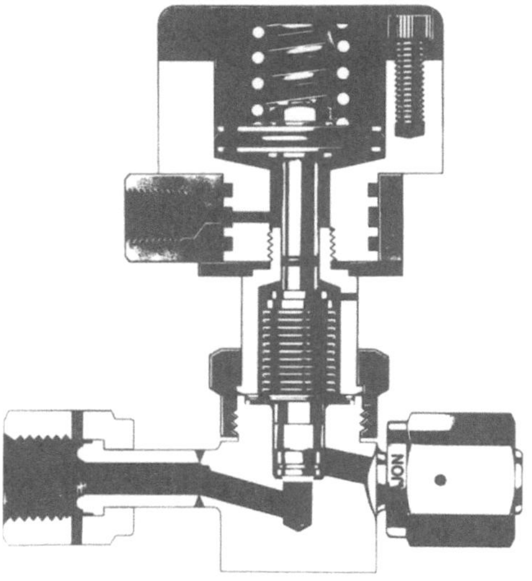

Figure 8. The BK valve illustrating a conventional bellows valve design with pressure inside the bellows and internal guiding. Nupro Model SS-4BKV51-1C).

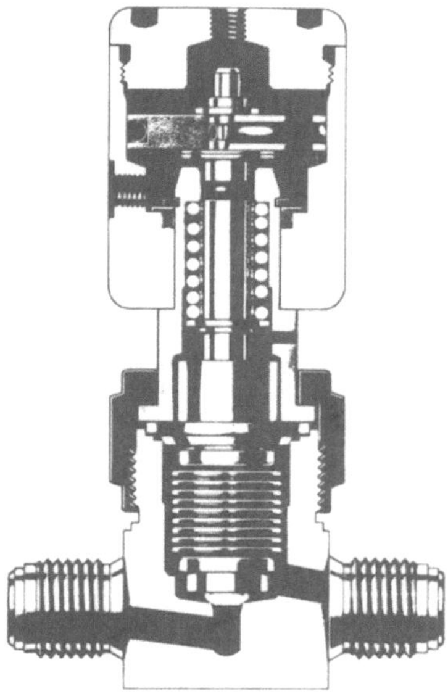

Figure 9. The BN valve illustrating an inverted bellows valve design with pressure on the outside of the bellows and external guiding. (Nupro Model SS-BNV51-C).

STATIC TESTS

Method

Each static test run began with a background measurement of the system, without a sample valve in place, using the following procedure:

For the aerosol system:
- Flush for 15 minutes at 150 slpm.
- Reduce flow to the planned test rate.

For the in situ system:
- Flush for 15 minutes at 840 slpm (twice the maximum counting rate).
- Reduce flow to 420 slpm (maximum counting rate).

For both systems:
- Measure the background count for 20 minutes.
- Maintain a zero count for the full 20 minutes.
- Install test valves and begin static tests.

Two static tests were run in the aerosol system at 29 and 121 slpm, providing safety factors of 1.5:1 and 6:1 based on test-to-operating flow ratios. One static test was run on the in situ system at 280 slpm, providing a 14:1 safety factor. Each test was repeated four times for the two valve designs cleaned by both methods.

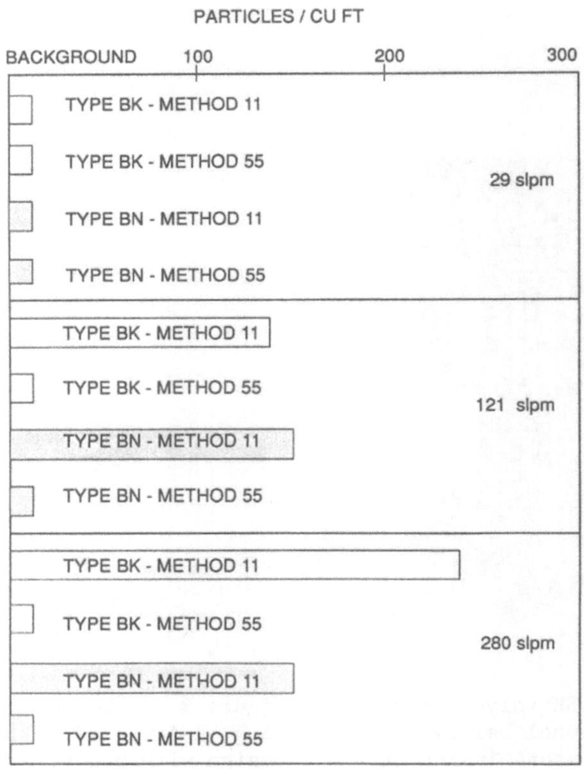


Figure 10. Static tests results at three flow rates.

Table I. Results of static tests at three flow rates.

| Method | Flow Rate (slpm) | Valve Type, Cleaning Method | Particles per Cubic Foot * | Number of Samples |
|---|---|---|---|---|
| Aerosol | 29 | Type BK: | | |
| | | Method 11 | Background | 10 |
| | | Method 55 | Background | 10 |
| | | Type BN: | | |
| | | Method 11 | Background | 10 |
| | | Method 55 | Background | 10 |
| Aerosol | 121 | Type BK: | | |
| | | Method 11 | 140 | 10 |
| | | Method 55 | Background | 10 |
| | | Type BN: | | |
| | | Method 11 | 150 | 10 |
| | | Method 55 | Background | 10 |
| In situ | 280 | Type BK: | | |
| | | Method 11 | 240 | 10 |
| | | Method 55 | Background | 10 |
| | | Type BN: | | |
| | | Method 11 | 150 | 10 |
| | | Method 55 | Background | 8 |

* Aerosol probe method measures 0.1-3 μm particles; in situ method measures 0.3-5 μm particles.

Results

As shown in Figure 10 and Table I, the aerosol system is unable to discriminate between the two valve designs (BK and BN) at either high or low flows. At 121 slpm, all samples cleaned by Method 11 show a similar increase in particle count. However, when flow is more than doubled to 280 slpm in the in situ system the BK valve shows a 70% increase in particle count. The BN valve shows no increase because its inverted bellows design makes it inherently easier to clean.

All samples cleaned by Method 55 remain at background level regardless of flow rate. Comparisons and differences were verified by testing for location of two independent samples using Student's t-distribution with a 97.5% confidence level.

## DYNAMIC TESTS

### Method

Each dynamic test began with background measurements of the system and the sample valve using the following procedure:

- Flush for 15 minutes at 840 slpm.
- Reduce flow to 420 slpm (maximum counting rate for the window).

- Measure the background count for 20 minutes.
- Maintain a zero count for the full 20 minutes.
- Install the test valve and repeat a static test to check the cleanliness level with continuous flow.
- Maintain the nitrogen flow at 420 slpm.
- Maintain a zero count for at least 10 minutes.
- Begin cycle testing.

The cycle test sequence, illustrated in Figure 11, is:

- Cycle open-shut once per second, 30 times, for 30 seconds.
- Open 30 seconds.
- Record the first 1-minute count.
- Flush 1 minute.
- Record the second 1-minute count.
- Repeat 15 times for 30 minutes, 450 cycles.
- Record the cumulative count after 30 minutes.

Counts taken at 10-second intervals (minimum available) show a rapid rise in particles during the 30 seconds of cycling and a rapid decline to background during the next 30 seconds. The second minute of open flow ensures that the count returns to background each time. The results are compared on the basis of the 30 minute cumulative count in order to have a significant count for even the cleanest valve.

Results

Dynamic testing measures the particles generated during operation. While cycling, the valve slams closed and reopens rapidly to shed particles by impact. Internal guiding surfaces can rub and generate particles. The results illustrated in Figure 12 and Table II reveal major differences between the conventional and inverted bellows valve designs and the two cleaning methods.

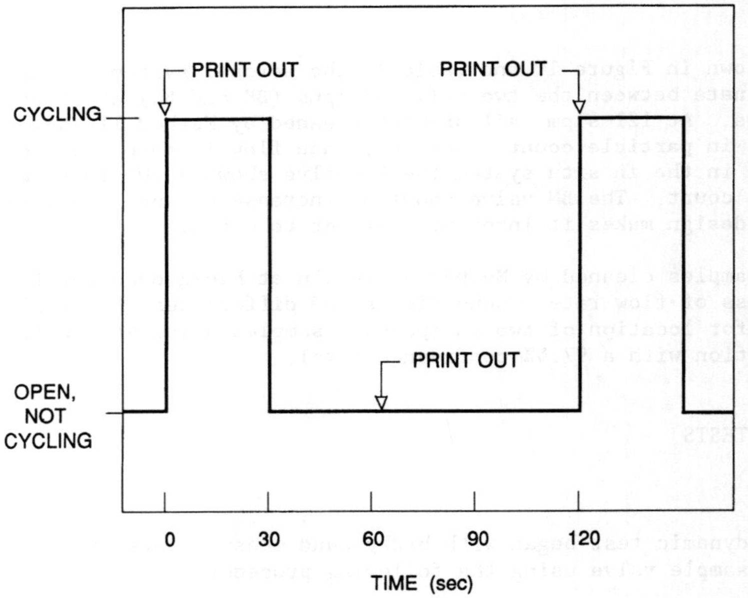


Figure 11. Operating pattern used for dynamic cycle tests. The valves are cycled once per second for 30 seconds during each two minute period which is repeated 15 times for 450 cycles.

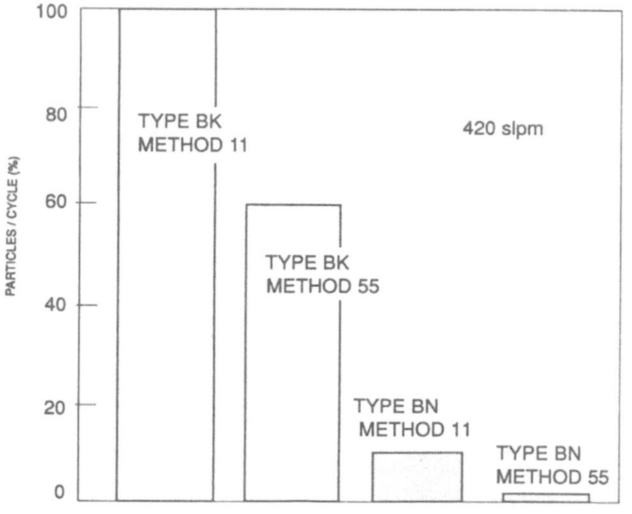


Figure 12. Dynamic test results for two valves and two cleaning methods.

The BK valve shows the highest particle count when cleaned by Method 11. Cleaning by Method 55 reduces this count by 40%, but does not reduce it to zero as in the static tests. This is because the BK bellows valve is a conventional bellows valve design with internal guiding which generates particles throughout the test.

By contrast, the BN valve shows 90% fewer particles than the BK valve when cleaned by Method 11 and 98% fewer when cleaned by Method 55. This is because the BN valve, with an inverted bellows and external guiding, does not generate particles in the system during operation. These comparisons were verified by Student's t-distribution testing for location of two independent samples, with a confidence level of 97.5%.

Table II. Relative particle counts for dynamic tests of two valve designs and two cleaning methods

| Valve Type, Cleaning Method | Particle Count * (% Increase Relative to Background) | Number of Samples |
|---|---|---|
| **Type BK:** | | |
| Method 11 | 100 | 10 |
| Method 55 | 60 | 10 |
| **Type BN:** | | |
| Method 11 | 10 | 10 |
| Method 55 | 2 | 10 |

* Method measures particles in 0.3-5 μm range.

## CONCLUSIONS

Several conclusions are apparent from the results of the static and dynamic tests with the aerosol and in situ counting systems.

- Low velocity tests can not evaluate high-purity cleaning methods.
- High velocity tests are perceptive in evaluating high precision cleaning methods.
- Static tests can not distinguish between valve designs if the valves are cleaned to a sufficiently high standard.
- Dynamic tests reveal that the valve design governs particle generation during operation.
- Ultrapure cleaning is valuable only if the valve does not generate particles while cycling.
- The in situ method is more useful than the aerosol method in evaluating cleanliness of valves because it makes it possible to test at high velocity.

### Specifications

With these conclusions in mind, consider the meaning of such common gas system specifications as "zero particles 0.1 $\mu$m in size, at 56 slpm (2 scfm) gas flow". The BK valve cleaned by Method 11 easily meets that specification but does not approach the goal of a system with the highest possible cleanliness level. Likewise, to specify dynamic cleanliness of "...zero particles per cycle" without any stipulation of flow rate, cycle rate, or number of cycles provides the user no assurance of achieving the stated goal. The use of a "cleanliness safety factor" based on particle counts at a known gas velocity well above normal will ensure that specifications achieve the desired results.

### Safety Factors

Safety factors are fundamental to sound engineering throughout system design. Material stresses, pressure ratings, leak integrity, and other forms of stress that lead to failure are protected by safety factors. Many unknown stresses can cause particle shedding in gas systems and clearly lead to product failure. The gas velocity is not constant but subject to surges of unknown magnitude during start-up, purging, or shutdown. Piping systems are never static, but constantly exposed to shock and vibration from machinery, traffic, gas cylinder changes, equipment maintenance, and people working nearby. These are all real-world stresses that cannot be quantified but will generate particles. A cleanliness safety factor based on velocity allows for such unknowns and ensures an ultraclean system.

## STANDARD TEST METHOD

The first objective was to identify a test method capable of detecting differences between valve designs and cleaning methods which achieve ultrahigh cleanliness levels. High-velocity tests performed with an in situ particle counter met that objective. The second objective was to define a qualification test method which meets four basic criteria: produce quantitative results, be repeatable, be reproducible, and be non-destructive to the product tested.

Quantitative Capability

Modern particle counting instruments readily provide a numerical value of particles per unit volume and size distribution, making it easy to obtain quantitative results.

Repeatability

Particle counting tests of components comparing different products and cleaning processes are described in numerous other papers.[13] Few have used a large enough sample of the same product to measure the repeatability of the test method, the product, or the cleaning process. The large number of samples tested in this program plus the statistical analysis demonstrated that both methods studied yield repeatable results.

Reproducibility

Most particle counting studies of components published so far have ranked the several components tested but none have reproduced the same ranking and quantitative values by independent researchers. We invite others to evaluate the reproducibility of the in situ method by a series of round-robin tests, as required of all ASTM standard test methods [14].

Non-destructive Capability

Counting particles in the air of a cleanroom is nondestructive. Air enters and leaves the counter unchanged; the counter makes the room neither more nor less clean. This is not true when counting particles on a surface. The act of removing particles from the object to count them makes the surface cleaner and changes the article being measured. Consequently, repeatability and reproducibility can not be demonstrated by testing the same sample repeatedly. Therefore, it is imperative to test enough like samples to obtain statistically valid results.

Proposed Method

Based on these results we developed a standard test to certify finished products for shipment. A statistical sample of valves from a production lot is tested by the in situ method at 300 m/sec. (1000 ft./sec.) velocity nitrogen flow for 10 minutes with the valve open and static. To be acceptable each valve must add no particles above background during the test.

This test method provides the semiconductor industry with a rational basis for specifying the cleanliness of gas system components. The method has several advantages:

- The equipment and instruments are standard, readily available commercial products.
- Prompt, real-time results are produced.
- The system user can confirm the results.
- Quality control tests of production lots are practical.
- New product designs can be qualified.
- Entire systems can be qualified.

The use of this test method will assure semiconductor producers that their process gas system will be a minimal contributor to product defects.

Further Work

This research has demonstrated that high velocity gas flow is required to detect sub-micron sized particles in gas system components. The critical velocity required to remove particles of known size and material from a surface is unknown and requires further research. The velocity within the viscous sublayer which affects particles of known size must be determined. With that information a Safety Factor can be established based on operating flow rates and critical particle size.

REFERENCES

1. R.J. Miller, D.W. Cooper, H.S. Nagaraj, B.L. Owens, M.H. Peters, H.L. Wolfe, and J.J. Wu, Mechanisms of contaminant particle production, migration and adhesion, J. Vacuum Sci. Technol., 6(3), 2097-2102 (1988).

2. D.A. Brandreth and R.E. Johnson , Removal of particulates from optical lenses, The Optical Index's J. Ophthalmic Dispensing, 54(1), (1979).

3a. A. Khilnani and D. Matsuhiro, Adhesion forces in particle removal from wafer surfaces, Microcontamination, 4(4), 28-30 (1986).

3b. K.L. Mittal, editor,"Particles on Surfaces 1: Detection, Adhesion and Removal,"Plenum Press, New York (1988).

3c. K.L. Mittal, editor,"Particles on Surfaces 2: Detection, Adhesion and Removal,"Plenum Press, New York (1989).

4. M. Corn, The adhesion of solid particles to solid surfaces. I.A review, J. Air Pollution Control Assoc., 11(11), 523-528 (1961).

5. M. Corn, The adhesion of solid particles to solid surfaces, II., J. Air Pollution Control Assoc., 11(12), 566-584 (1961).

6. J.A. Roberson and C.T. Crowe, "Engineering Fluid Mechanics", Houghton Mifflin Company, Boston, 1980.

7. R.L. Panton, "Incompressible Flow", pp. 29-450, 561-637, 761-763, John Wiley, New York, 1984.

8. "SEMI Specifications for Gases", SEMI Designation C3-86, Semiconductor Equipment and Materials International, Mountain View, CA, 1987.

9. W.T. McDermott, A. Schwarz, and R.C. Ockovic, Particle deposition in pressurized gas sampling systems, Proceedings Institute of Environmental Sciences, pp. 392-399, Mount Prospect, IL, 1987.

10. C.E. Nowakowski and J.V. Martinez de Pinillos, In-situ monitoring of particles in gases, Proceedings Microcontamination Conference, pp. 160-176, Santa Clara, CA, 1985.

11. "NC-55: Ultra-Clean Production Specification", Rev. B, Nov. 13, 1989, Nupro Company, Willoughby, Ohio 44094.

12. "NC-11: Special Cleaning and Packaging Specification", Rev. L, Jan. 9, 1987, Nupro Company, Willoughby, Ohio 44094.

13a. T. Govers, Ph D, G. Torelli, D. Gary and C. di Giulio, Department for Analysis and Equipment Development, L'Air Liquide, Claude-Delorme Research Center, Jouy-en-Josas, France.

13b. D. Jensen, Senior Mechanical Engineer, ASM America, Inc., Phoenix, Arizona, and S. Goldsmith, DIrector of Engineering, IBR, Inc., Ann Arbor, Michigan.

14. "Standard Practice for Conducting an Interlaboratory Study to Determine The Precision of a Test Method", ASTM Designation E691, ASTM, Philadelphia, 1987.

# REAL-TIME MEASUREMENT OF PARTICULATE LEVELS IN GASES IN A PRODUCTION DIFFUSION ENVIRONMENT

Bruce Huling
Motorola Inc.
2200 West Broadway Rd.
Mesa, Arizona 85202

Saied Tousi and
Barry Gotlinsky
Pall Corporation
30 Sea Cliff Avenue
Glen Cove, New York 11542

It has been shown that the contamination levels of the process fluids have a direct impact on production yield and system performance. It has also been shown that filtration of such fluids can be an effective means of quality and performance improvements. However, no comprehensive study has been performed to characterize system performance and the extent of improvements for a properly maintained fluid distribution system.

In this study the particulate contamination levels of an actual production environment were closely monitored. Using a laser particle counter, at various locations, the particulate contamination of the gas distribution system was studied. Based on the results of this study, a new replacement schedule has been implemented at this site (Motorola Wafer Manufacturing Division, Mesa, Arizona) which has resulted in substantial savings. In addition, since the changes have been made, there has been no yield loss due to equipment failure. During this study, the performance of used and new filters as well as the cleanliness of the gas delivery systems were monitored and the results are reported.

## INTRODUCTION

Particulate contamination of process fluids directly affects the product quality, yield, and reliability in today's semiconductor wafer fabrication areas. As the need for better device performance increases, integrated circuits with higher complexity and shrinking geometries are needed[1]. As the industry's requirements for performance have become more stringent, attention has been focused on contamination issues relating to equipment construction, fabrication, and maintenance, as well as the

*Particles in Gases and Liquids 2,* Edited by K.L. Mittal
Plenum Press, New York, 1990

process fluids. As the need to reduce or eliminate such contamination becomes vital, filtration of process fluids becomes ever more important, as filtration is the only reliable method of removing particulate contamination from process lines where all sources of generation of contaminants can not be eliminated. Even where components of the system can be improved so that they do not contribute to contamination, filtration serves as the tool to eliminate contaminants generated from unexpected process events[2].

The technological advances of the semiconductor industry have placed strict requirements on the materials used to prevent contamination of the process fluids. Specifically, the filters are required not only to remove all particulate contamination above a certain size, but, also, to not add any contaminants to the process fluids. For instance, if a filter sheds or releases particles, then the process fluid can become contaminated to a greater degree than before the fluid was filtered. Such contamination will have a detrimental effect on device yields, as well as the maintenance of the process equipment, necessitating expensive clean-up of the equipment and down-time of the process.

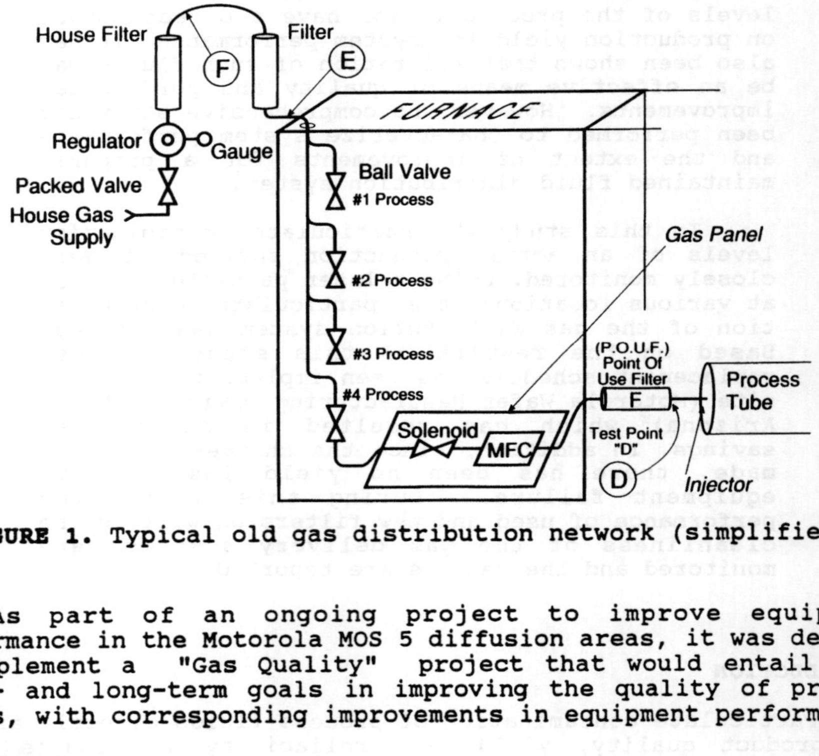


**FIGURE 1.** Typical old gas distribution network (simplified).

As part of an ongoing project to improve equipment performance in the Motorola MOS 5 diffusion areas, it was decided to implement a "Gas Quality" project that would entail both short- and long-term goals in improving the quality of process gasses, with corresponding improvements in equipment performance.

**BACKGROUND**

At the outset of the project the short-term goal was the characterization of the current diffusion furnace gas delivery systems. These systems are approximately eight years old (Fig. 1 and 2) and were scheduled to be replaced with new, state-of-the-art gas distribution systems (Fig. 3). By examining the performance of the "old systems" it was planned to determine the current level of gas quality delivered to the product in the processing environment. Initial plans called for the study of particulate as well as gaseous impurity levels. These observations were to be used as a "base line" for the newer systems to be compared against. This paper will address only the results of the particulate contamination study. The gaseous contamination issues will be reported later. As part of this investigation, different sources of particulate contamination were examined.

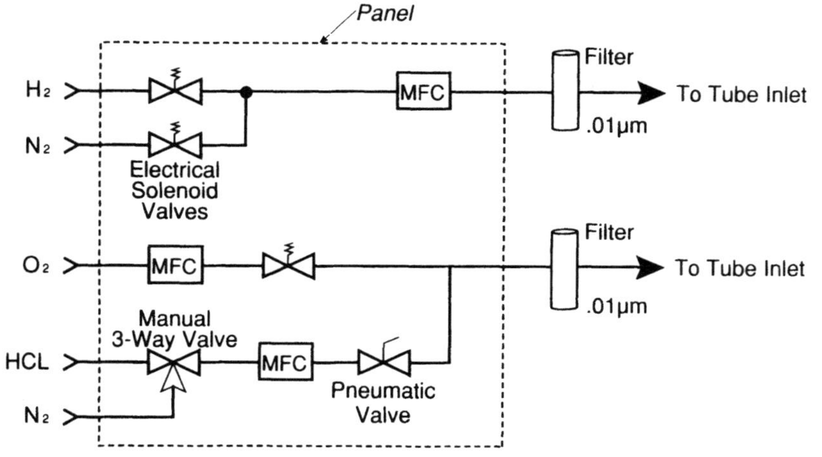


**FIGURE 2.** Old gas delivery panel (oxidation).

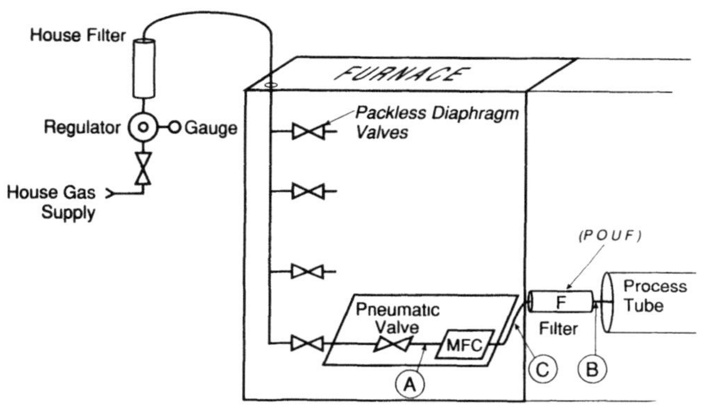


**FIGURE 3.** Typical new gas distribution network (simplified).

This investigation would aid in providing insight into the areas requiring improvement, and the long-term goal: <u>control</u> of the processing environments. The long term goal for the Gas Quality analysis in the diffusion area was not only to evaluate and improve the gas delivery systems, but to also provide a means to <u>control</u> the systems performance over their useful life in the production environment with respect to particulate contamination. <u>Control</u> would manifest itself in many ways, from detailed preventive maintenance schedules, service procedures, and particulate control checks, to quick identification of sources of deviation from the norm of gas quality using particle measurement equipment on a routine basis.

Obviously, without base line performance data, it would be impossible to evaluate the performance of future gas delivery systems. This data base would also aid in determining the frequency at which the system would require servicing (filter changes, system purges, service techniques, component replacement, etc.) to maintain an accepted and established level of performance. A ramification of this evaluation would be a better understanding of actual filter life in real-time wafer fabrication areas using state-of-the-art VLSI, MOS wafer fabrication methods and equipment.

Prior to purchasing new wafer fabrication equipment, it was recognized that proper evaluation was necessary to calculate a return-on-investment on a gas delivery system. Significant amounts of time and money can be wasted if the end result were not a system that performed better than the existing systems. The average gas delivery system, excluding the system outside of the diffusion furnace source cabinet, costs approximately $10,000 each. In a wafer fabrication area with over 60 systems this corresponds to a capital investment of over $600,000 (excluding filter costs). Filter changes were also to be examined since costs at different frequencies of filter change can increase rather quickly.

> For example: 14 filters per stack furnace x 15 stacks x $200 per filter x 2 changes per year = $84,000 per year. If the filters are changed once a year the cost is reduced by $42,000 (excluding labor costs).

It should be noted that additional savings are realized since a corresponding down-time is associated with changing of the filters and bringing the systems back to production status. However, the single most important concept in contamination control in gas delivery systems is to provide a known acceptable quality of gasses to the product process environments.

**TEST PROCEDURE AND FACILITY TYPE**

A PMS LPC101-HP laser optical particle counter manufactured by Particle Measuring Systems (Boulder, CO) was used to monitor the particles in the gas stream. The counter monitors the concentration of particles larger than 0.1 µm in a gas line. The use of the High Pressure counter enabled the direct connection of the counter to the gas lines without a need for pressure reducing devices. The counter samples at a flow rate of 0.1 CFM. The inlet pressure of the filters ranged from 30 to 60 psi. At various points of the system, the counter was connected to the gas lines and flow at line pressure was initiated to the counter. A typical sampling time of 30-60 minutes was used for each test. To study filter performance under pulse conditions, the distribution system was exercised for some of the filters. Individual test conditions are listed in Tables I and II.

To study the performance of filters under simulated mechanical shock, using a wrench, the body of the test filter was struck by hand, 200 times a minute for ten minutes. Even though it was attempted to strike the filters with consistent force, it is understood that these tests are not precisely reproducible.

The experiments were conducted in a real-time manufacturing environment of a Motorola MOS VLSI front-end diffusion area. The furnace systems were originally installed in the late 1970's and have been in continuous operation since start-up, essentially unmodified since construction. The systems are maintained by a team of equipment services technicians of Motorola.

Table I. Particle Counts in the Old System. *

| TEST # | LOCATION | DESCRIPTION | FLOW CONDITION | INITIAL PARTICLE COUNT > 0.1μm per $ft^3$ |
|---|---|---|---|---|
| 1 | - | Background | -- | 0 |
| 2 | A | Inside the gas panel | Steady | 0 |
| 3 | A | Inside the gas panel | Valve cycled | 25115 |
| 4 | B | House filter (pleated) | Steady with mechanical shock | 0 |
| 5 | C | Entry to P.O.U.F. (pleated) | Steady | 0 |
| 6 | C | Entry to P.O.U.F. | MFC ** cycled (1 cycle/min) | 150 |
| 7 | D | Downstream of P.O.U.F. (pleated) | MFC ** cycled (1 cycle/min) | 0 |
| 8 | E | Downstream of gas panel entry filter (stacked disc) | Steady | 58 |
| 9 | C | Entry to P.O.U.F. | MFC *** cycled | 3190 |
| 10 | D | Downstream of P.O.U.F. (pleated) | MFC *** cycled | 0 |
| 11 | F | House and entry to panel | Steady | 496 |
| 12 | D | Downstream of P.O.U.F. (stacked) | Steady w/mechanical shock | 97 |

* See Figures 1, 2 and 5
** Tylon MFC
*** Unit MFC

Table II. Particle Counts in the New System. *

| TEST # | LOCATION | DESCRIPTION | FLOW CONDITION | INITIAL PARTICLE COUNT $> 0.1 \mu m$ per $ft^3$ |
|---|---|---|---|---|
| 1 | -- | Background | Steady | 0 |
| 2 | A | Upstream of MFC | Steady | 99 |
| 3 | A | Inside the gas panel | Valve cycled (one cycle/sec) | 386 |
| 4 | B | Downstream of P.O.U.F. (pleated) | Steady | 0 |
| 5 | A | Repeat of test # 2 after flushing | Steady | 6 |
| 6 | A | Regulator | Steady with Mechanical Shock | 8 |
| 7 | C | Upstream of P.O.U.F. Filter (pleated) | Steady | 2000** |
| 8 | B | Downstream of P.O.U.F. (pleated) | Steady | 0 |

* See Figures 3 and 4
** Brand New Line

Prototypes of the "future" systems were built by a joint effort between the equipment engineering group of Motorola and Flopure (San Antonio, TX), utilizing many of the newest fabrication and design techniques and components. All components, including the mass flow controllers, isolation valves, and fittings were carefully selected for the system, based on the criteria outlined in a publication by Jensen[5], as well as internal Motorola specifications and experimental results. Chemically polished tubing was used, incorporating a number of cleaning and assembly techniques to optimize the surface preparation. Automatic orbital welding was also exclusively used for all welds, with fittings of the automatic weld type. On certain welds, Flopure used a proprietary procedure that is believed to significantly improve the joint quality. Packless diaphragm valves were used for manual operation, with air-actuated pneumatic valves used for gas selection. All systems were assembled in Class 100 clean room areas with proper clean room operation procedures.

Prior to each particle counting test, the background counts of the counter were checked to ensure the validity of the experimental results. It was also attempted to perform the particle counting experiments as close to established procedures as possible[6].

**RESULTS AND DISCUSSION**

Tables I and II summarize the results for the new and the old systems (see Fig. 1-4 for test locations). It should be noted that the test times are shorter than those of similar experiments performed in laboratories. This is a natural limitation of performing experiments in a product manufacturing environment. In addition, only the initial particle counts are reported since they represent the worst case scenario (practically in every case the initial particle counts were also the highest). However, the trends concluded from these results are useful and comparisons can be made among systems with varying conditions and components.

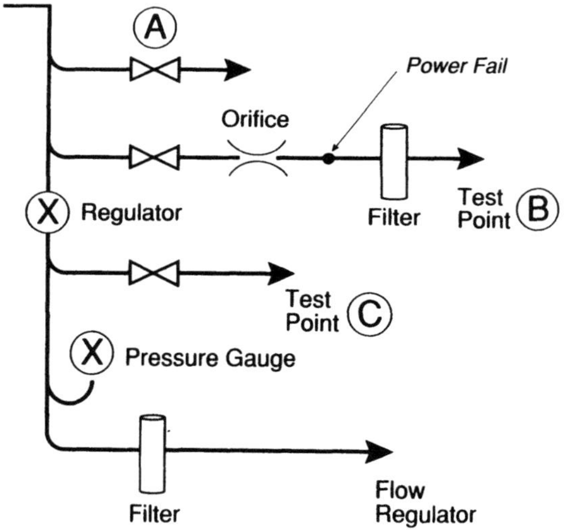


**FIGURE 4.** New gas delivery system.

Table I summarizes the test results for the "old system". While characterizing this system, it was surprising to find that minimal particulate contamination was occurring at the gas injection point to the process environment. That is, **no particles were detected by the particle counter.** This was measured at the outlet of the point-of-use (POU) filter (see test numbers 7 and 10 in Table I). These data points became the focus of one of the most important observations in this experiment, which will be discussed later. Upstream of the filter, it was also interesting to find that only few or no particles were detected by the counter during steady state gas flow. This was surprising on systems that had been in service continuously for more than eight years, considering the use of standard 316 stainless steel and copper tubing, pipe fittings, teflon tape joints, electrical solenoid valves, socket welds, hand welds, oxidized weld sites, leak integrities not exceeding $10^{-4}$ atm-cc/sec in some cases, etc. However, as can be seen in test numbers 3, 6, and 9, when the systems were pulsed the particle counts would significantly rise. The most significant contributors to particle counts were the ball valves used, with their resultant clean out period being quite long. However, with the installation of POU filters, the gross contamination generated by the gas delivery system becomes a moot point as the POU filters in every case produced **zero** outlet counts!

It should also be noted that the mechanical shocks had no effect on the performance of POU filters with pleated structure (see test number 4). However, under similar test conditions, the POU filters which use stacked disc configuration, showed relatively high counts (see test number 12).

Another interesting portion of the study conducted was the observations regarding filter performance longevity. Currently Pall Gaskleen filters, which use a pleated structure, are used exclusively on all diffusion furnace gas delivery systems at this site (see figures 1 and 3). Some filters have been in service for more than three years, and yet have retained the performance of new out-of-the-box filters. As a result of the data obtained, a filter change schedule of fourteen months was selected as a frequency that would insure integrity and performance. (This frequency also coincides with the major plant preventative maintenance schedule). A cost savings of over $40,000 is realized over the prior schedule.

Table II summarizes the test results for the new system. The results for these prototype systems were less dramatic, but still noteworthy. The new gas delivery systems realized an improvement in contamination levels over the old systems. The most obvious feature was the significant reduction of particulate contaminants occurring during valve cycling. It is estimated that with these systems, the filter change frequency could easily be stretched to two years with no detrimental effects on the performance of the system (this corresponds to a three year savings of approximately $100K if instituted). However, these figures are applicable to oxidation and doping atmospheric furnaces only. Further experiments are scheduled to study CVD systems.

The location at which the filters should be installed in the gas delivery system is another topic of interest. Two locations were used with the intention of providing filtration prior to critical components: mainly as a first line of defense against particulate contamination of the gas jungle, and to further reduce the effects of particulate contaminants on component reliability and performance (i.e., valves, sensors, etc.). The other consideration is the utilization of "point-of-use" filters, located prior to the gas entry points to the process environment. These filters are at the most critical location, as it is here that all particles of interest should be stopped.

We also had interest in understanding the "turn on" characteristic of a new jungle, or system: particle counts would initially be relatively high on a newly started-up system, but rapidly dropping in a short time. Prior to this experiment, it was estimated that it would take a large number of cycles, and at least 24 hours of continuous use, to reduce particle counts to acceptable levels, but our experiments indicated that after system start-up, approximately 1 hour would provide adequate time for the particle counts to drop to **zero**! Using these data, it was decided that after component replacement, or upon system start-up, a four hour cycling procedure would be more than adequate for providing a particle-free start-up condition. To facilitate this procedure, digital controllers were utilized with software generated to perform an extended "hands off" cycling and purge flow. This program also serves as a check of the system from an operational reliability standpoint (i.e. failure analysis). Also at the end of this purge period, a particle counter can be used to monitor specific points of the system to verify the particle count and to ensure proper operation of the system prior to involving either a "test" run or actual product material.

**SUMMARY AND CONCLUSIONS**

After this study, a better understanding of the behavior of components was achieved. Hence, it is possible to reduce the number of particle monitoring tests needed to characterize a gas delivery system. The pulsing and vibration tests were extremely interesting. In the new system, little or no effect on the particle counts could be detected due to mechanical or flow shocks to the system. In addition, the systems implemented to date have sustained minimal reliability problems which indicate that the systems reliability has increased dramatically over the old systems. In the new system, the number of component failures in the gas delivery system has been reduced dramatically (as indicated by the systems up-time and reliability performance).

Some of the test points were downstream of the filters that had been in use for more than three years. It appears that the filters had maintained the performance of out-of-the-box filters, indicating that preventive maintenance schedules can be adjusted to provide optimum systems performance, without incurring unnecessary down-time and component replacement costs. Filter changes can be scheduled confidently with performance known at extended intervals to be satisfactory so that even in the event that a filter change date is accidentally missed, the system performance would not be in jeopardy. (Adopting such a schedule in this front-end will result in cost savings of at least $40K per year, excluding labor and down-time costs).

It is clearly concluded that point-of-use filtration is a critical and necessary design feature of diffusion gas delivery systems. In our study, the gas had its highest cleanliness level downstream of the POU filters. The use of POU filtration eliminates any chances of particles entering the furnaces.

Reduced equipment start-up times can be realized with a base line set up for equipment contamination levels, with confidence that the particulate levels are acceptable after a relatively short duration. Equipment for testing high pressure gas line particle levels is imperative for the characterization, sustaining, and understanding the performance of gas delivery systems.

**ACKNOWLEDGMENTS**

We would like to thank Roger Gregoire and John Wheeler of Flopure and Joe Barto of Process Technology for the many hours spent in consultation on the design and manufacture of the new systems. Their valuable comments are deeply appreciated. We also thank the many people involved at PMS for the loan of the particle counter used during this study. Their willingness to explore particle counting in a manufacturing environment enabled us to accurately and reliably make measurements where few have been made before.

Finally, we thank Motorola Inc. and Pall Corp. for their permission to publish the results of our research.

**REFERENCES**

1. S. U. Gunawardena, U. Kaempf, B. Tullis, and J. Vietor, SMIF and its impact on cleanroom automation, Microcontamination, 3, (September 1985).

2. M. W. Mittelman, Contamination control activities in today's semiconductor market- An interview with Robert W. Martin, Microcontamination, 5, (May 1987).

3. S. D. Cheung, and R. P. Roberge, An inside look at measuring particles in process equipment, Microcontamination, 5, (May 1987).

4. R. N. Castellano, The need for analyzing contamination in process equipment, Microcontamination, 5, (September 1987).

5. D. Jensen, Reducing microcontamination generating and entrapment within high-purity gas delivery systems, Microcontamination, 5, (May 1987).

6. S. Tousi, and B. Gotlinsky, High purity gas filtration, Solid State Technol., 32, (March 1989).

# IN-SITU PARTICLE MONITORING IN A PLASMA ETCHER

John Gregg and Peter G. Borden

High Yield Technology, Inc.
762 Palomar Avenue
Sunnyvale, CA 94086

A particle sensor appropriate for use with plasma etchers used in VLSI processes has been demonstrated. The sensor uses laser light scattering to detect particles, with 0.3 µm diameter being the smallest detectable particle size, and is protected from attack by reactive halogens. Correlation between sensor particle count levels and wafer surface particle counts is demonstrated. Count rates are sufficient to allow use in both process optimization and process control applications. Examples of count rates as a function of process parameters are shown. Both short-term episodic events and long-term trends related to etcher operation are seen during production monitoring.

## INTRODUCTION

It is generally recognized today that one of the largest sources of VLSI yield loss is particle contamination added to wafers as they pass through processing equipment[1,2,3]. This has motivated considerable work to clean up equipment by improvements such as "soft" pumping and venting, modified wafer handlers, and in-situ cleans[4,5]. Nevertheless, newer processes are presenting even greater demands for particle control as line widths shrink and die sizes increase.

The rate of progress in reducing particle levels within equipment has been somewhat limited by the measurement tools available. The measurement tool in most common use is the surface defect scanner, which counts particles on a wafer surface. An accepted technique used to characterize equipment is to count the particles on a wafer before and after it passes through a process tool. This gives a measure called the particles per wafer pass[6]. While this technique is very powerful, and has led to major reductions in the particle levels within process equipment, it suffers from two basic limitations. First, it generally uses unpatterned wafers, since a pattern or surface roughness will mask small particles. In many cases, this makes it difficult to see particle contributions while the process is actually running, since the process often modifies the front surface of the wafer. Also, the effect of running product wafers must be inferred, which may be difficult since the product wafer usually has a significantly different surface from the monitor wafer. Second, the surface scanner looks at the sum of particles added in a complete pass through the process tool, so that it is difficult to locate particle sources within the tool itself, or to determine where in a sequence of steps particles have been added.

These drawbacks are even more pronounced when surface scanning is used for statistical process control during production. Through in-situ monitoring of particle levels during production in ion implanters, sputter coaters, and loadlocks of other types of equipment, it has been empirically observed that particle generating events are often episodic, lasting a few minutes to an hour, and are of higher level when product wafers are run than when monitor wafers are run[7,8,9]. The episodic nature probably arises because particles released from a generating event such as improper wafer handling are flushed out through repeated pump and vent cycles, or carried out by wafers passing through the system. Particle levels rise when product wafers are run because product wafers usually have been exposed to considerable handling and processing, and are therefore more likely sources of particles than virgin wafers commonly used as monitors.

For these reasons, it can be beneficial to use real-time, in-situ monitors to characterize process equipment particle generation. These monitors, when installed properly, are capable of reporting changes in the relative particle levels as a process occurs, allowing immediate identification of active sources. They are also capable of observing particle levels during production. Sensors based on laser light scattering, and their application in ion implanters, sputter coaters and airlocks have been reported[5,7,8,9].

An application that has been more difficult to address is oxide etching. This is because the process chamber and effluent gasses contain reactive fluorine species that can attack sensor optics. The sensor described below has been protected through design improvements and the use of special coatings. As will be shown here, it has performed in this environment for several months without degradation. This has allowed it to be used to characterize and monitor production oxide etch processes. A number of experiments will be detailed to show how the sensor has been used to characterize particle levels as a function of process variations, observe particle levels over short and long periods of production, and demonstrate that particle levels seen with an in-situ sensor correlate to wafer surface particle levels.

## SENSOR DESCRIPTION

A particle sensor for a plasma etcher must work both over a wide range of pressures and in the presence of reactive halogen species such as fluorine, chlorine and bromine. A sensor appropriate for these environments (High Yield Technology model 20) is shown in figure 1. A beam of laser light projects between two photocells. As particles pass through this beam they scatter light to photocells, creating electrical signals that are used to count and size the particles. The detection cross-section for this sensor has been measured using polystyrene latex spheres. It is 12 $mm^2$ for 0.5 µm diameter particles, 6 $mm^2$ for 0.38 µm particles, and falls to zero for 0.28 µm particles.

Because this sensor relies on free particle motion, it works independent of pressure, and can be used from atmosphere to hard vacuum. Proprietary coatings have been placed on the optical surfaces to protect them from etching. As will be seen below, this allows the sensor to remain stable after several months of exposure to low pressure gas flows with fluorine species present.

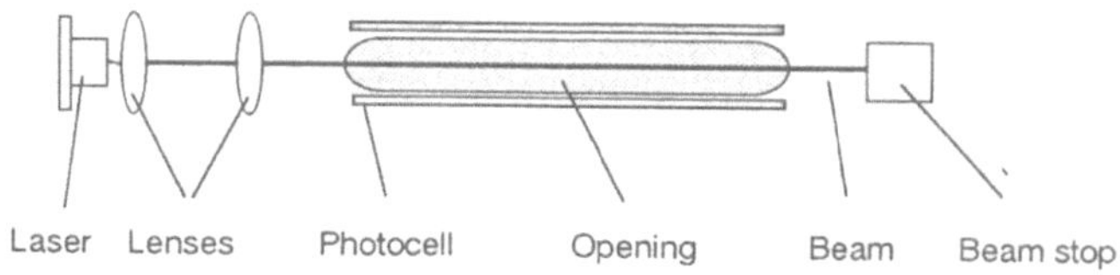


Fig. 1. Detailed drawing of the sensor used in the etcher exhaust line. A collimated laser beam passes through an opening in the sensor through which particles pass. Photocells on both sides of the opening collect light scattered by particles.

Installation in a Single Wafer Etcher

A typical single wafer etcher is shown in figure 2. Wafers are set on a cassette loader and passed one at a time into an input loadlock. Here they are pumped down and passed into the process chamber. After etching, they move to the exit loadlock, where they are vented to air. Finally, they are sent to another cassette. With this configuration, the primary sources of contamination are expected to be stirring up of particles during pumping and venting in the airlocks, and process related contamination in the process chamber.

One sensor mounts in the exhaust line of the etcher, between the process chamber and the throttle valve. This is accomplished by forming the base of the sensor in the shape of a KF-40 flange and placing a tee section of vacuum pipe on the pump line. The sensor then plugs into the leg of the tee, allowing easy installation and removal.

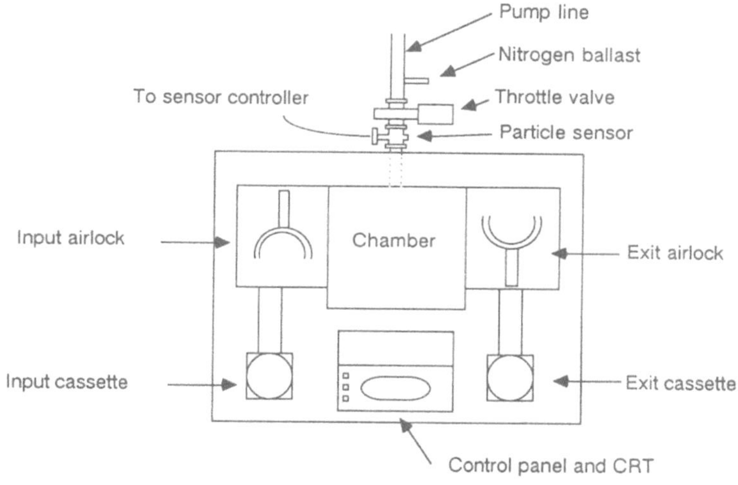


Fig. 2. Top view of a typical single wafer etcher, showing the airlocks, process chamber, and pump line. Particle sensors are mounted in the airlock covers and the exhaust line between the chamber and the throttle valve.

Particles detected with the sensor in this location are carried from the process to the sensor in the process gas flow. As shown in figure 3, these gasses pass through the top electrode (the cathode), over the wafer, and out the exhaust line. A convenient rule of thumb is that a one µm particle falls about 1 cm in 1 second at a pressure of 1 Torr. This can be verified using settling velocity calculations[10]. At typical process gas flow rates and a process pressure of 2 Torr, the gas velocity at the sensor is about 1 m/sec and the gas residence time is less than one second. Thus, it is evident that particles less than or about equal to one µm in diameter will be carried from the region above the wafer to the sensor with a negligible delay.

A second sensor was mounted on the cover plate of the entrance loadlock. This uses multiple laser beam passes (High Yield Technology model 15A). It has a detection cross-section of 35 $mm^2$ at one µm, with the cross-section dropping to zero at 0.5 µm.

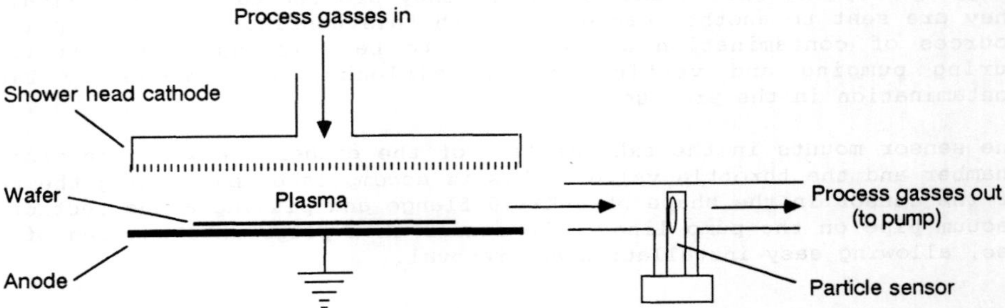


Fig. 3. Configuration of the etcher process chamber. The wafer sits on the anode. The cathode is perforated to allow process gasses to bathe the wafer. These flow out the cathode, over the cathode, and into the exhaust line, carrying particles from the process to the sensor.

## Particle Count Levels for Single Wafers

Particle count levels are highly dependent on a number of factors, including the process chemistry, process sequence, and the presence or absence of episodic excursions due to process problems. Figures 4 and 5 show particle counts from the airlock and process chamber, respectively, for a sequence of five wafers passing through the etcher while it ran a Freon based process. Note that the particle events are staggered, conforming to the operating sequence of the etcher. In the airlock, the first short spike is the pumpdown, and the larger spike is the vent. In the process chamber, the counts are seen to be fairly low initially, but rise to much higher levels as etching proceeds.

This gradual rise in time is strong evidence that the sensor is detecting real particles, rather than plasma emissions or electrical interference. Interference would be expected to cause a constant count level over the whole process. To confirm this, a clear cover was placed over the sensor. This lets in light and electrical noise, but not particles. This caused the particle count rate to drop to zero.

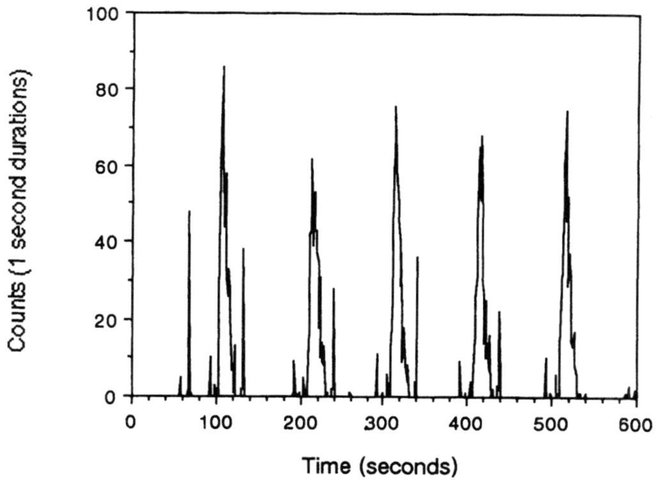


Fig. 4. Particle counts seen in the airlock during production processing of five wafers.

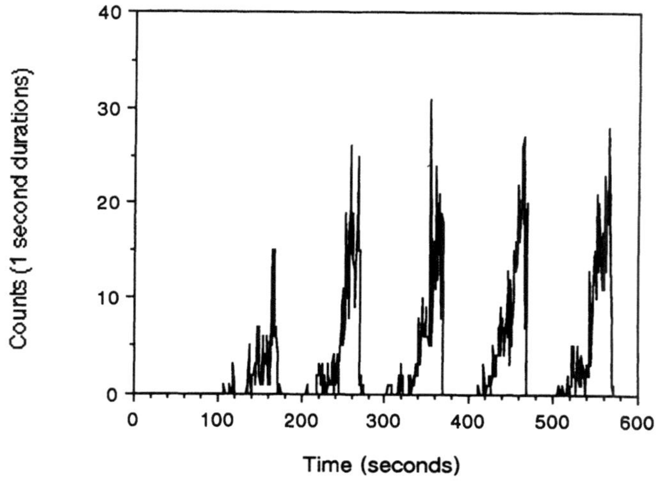


Fig. 5. Particle counts seen in the exhaust line during production processing of five wafers.

Correlation of Particle Counts to Wafer Surface Counts

Some fraction of the small particles entrained in the process gas flow is expected to deposit on the wafer. If this fraction is constant, it is expected that particle densities measured in the exhaust line will correlate to particle deposition rates on the wafer. To confirm this, the etcher was run under a variety of process conditions, both normal and extreme, so that a wide range of particle counts could be obtained. Wafer surface counts were obtained using a Tencor 4500 surface scanner, and related to particle counts obtained in the exhaust line. Airlock particle counts were not measured.

Photoresist coated wafers were used in order to prevent etching of the silicon surface. This etching roughens the surface, so that it is no more possible to count surface particles with the surface scanner. A problem with this technique is that the surface scanner sensitivity is a function of the resist thickness, especially for small particles[11]. This is because the resist acts like an antireflection coating. Most etch process will etch back the resist to a certain extent. Data points in which the resist is etched significantly were rejected because of this effect.

The resulting correlation is shown in figure 6. The cluster of points near the origin represents normal operating conditions; the other points represent extreme conditions. The correlation coefficient of 0.90 shows that most of the variation of the data can be explained by assuming a linear relation between exhaust line particle density and wafer deposition rate. The shift in the origin is probably due to particles generated in the airlocks or to wafer handling.

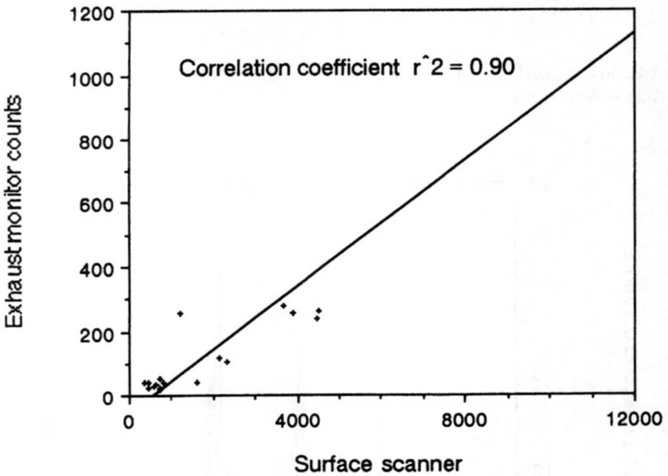


Fig. 6 Correlation between surface scanner and exhaust line monitor particle counts.

Process Optimization

The sensor can be used for both process monitoring and process optimization. For the latter case, since the particle levels can be measured in real-time, it is possible to correlate changes in process conditions to changes in particle levels. An example of this is shown in figure 7, in which particle counts during the etching of a single wafer using an $SF_6$ process are shown. The gap spacing between the electrodes has been changed, and the particle level is seen to vary inversely with this spacing. The reason for this is probably that the intensity of the plasma varies inversely with the spacing, and the more intense plasma sputters more material from the cathode.

An etch will often require a change in gas mixtures. For example, one mixture may be used initially to provide a high rate and low selectivity. The mixture may then be changed for higher selectivity as the bottom of the film being etched is reached, or to etch away residues. Figure 8 shows particle counts during etching with a freon-based process. Near the end of the process, the gas mixture is changed. The introduction of the new gasses appears as a change in particle levels while the pressure is being stabilized.

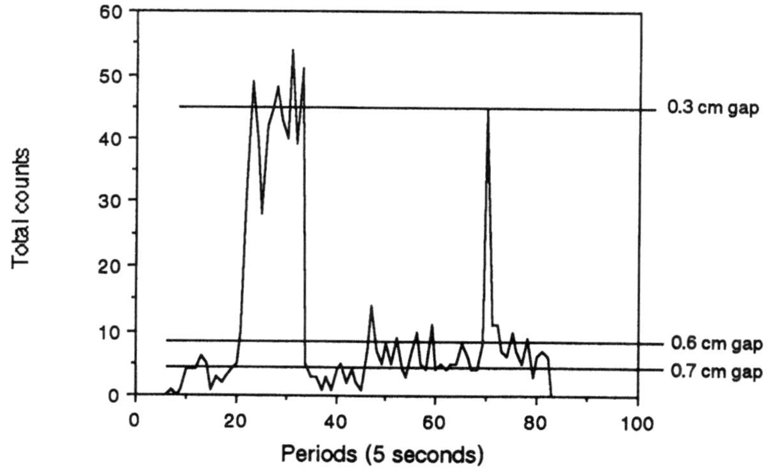


Fig. 7. Particle counts in 5 second periods as a function of gap spacing during the etching of a single wafer using a sulfur based chemistry. The gap spacing is programmed to change to several values during the etching of the wafer. As the spacing increases (decreases) the particle count decreases (increases).

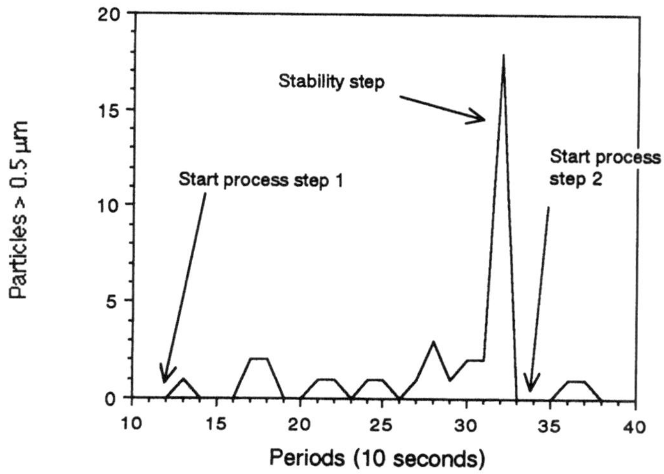


Fig. 8. Particle counts as a function of time over 300 seconds during the etching of a single wafer using a freon based chemistry. Note the high counts as the gas pressure stabilizes at the start of step 2.

It is important to note that particles are only seen when the process runs. Almost no particles are seen when the plasma is off and process gasses flow through the etcher. This, along with the good correlation between surface and monitor particle counts, suggests that the dominant source of particles is plasma processing, not the machine background.

## Particle Counts in Production

The sensor can also be used to monitor particle levels during production, so that statistical process control techniques can be applied to maintain process stability. This type of monitoring will show long-term trends related to etcher operation, and short-term trends related to episodic out-of-control events. The long-term trends can be found using periodic sampling of the data, but the short-term trends must be found using continuous monitoring.

An example of long-term trending is shown in figure 9, in which particle levels over 70 days are shown. At the end of the 70 day period, the sensor was removed and its calibration checked. It was found to be the same as at installation. When the sensor was installed, the etcher process was switched from $CF_4$ to $SF_6$. The latter process is considerably cleaner because it does not have associated polymer generation. Note that the particle baseline gradually dropped by about a factor of five over the two weeks following the process change. At 54 days a quarterly preventive maintenance (PM) was performed. The particle baseline after the PM is double that seen before the PM. Finally, at 69 days the cathode was changed, and the particle levels rose to very high levels. This is consistent with the fact that the cathode is the electrode through which the process gasses pass. A new cathode will probably have loose particles and rough points that are prone to attack by the plasma.

For short-term trending, it is best to measure particle levels in units associated with machine operation, such as particles per wafer or, for constant lot size, particles per lot. Examples of these data are shown in figures 10 and 11. Figure 10 shows particles per wafer for about one day of production. A slight downward trend is seen until a wafer breaks at the end of the monitoring period. Figure 11 shows particles per lot for

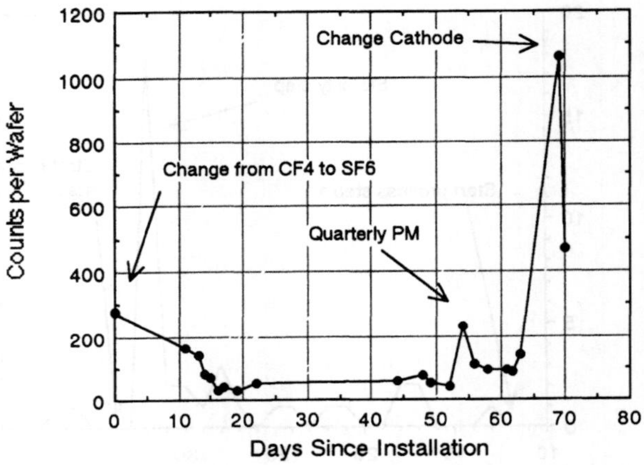


Fig. 9. Particle counts during a 70 day period, showing a correlation between particle levels and etcher operation.

two days of production. Again, the process is generally stable, but episodic out-of-control conditions are seen at the start and end of the monitoring period.

CONCLUSIONS

Certain conclusions about particle generation in an oxide etcher are evident from these data. It appears that the dominant source of particles in the process chamber is attack of the carbon cathode by the plasma. This is consistent with the fact that significant particle levels are seen only when the plasma is on, that a cathode change results in a very large increase in particle levels, and that a carbon residue is found in the exhaust line. It is possible, however, that these particles may be due to other effects, such as condensation[12]. Good correlation between wafer surface and exhaust line particle levels is obtained.

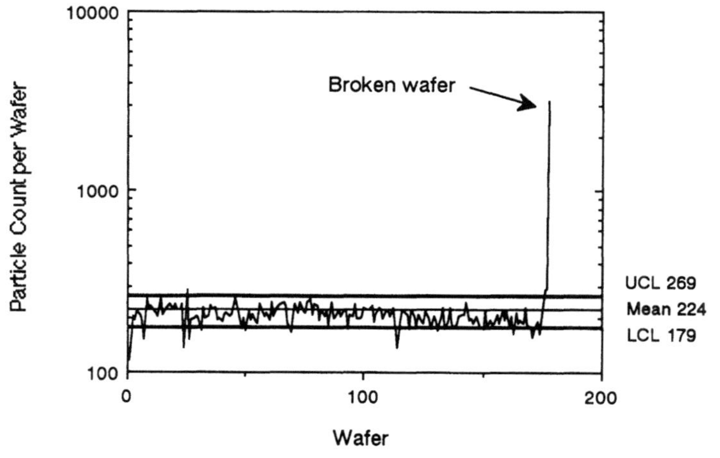


Fig. 10. Particle counts per wafer over a day of operation. Note the increase when the wafer breaks.

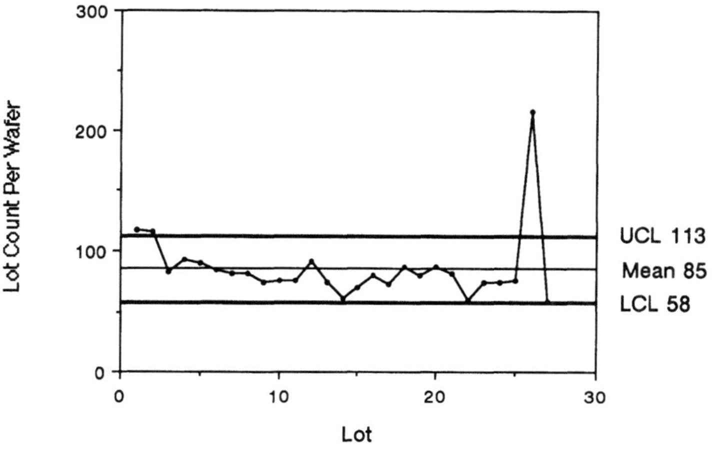


Fig. 11. Particle counts per lot over two days of operation, showing episodic out-of-control events at the start and end of the monitoring period.

It is also seen that the particle level is a strong function of process conditions, the service history of the etcher, and episodic events that can only be explained by tracking the history of individual lots. The process conditions and service history cause a baseline level that varies slowly and can be tracked with sampling inspection. In-situ monitoring facilitates tracking this baseline since it is readily automated. Episodic events, however, are short and can only be tracked with continuous monitoring. This is where the monitoring techniques described above have the potential to offer a significant benefit, since they can provide an immediate warning of out-of-control conditions. This allows corrective action to be taken quickly, minimizing yield loss.

REFERENCES


1. T. Fukumoto and T. Omori, Total contamination control technology in VLSI wafer fabrication process, Semicon West Technical Proceedings, p. 3, 1988.

2. R. W. Martin, Defect density management, Proceedings of the International Committee of Contamination Control Societies (ICCCS) meeting, p. 1, Los Angeles, Sept. 1988.

3. S. D. Cheung in "Particles in Gasses and Liquids 1," K. L. Mittal, editor, pp. 167-173, Plenum Press, New York, 1989.

4. T. C. Smith, this volume.

5. P. G. Borden in "Particles in Gasses and Liquids 1," K. L. Mittal, editor, Plenum Press, New York, 1989.

6. B. J. Tullis, A method of measuring and specifying particle contamination by process equipment, Microcontamination, 67 (November 1985).

7. P. G. Borden and L. A. Larson, Benefits of real-time, in situ particle monitoring in production medium current implantation, IEEE Transactions on Semiconductor Manufacturing, 2(4),141, (1989).

8. W. Weisenberger, R. Van Moorleghem, P. Borden and W. Knodle, Real-time, in-situ particle monitoring in a high current ion implantation production bay, Proceedings of the 7th International Conference on Ion Implantation Technology, p. 163, Kyoto, June 1988.

9. P. Borden and W. Knodle, Monitoring particles in vacuum equipment, Proceedings of the 9th International Symposium on Contamination Control, p. 204, Los Angeles, Sept.1988.

10. W. C. Hinds, "Aerosol Technology," Ch. 3, Wiley Interscience, New York, 1982.

11. B. Fishkin and E. J. Baker, Particle performance evaluation of CVD and epitaxial processes and equipment," Proceedings of the 1988 Annual Meeting of the Institute of Environmental Sciences, p. 517.

12. G. S. Selwyn, J. Singh and R. S. Bennett, J. Vac. Sci. Technol. A7(4), 2758, (Jul/Aug 1989).

PROTECTED PARTICLE COLLECTION FROM GAS STREAMS FOR CHARACTERIZATION BY ANALYTICAL ELECTRON MICROSCOPY

M. R. Anewalt, Air Products and Chemicals, Inc., 7201 Hamilton Blvd., Allentown, PA 18195

P. S. Schabes-Retchkiman, On leave from Instituto de Fisica, Universidad Nacional Autonoma de Mexico P.O. Box 20-364, 01000 Mexico D.F.

C. E. Lyman, Department of Materials Science and Engineering, Lehigh University, Bethlehem, PA 18015

Submicron particles are known to cause yield losses in the manufacture of electronic devices. Analytical electron microscopy (AEM) is a powerful tool for characterizing particles by size, morphology and composition. Data obtained can be used to identify particle sources. Major problems include particle collection and contamination control during sampling and transfer into the AEM. A controlled atmosphere specimen holder was used to collect particles from a gas stream, to isolate them from ambient contamination and to transfer them into the specimen chamber of the AEM. The collecting grid was examined in the AEM before particle collection to verify its cleanliness and provide background data. The grid retracts into an isolation chamber inside the specimen holder, thus protecting it from extraneous contamination during transfer to the AEM from the sampling site. The AEM combines a number of techniques to characterize particles by size, morphology, composition and crystal structure. Elemental analysis of individual particles and elemental images of high resolution are obtained by X-ray Emission Spectroscopy using the characteristic x-rays to identify specific elements. Examination of a large number of particles provides a histogram of the types of particles present.

## INTRODUCTION

Contamination control is of critical importance in the semiconductor industry to control device yield losses. As the purity of gas supplies becomes more stringent, particle characterization and source identification take on increased importance. Submicron contaminants in ultrapure gas supplies present a particular challenge to the analyst. This challenge is met by the recent technique known as Analytical Electron Microscopy (AEM).

*Particles in Gases and Liquids 2,* Edited by K.L. Mittal
Plenum Press, New York, 1990

Energy Dispersive X-ray Spectrometry (EDS) in the Scanning Electron Microscope (SEM) or Electron Probe Microanalyzer (EMPA) has been valuable in the study of particles of size >1 µm. Particle sampling for this conventional technique usually involves filtering the gas through a membrane filter. The filter is loaded into a filter housing in a clean atmosphere, sealed and transported to the sampling site. After sampling, the housing is again sealed and transported to the laboratory. Ideally, the laboratory is in a clean environment. The filter is removed from the filter housing and examined by SEM/EDS. A blank control filter is also carried through every step along with the sample filter, except that gas flow is limited only to the rapid opening and closing of the supply valve. This blank will contain particles introduced during handling. Statistically significant particle population differences must be found between sample and blank filters in order for true identification of gas contaminants.

Collection and analysis of particles in this manner has several limitations. Examination of filters can be time consuming, since the filter area exposed is usually on the order of several $cm^2$ and examination fields of view are by comparison small (several hundred $\mu m^2$). The smallest particle that can be reliably analyzed for elements by SEM/EDS or electron microprobe is about 1 µm. An isolated submicron particle may be below the minimum detection limit by volume dilution. It may also be difficult to recognize particles <0.1µm due to image resolution restrictions. Despite the limitations of SEM/EDS analysis of filter samples, this method has a valuable place in characterization of large particles. Submicron particle characterization, however, can be difficult, if not impossible, by the SEM/EDS technique.

Analytical electron microscopy is perhaps the most powerful analytical technique for simultaneously assessing submicron particle morphology and size through the several high resolution imaging modes available, particle composition by x-ray spectrometry, and particle crystal structure by both selected area electron diffraction and convergent beam electron diffraction.[1,2] AEM can be performed in either a dedicated scanning transmission electron microscope (STEM)[3,4] or in a conventional transmission electron microscope (TEM) converted to STEM operation and generally termed a TEM/STEM. Regions of particles on the order of 10 nm in diameter or less can be fully characterized as to elemental composition and crystal structure.

Analytical electron microscopy is thus a very powerful technique for the characterization of extremely small particles in gas streams. Obtaining a sample and controlling background particles is difficult, however, since typical laboratory air contains many submicron particles. In this paper we describe a special specimen holder used for sampling and for sample transport. We present results showing an array of analytical techniques that we have found valuable in submicron particle characterization and source identification.

## EXPERIMENTAL

The recently developed Gatan* vacuum-transfer electron microscope specimen holder is the basic component for both sampling and sample transport. The sample holder is designed for specimen transfer to and from the TEM/STEM either under vacuum or in a controlled gaseous atmosphere. The tip of the device, in which the specimen grid is located, can be withdrawn into the holder shaft and sealed. The holder is shown in Figures 1 and 2.

*Gatan, Inc., 780 Commonwealth Dr., Warrendale, PA 15086

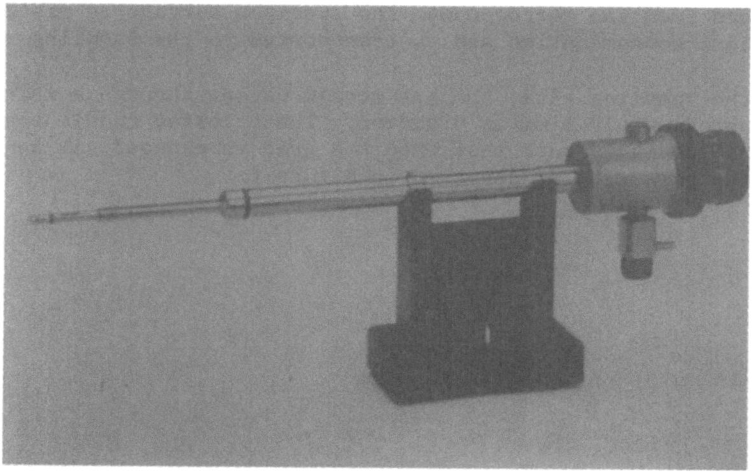

Figure 1. The Gatan vacuum-transfer electron microscope specimen transfer holder used for both sampling and sample transport.

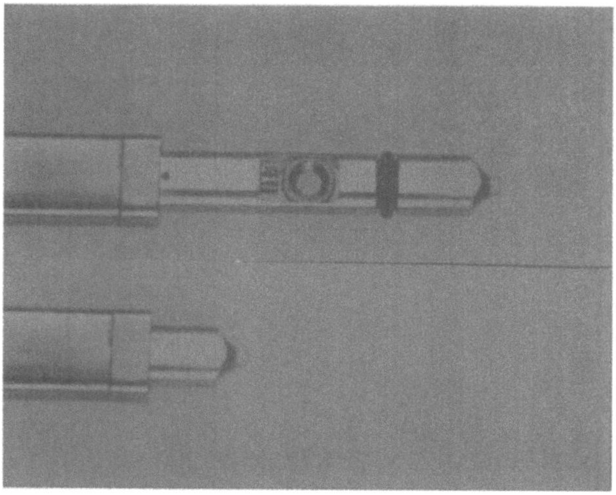

Figure 2. The tip of the transfer holder is extended (top) during gas sampling, and retracted (bottom) to protect the specimen between sampling and characterization.

Prior to sampling the gas stream, a clean, carbon coated beryllium grid is inserted in the holder and is thoroughly examined in the AEM, searching for contamination. The cleanliness of the sampling surface is demonstrated in the bright field image and x-ray spectrum, Figure 3. After ascertaining the absence of extraneous particles, the blank specimen grid is retracted into the isolation chamber of the holder and removed from the microscope. The specimen surface is protected from ambient contamination and is transported to the sampling site.

At the sampling site, the gas stream passes through a shielded nozzle approximately 1 mm in diameter. The transfer rod is positioned in the shielded area such that when the grid is exposed, it is within 2 mm of the gas exit, as depicted in Figure 4.

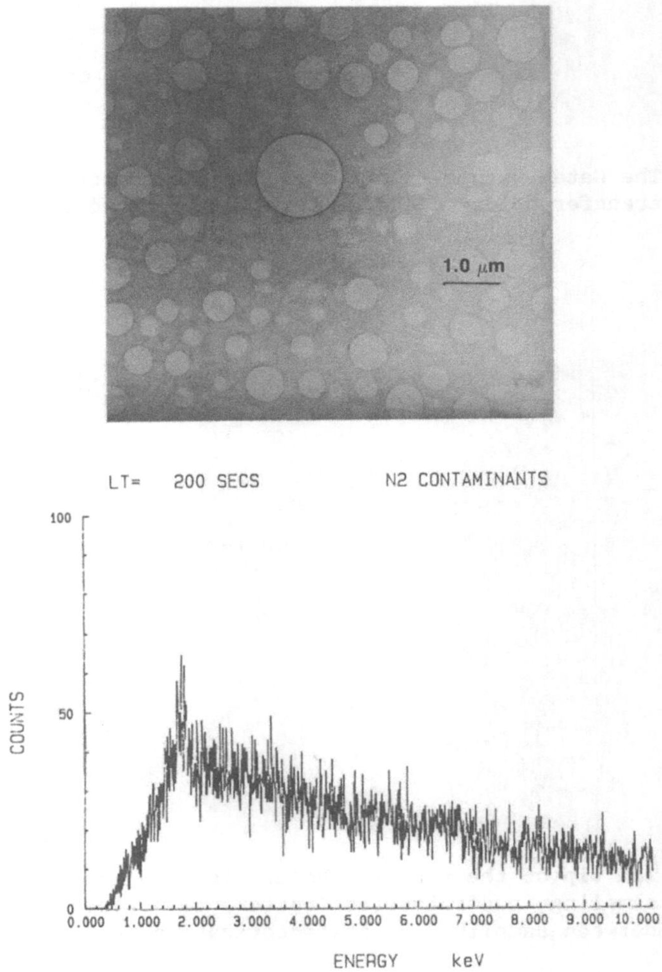


Figure 3. Clean Be, carbon coated grid, bright field, and x-ray spectrum from the carbon film. The cleanliness of the sampling surface is verified.

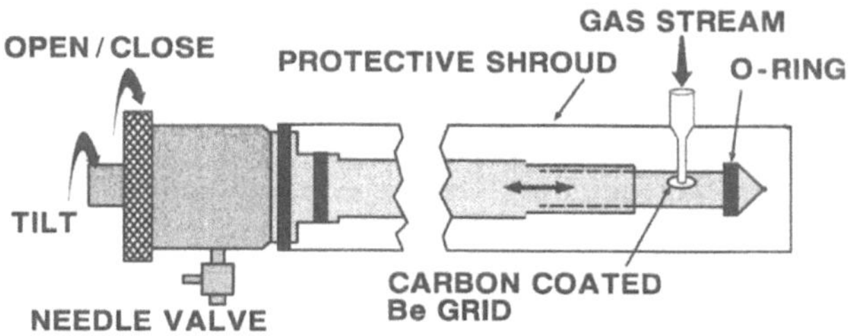


Figure 4. Schematic of sampling setup. The gas to be sampled flows over the specimen grid. The holder is protected from ambient contamination by a shroud.

The sampling grid remains retracted into the holder while the rod is locked into position. Valves controlling gas flow are opened and the shielded region is purged for several minutes. After the purge, the grid holding section of the rod is extended to the proper prealigned position. The grid is then exposed to the gas stream. After the sampling period, the sample grid is retracted into the holder shaft and sealed. Once again, the sample grid is protected from ambient contamination, and is securely transferred to the AEM laboratory.

In the laboratory, the sealed transfer rod is introduced into the AEM and prepumped to about $10^{-1}$ Pa before extending the sample grid section of the rod into the $10^{-5}$ Pa vacuum of the instrument. Particles isolated on the grid are then analyzed.

Results reported herein were obtained from sampling gas for 48 hours from a laboratory house nitrogen system without filtration or special care taken when installing the piping. Most AEM was performed at 120 kV on a Philips EM400T $LaB_6$ gun system with an EDAX Si(Li) x-ray detector employing a Be window. Bright field micrographs were obtained along with selected area diffraction patterns. X-ray emission spectra were also obtained from the particles observed, in both TEM and STEM modes.

X-ray elemental imaging was performed on a Vacuum Generators 100 kV HB-501 field-emission STEM with a magnetic gun-lens. This system was fitted with a windowless Si(Li) detector manufactured by Link Systems. The electron beam was controlled by the digital scan generator in a Link Systems AN10000 x-ray analyzer.[3,4] Typical x-ray images contained 128 x 128 pixels with a dwell time for each pixel of 100 ms. Thus, the length of time to acquire an x-ray image was nearly a half hour.

RESULTS

In Figure 3 a clean carbon support film before collection is shown along with the x-ray spectrum from the carbon film. After exposure to house nitrogen, the carbon film has collected a number of particles. Figure 5 shows that submicron particles are adequately separated for individual examination. The EDS spectrum of the support film away from the particles does not contain characteristic peaks and is similar to that in Figure 3. In Figure 6, a bright field image of a 0.4µm particle is shown along with the energy dispersive x-ray (EDS) spectrum and selected area electron diffraction pattern (SAED). The x-ray spectrum indicates Al as the major element with atomic number greater than 11 (sodium) that can be detected through the Be window of the detector. The SAED shows a semi-hexagonal spot pattern. The d-values measured from the pattern closely match those for $\gamma$-$Al_2O_3$.

Figure 7 displays two submicron particles, separated only by approximately 2.5 µm, with associated EDS spectra. The two particles are shown to be different compositionally. One is believed to be attributed to leaded brass and the other to aluminum or alumina (cf. Figure 6). Brass valves, aluminum storage tanks and aluminum vaporizer fins are components of the nitrogen supply system.

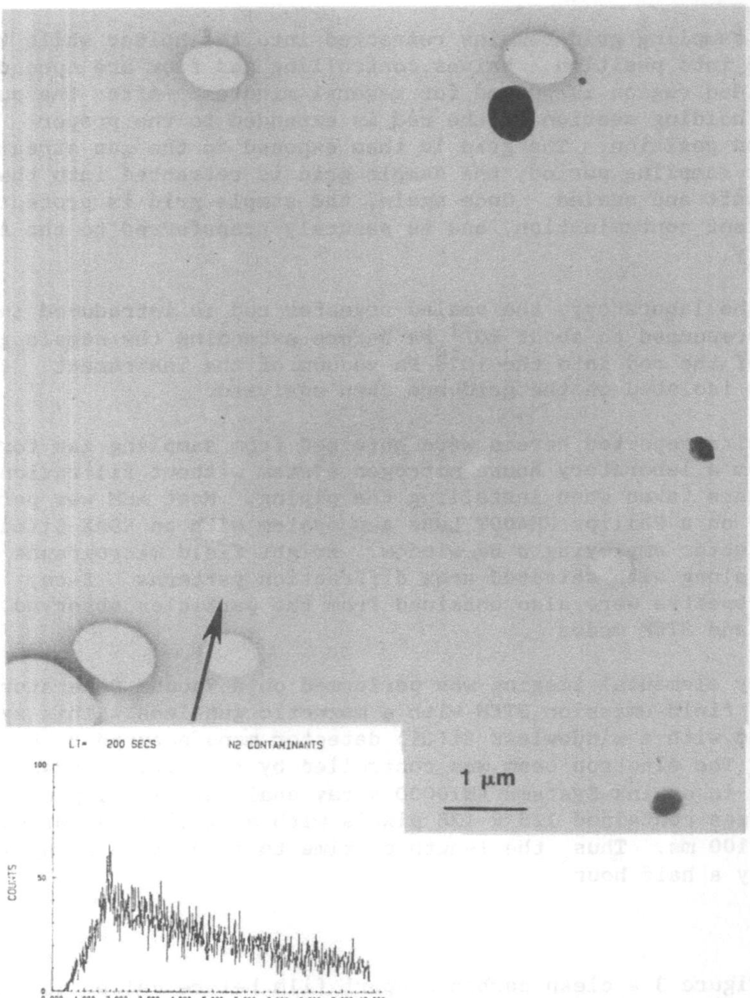


**Figure 5. General view and background EDS spectrum of particles collected from nitrogen.**

In Figure 8, we present a different particle with another morphology. A bright field image is shown together with the corresponding EDS spectrum which clearly shows the presence of Fe, Mn, Cu and Zn. This is probably a low carbon steel together with brass. Assuming relative concentrations, the steel contains approximately 98% Fe and 2% Mn. The Cu, Zn relative concentration is roughly 60% Cu, 40% Zn. This particle probably came from piping or a valve.

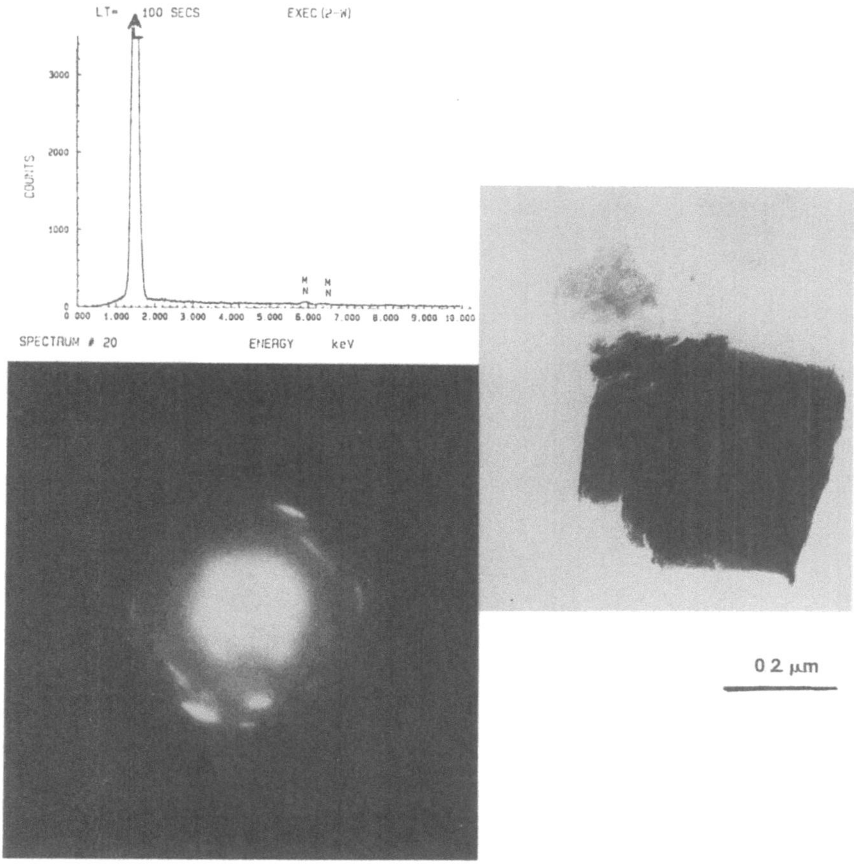


| d(Å) | $\gamma$-$Al_2O_3$ | * |
|---|---|---|
| 4.55 | 4.56 $(\bar{1}11)$ | |
| 2.39 | 2.39 | |
| 1.98 | 1.98 | |
| 1.53 | 1.52 | |
| 1.39 | 1.40 | |
| 1.13 | 1.14 | * |

Figure 6. Bright field image, EDS spectrum and selected area electron diffraction pattern from alumina particle collected from the gas stream. The full identification as γ-alumina is possible only because of the electron diffraction data.

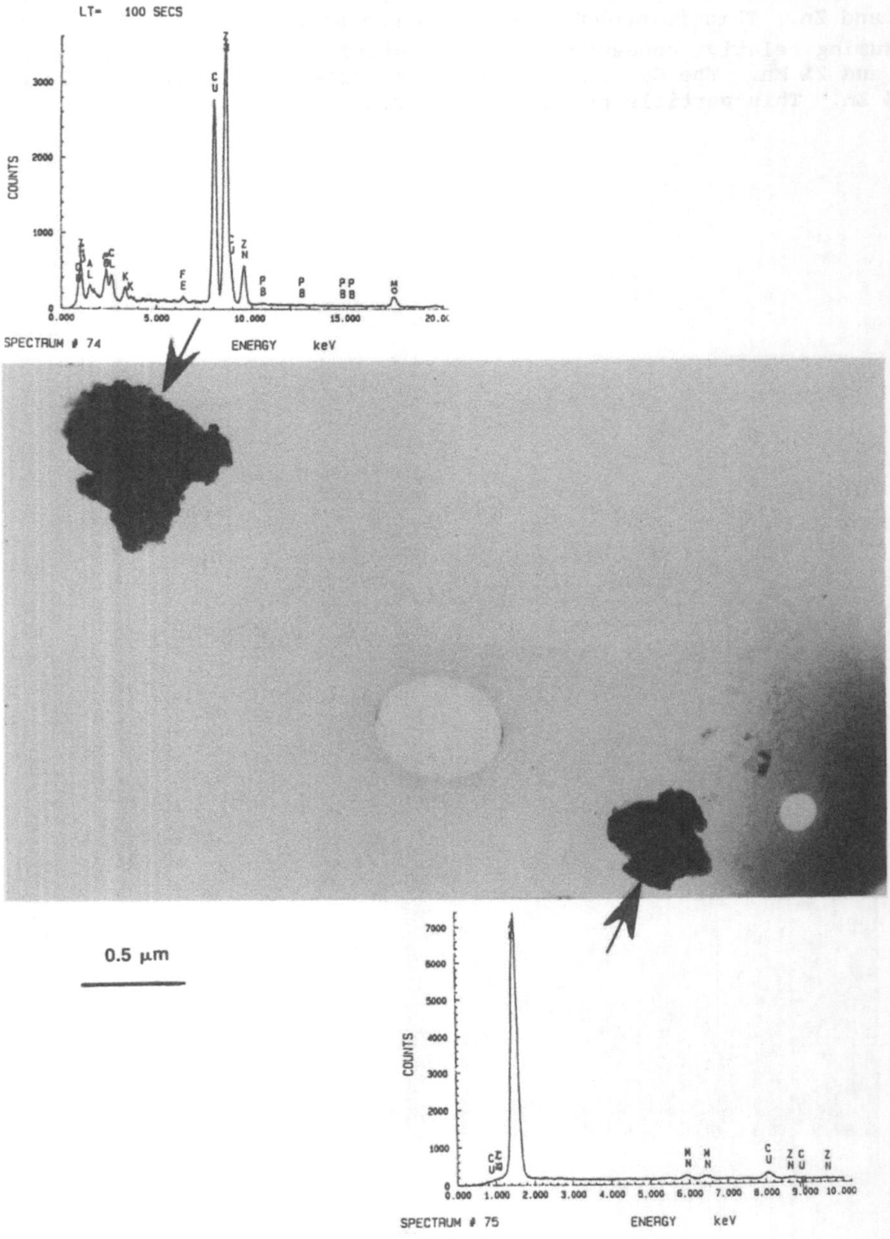


**Figure 7. Bright field image of two submicron particles and their corresponding EDS spectra.**

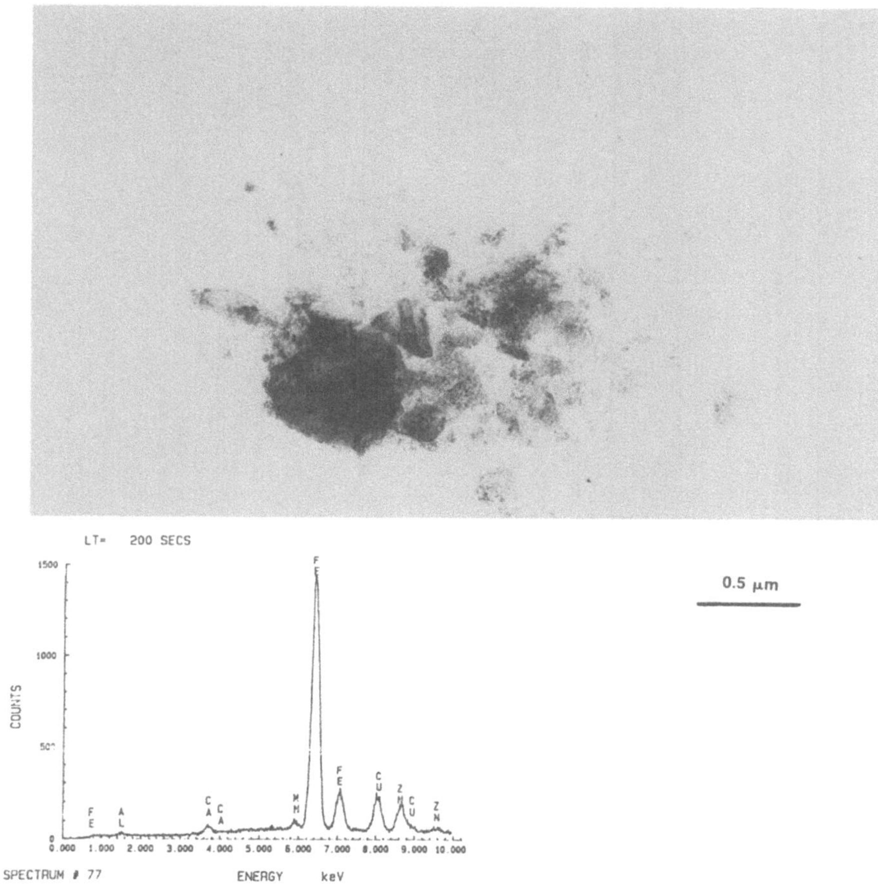


**Figure 8.** Bright field image and EDS spectrum of particle collected from the gas. Fe plus approximately 2 percent Mn may indicate the particle is from a carbon steel. The presence of Cu and Zn indicates brass. This particle probably came from a pipe fitting or valve.

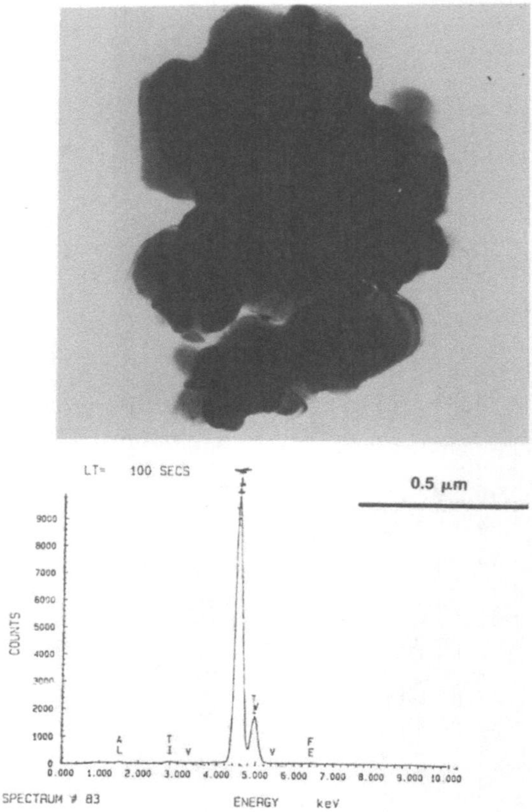


Figure 9. Bright field image and EDS spectrum of Ti particle collected from the gas stream. The morphology is characteristic of titanium dioxide.

In Figure 9, a bright field image and EDS spectrum of another particle are presented. The morphology of the particle is characteristic of $TiO_2$, a paint pigment. EDS results are consistent with this assignment.

X-ray images showing the distribution of a particular element can be obtained by moving the electron beam in a digitally stepped scan. Characteristic x-ray intensities are monitored for elements of interest at each step or pixel. The relative intensities for an element are displayed on a brightness scale creating an x-ray image showing the distribution of the element.

In Figures 10 and 11, an annular dark field STEM image of a particle field is presented along with the oxygen, aluminum, titanium and calcium x-ray images. It has been estimated[3] that for the image to have any significance, the number of counts above background for any given pixel should be greater than 7. The oxygen Kα X-ray image indicates most particles contain varying amounts of oxygen; the maximum number of counts is 48. In the case of aluminum, the maximum number of counts is 8, so the signals depicted are barely significant. The particle labeled 4 in the dark field image is not

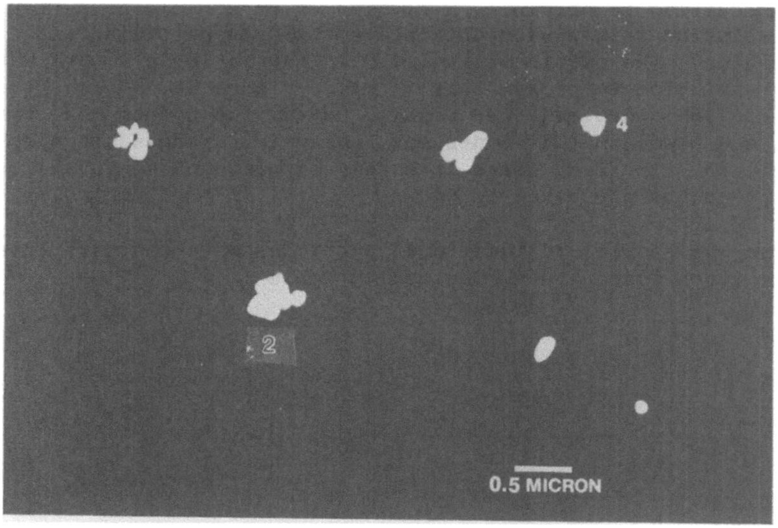


Figure 10. Annular dark field STEM image of particles collected from gas on carbon support. The corresponding x-ray images are shown in Figure 11.

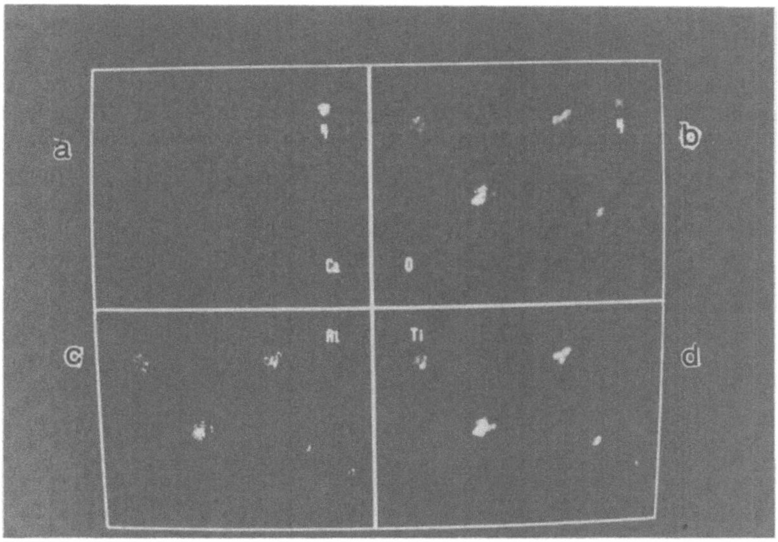


Figure 11. (a) Calcium x-ray image, (b) Oxygen x-ray image, (c) Aluminum x-ray image, (d) Titanium x-ray image. The corresponding annular dark field STEM image is shown in Figure 10.

seen in the aluminum image, and hence this particle does not contain Al. The calcium image in Figure 11 clearly shows that this particle contains calcium, probably as a calcium compound.

In addition to obtaining x-ray images of a field of particles, high resolution images were obtained for single particles and agglomerates. The particle labeled 2 in Figure 10 is imaged with higher resolution (approximately 5 nm) in Figure 12. Corresponding titanium (196 counts max. per pixel), oxygen (34 counts max. per pixel), and aluminum (18 counts max. per pixel) images are displayed in Figure 13. The S-Mo image is barely significant since the maximum number of counts per pixel is 9.

From one sample, 86 individual particles were characterized. Particle types found are summarized in Table I.

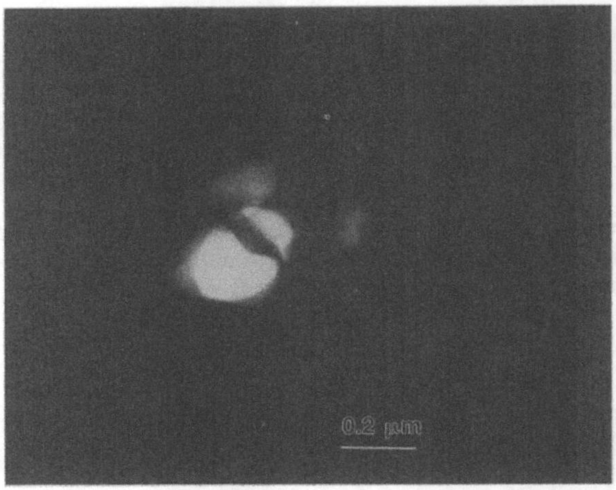


Figure 12. STEM dark field image of particle 2 in Figure 10. Corresponding x-ray images are shown in Figure 13.

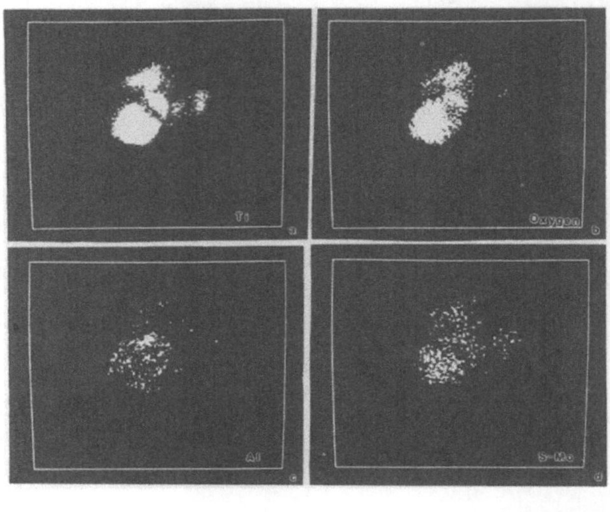


Figure 13. X-ray images of particle in Figure 12: (a) titanium, (b) oxygen, (c) aluminum, (d) sulfur/molybdenum.

Table I. Summary of AEM Identification of Particles Collected from Nitrogen Gas Stream.

| Particle Type | Number | Size Range (μm) | Mean Size | Remarks |
|---|---|---|---|---|
| Al | 30 | 0.1-0.8 | 0.5 | Electron diffraction gives $\gamma$-$Al_2O_3$ |
| Ca | 5 | 0.1-0.5 | 0.3 | Alone or with Cl, Cu |
| Ti | 5 | 0.1-0.8 | 0.5 | Probably oxides |
| Fe<br>Fe-Mn<br>Fe-Cr | 7<br>3<br>2 | 0.3-1 | 0.6 | Often found with Cu-Zn |
| Pb | 5 | 0.3-0.4 | 0.4 | |
| Cu<br>Cu-Zn | 5 | 0.2-0.8 | 0.3 | From 100% Cu to a 50% Cu-Zn found also with Fe-Mn |
| S-Mo | 16 | 0.05-0.5 | 0.2 | Most contain oxygen, a few with Cl, K |
| Si | 6 | 0.5-0.6 | 0.5 | Probably oxides |
| K | 2 | 0.2 | 0.2 | Appears also with Cl |

## DISCUSSION

In the sample considered here, about a dozen different types of particles were identified. The above results show the enormous capability of the method described here for capture, detection and analysis of submicron particles in a gas without ambient contamination. Unlike other analytical methods, AEM can analyze submicron particles by two complementary techniques: X-ray spectrometry and electron diffraction.

Using the sampling technique described here, contamination producing a background of particles from sample handling is virtually eliminated. The analyst surveys the collection surface before sampling to assure cleanliness. With the grid secured within the transfer rod, the apparatus can be safely moved to and from the sampling site. In the case described here, the sampling site is nearly 20 miles away from the AEM laboratory. The authors are confident of the protected chamber seal. Clean grids have been shown to remain clean until exposed at the sampling site. The gas stream sampled was a general purpose laboratory house nitrogen supply. It is not intended for high purity use. From microcharacterization, many of the particles examined appear to be related to the materials of construction of the gas supply system. These materials include aluminum, copper, brass, carbon steel and stainless steel. Molybdenum disulfide may have been used in certain components as a lubricant. Titanium oxides may have been introduced into the system from paint. Common dust and minerals may be the source of calcium, silicon and potassium containing species.

Had these same particles been isolated on a filter for SEM/EDS examination, on an individual basis they would be small relative to the x-ray emission volume. Consequently, insufficient signal would be available for comparable characterization.

Because less than 100 particles were analyzed in the AEM, great statistical significance in the relative proportion of the particles should not be inferred. There may even be particle types that were missed by this manually performed analysis. Automated analyses of this nature, however, will be soon possible. By analyzing numbers of particles in excess of 1000, a more reliable overall analysis can be made.

## CONCLUSIONS

The use of a retractable specimen holder for collection, isolation and observation of submicron particles in $N_2$ gas has been shown to be very effective. Particulate background from ambient contamination can be virtually eliminated. Particles with different morphologies and different compositions have been individually characterized by AEM. Elemental images with 5nm resolution can be achieved. The unique advantages of analytical electron microscopy, in particular electron diffraction and x-ray emission spectrometry, allow complete characterization of individual particles. The AEM allows the determination of relationships between chemistry, particle size and morphology. In some cases, this information gives clues to the origin of the particles.

## ACKNOWLEDGEMENTS

The authors are grateful for the assistance of J. R. Michael and the Bethlehem Steel Corporation for use of the VG HB-501. We also thank Air Products and Chemicals, Inc., and Lehigh University for support and permission to publish this work.

## REFERENCES

1. D. C. Joy, A. D. Romig, and J. I. Goldstein, "Principles of Analytical Electron Microscopy," Plenum Press, New York, 1986.
2. D. B. Williams, "Practical Analytical Electron Microscopy in Materials Science," Philips Electron Optics Publishing Group, Mahway, NJ, 1984.
3. C. E. Lyman, Ultramicroscopy, 20, 119 (1986).
4. C. E. Lyman, H. G. Stenger, Jr., and J. R. Michael, Ultramicroscopy, 22, 129 (1987).

# PHOTOACOUSTIC DETECTION OF RADIATION ABSORBING PARTICLES IN GASES

D.A. Jackson, K.B. Tennal, M.K. Mazumder, J.D. Wilson, and B.A. Ragent*

University of Arkansas at Little Rock
Department of Electronics and Instrumentation
2801 S. University, Little Rock, Arkansas 72204 and
*NASA Ames Research Center, Moffett Field, CA 94035

Photoacoustic detection methods have been successfully applied to monitoring trace contaminants in gases. The major focus of this study was the improvement of the response characteristics of photoacoustic detection systems which are sensitive to noise sources arising from: 1) vibration, 2) ambient acoustic noise, and 3) contamination of the optical windows of the photoacoustic devices. In this study one commercially available non-resonant spectrophone (Burleigh PAS-100) and two laboratory made spectrophones - one operating in a Helmholtz resonant mode and the other in a longitudinally resonant mode - were evaluated for their sensitivity for measuring light absorption by aerosols. The longitudinally resonant spectrophone was found to provide the greatest sensitivity of the three photoacoustic cells investigated. Using two microphones and applying noise subtraction techniques in the electronic signal processing, it was possible to reduce the noise contribution due to vibration by approximately 40 dB. Using an open-ended resonant cell in the longitudinal mode, the maximum sensitivity of $3.0\times10^{-7}m^{-1}$ was achieved in the laboratory experiment. With one-Hz bandwidth, typical detection limits were: 17 ppb for $NO_2$ in $N_2$ gas, and 0.39 $\mu g/m^3$ for 1.0 $\mu m$ diameter carbon particles.

## INTRODUCTION

Photoacoustic spectroscopy techniques[1] have been successfully used for monitoring both gaseous and particulate contaminants in air. The method provides near real-time measurements and has been applied to organic vapors, inorganic gases, solids, and particulate matter. The method is based upon the absorption of radiation by a sample and measurement of the absorbed energy released as heat that expands the gaseous sample and results in an increase in pressure. The incoming radiation (UV, VIS, or IR) is modulated so that the fluctuating pressure creates an acoustic signal. The acoustic signal intensity is proportional to the concentration of the absorbing species in the gas. The signal is detected by using a sensitive microphone. The sensitivity

*Particles in Gases and Liquids 2*, Edited by K.L. Mittal
Plenum Press, New York, 1990

of the system depends upon a number of factors: intensity of the incoming radiation, absorption cross-section of the species being measured[2], photoacoustic cell design, and sensitivity of the microphone. Since a microphone is used as a detector, the sensitivity is often limited by the ambient noise and vibration. A major focus of the study was the reduction of spectrophone sensitivity to external vibrations and acoustic noise. The impetus for the study was to determine the feasibility of using a spectrophone on board an aircraft. However, the results apply to noise sources such as those present in an industrial environment as well.

The insensitivity of the photoacoustic technique to scattered radiation permits one to obtain radiation absorption spectra on highly transparent media, such as powders, amorphous solids, gels, and colloids. The photoacoustic technique can thus be used in the study of adsorbed and chemisorbed molecular species and compounds on the surface of metals, semiconductors, and insulators. Photoacoustic spectroscopy studies in the infrared (IR) region have also been successfully performed[3] using a Fourier-transform photoacoustic spectrometer.

PHOTOACOUSTIC MEASUREMENT

The basic theory of photoacoustic detection is quite simple. When light energy is absorbed by a particle, molecules of the particle are excited to higher potential energy states. In general, this absorbed energy results in heating of the particle with subsequent transfer of the heat to the surrounding gas. If the sample is contained in a closed cell, this heating of the gas results in an increase in the pressure within the cell. If the light source is intensity modulated or chopped, then pressure variations will occur in the cell at the modulation frequency. These pressure variations can be detected with a microphone. The cell with windows and microphone is called either a photoacoustic cell or a spectrophone.[4-6] A schematic of a typical photoacoustic apparatus for aerosol concentration measurements is illustrated in Figure 1.

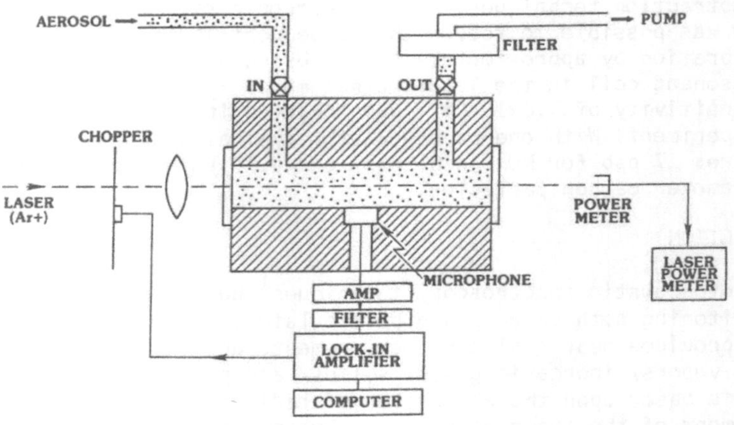


Figure 1. Schematic of photoacoustic measurement apparatus for aerosol light absorption studies.

The spectrophone provides a means of continuously monitoring ambient aerosol particulate concentrations that is far less time consuming than filter collection techniques. The time response of the spectrophone is on the order of seconds, with detection sensitivities of sub-ppb concentration levels.

Light extinction in a homogeneous medium is described by the Beer-Lambert Law.

$$I/I_o = \exp(-\beta_{ex}L) \quad (1)$$

where I is the light intensity at a distance L into the medium, $I_o$ is the incident or initial intensity, and $\beta_{ex}$ is the extinction coefficient for the medium at the specific frequency or frequencies of the incident light.

Extinction is due to absorption and scattering of the electromagnetic energy. The extinction coefficient is the linear sum of the absorption and scattering coefficients.

$$\beta_{ex} = \beta_{ab} + \beta_{sc}. \quad (2)$$

The extinction cross section ($\alpha_{ex}$) of molecular species or aerosol particles is defined as the hypothetical area normal to the incident radiation that would geometrically intercept the total amount of radiation actually extinguished by the particle.[2] The extinction coefficient of a monodisperse aerosol can be considered to be the product of the extinction cross section for individual elements ($\alpha_{ex}$) of the medium and the number concentration, N, of those elements.

$$\beta_{ex} = N\alpha_{ex}. \quad (3)$$

The amount of radiant power, $P_{ex}$, absorbed and scattered by the particle is given by,

$$P_{ex} = I_o\alpha_{ex}. \quad (4)$$

The extinction cross section can also be written as a linear summation of the absorption and scattering coefficients.

$$\alpha_{ex} = \alpha_{ab} + \alpha_{sc} \quad (5)$$

where the absorption and scattering cross sections, $\alpha_{ab}$ and $\alpha_{sc}$, are defined in the same manner as the extinction cross section.

For small beam attenuation and negligible scattering per path length of absorbed power, Eq. (2) reduces to

$$I/I_o = \exp(-\beta_{ab}L). \quad (6)$$

Normalizing $\beta_{ab}$ to the mass concentration, $M_c$, of particles in the gas gives the specific absorption coefficient $A_a = \beta_{ab}/M_c$.

When the light source is only weakly attenuated in the spectrophone, the microphone signal S is given by

$$S = R\ B\ A_a\ M_c\ W, \quad (7)$$

where W is the time averaged power of the light source, $A_a$ is the specific absorption coefficient with units of inverse length per unit mass concentration, $M_c$ is the mass concentration of the sample, R is the responsivity of the spectrophone which depends on the cell geometry and the microphone sensitivity, and B is the fraction of the absorbed optical energy that is converted into translation energy of the gas molecules. For fixed physical and chemical make-up of the particles and fixed excitation wavelength, $A_a$ is a constant. Hence, S is proportional to the mass concentration of particulates in the aerosol.

System sensitivity is defined by the absorption coefficient

$$\beta_{ab} = \alpha_{ab} C, \tag{8}$$

where $\alpha_{ab}$ is the absorption cross section of the species being measured, and C is the fractional concentration of the absorbing species.

In the case of gaseous absorption the system sensitivity, $\beta_{ab}$, in Eq. 8 can be written as

$$\beta_{ab} = \alpha_m C n_o, \tag{9}$$

where $\alpha_m$ is the molecular absorption cross section ($m^2$/molecule) of the species being measured, C is the fractional concentration of the absorbing species, and $n_o$ is the total gas density (molecule/$m^3$).

As a numerical example, consider one watt of chopped laser power at a wavelength of 488 nm, a 1 um diameter carbon particle, having an absorption cross section, $\alpha_{ab}$, of $1.05 \times 10^{-12} m^2$, and a system detection limit of $3 \times 10^{-7} m^{-1}$. This gives a detection limit corresponding to a particle concentration of about 0.39 $\mu g/m^3$ (0.011 $\mu g/ft^3$ or about 8400 particles/$ft^3$).

## EXPERIMENTAL

Three spectrophones were tested for their sensitivity in making light absorption measurements: 1) a Burleigh Instruments, Inc., PAS-100 non-resonant spectrophone, 2) a laboratory built Helmholtz resonant spectrophone, and 3) a laboratory built longitudinal mode resonant spectrophone. For one watt of average power from the 488 nm line of an argon-ion chopped laser beam the sensitivities for the three spectrophones were $6.5 \times 10^{-6} m^{-1}$, $2.5 \times 10^{-6} m^{-1}$, and $3.0 \times 10^{-7} m^{-1}$, respectively. The sensitivities are for a one second time constant on the lock-in amplifier and for a signal-to-noise ratio of one. The noise levels were measured under ambient laboratory conditions with no absorber in the spectrophone. Nitrogen dioxide, $NO_2$, was used for calibrating the spectrophones because it exhibits strong absorption of visible light and because absorption spectra and absorption coefficients were available from literature. The $NO_2$ concentration was monitored using the wet chemical reagent technique developed by Saltzman.[7]

The Burleigh spectrophone, which could be operated at a frequency between 50 and 2000 Hz, was tested for sensitivity to externally applied vibration and acoustic noise. Methods for nulling of the vibration signal and for isolation from acoustic noise were incorporated into the system. Light absorption measurements on aerosols were made using the Burleigh and longitudinally resonant spectrophones.

The main cylinder of the Burleigh spectrophone was modified by adding a second valve near the opposite end from the original valve and by mounting a second microphone (Knowles Model 1759) diametrically opposed to the original microphone. This microphone could be replaced by a plate-mounted accelerometer (Endevco Model 2220C). A Bruel and Kjaer PM vibration exciter type 4809 was used to subject the Burleigh spectrophone to both sinusoidal and broadband vibrations. The experimental arrangement used to test the Burleigh spectrophone is shown in Figure 2.

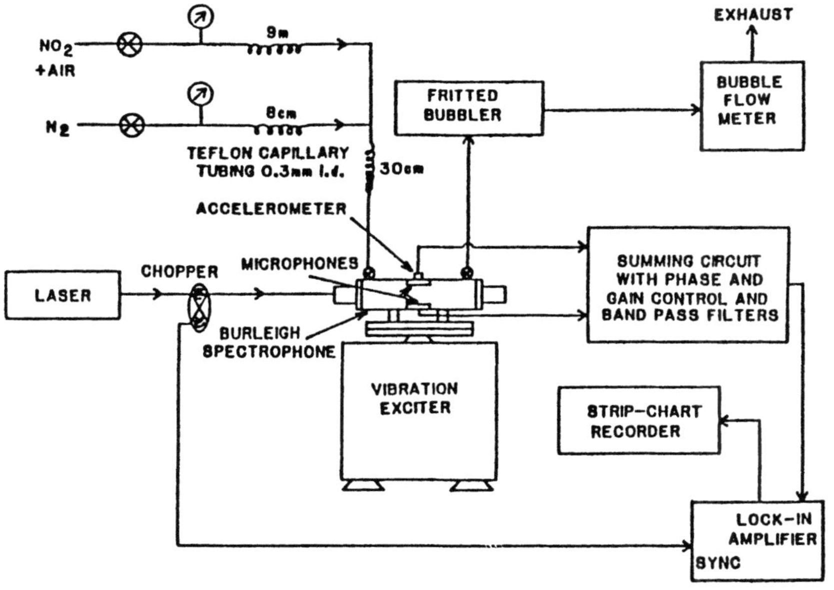


Figure 2. Experimental arrangement for making light absorption measurements and vibration measurements with the Burleigh spectrophone

The response of the spectrophone to external noise was examined by placing the spectrophone in an acoustic chamber and subjecting it to sound power levels of up to 108 dB. A GENRAD 1988 precision integrating sound level meter and analyzer was used to measure the sound power levels in the chamber. A Hewlett Packard spectrum analyzer model 186A was used to monitor the frequency spectra of vibrations due to sound power levels between 0 and 2000 Hz. The responsivity of the Burleigh spectrophone was found to be about 30 $\mu$V per watt of chopped laser power per ppm of $NO_2$ at a 100 Hz chopping frequency. The noise level with a one-Hz bandwidth on the lock-in amplifier was about 0.5 $\mu$V giving a detection limit of 17 ppb of $NO_2$ in $N_2$.

The Helmholtz resonant spectrophone is shown in Figure 3. It consists of two identical absorption chambers joined at their longitudinal midpoints by a connecting tube. At resonance frequency (242 Hz) the pressure in the two absorption chambers will vary approximately sinusoidally and with a 180-degrees phase difference. Because of this phase difference the two chambers can be driven by alternate half cycles of the same light source. It was therefore possible to use both halves of the chopped laser beam, thus increasing the photoacoustic signal. The responsivity was found to be 49 uV per watt of chopped laser power per ppm of $NO_2$. The signal increase of the Helmholtz over a single-drive system was about 50% rather than the 100% theoretically possible.

This was due to optical power loss in the chopper blade and at the extra mirrors required for the system. The resulting sensitivity is about 12 ppb of $NO_2$ in air for one watt of laser power, which is equivalent to approximately $6.5 \times 10^{-6} m^{-1}$.

The third spectrophone tested was a double-open-ended tube operated at its first longitudinal resonance. This optical arrangement permits elimination of optical window noise. The first longitudinal resonance was at 814 Hz. A Radio Shack 270-092A electret condenser microphone was mounted at the longitudinal midpoint of the tube. This tube was housed

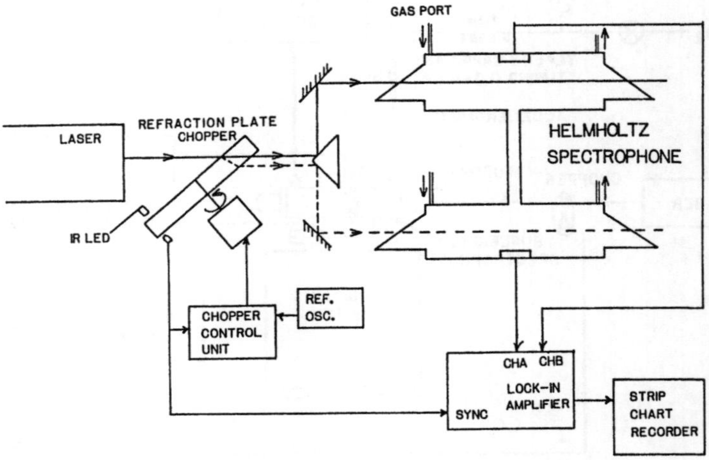


Figure 3. Experimental arrangement with the "Dual-Drive" Helmholtz resonant spectrophone using a slotted refraction plate chopper.

inside an acoustic isolation chamber. Figure 4 shows the experimental arrangement for making measurements of light absorption by aerosols

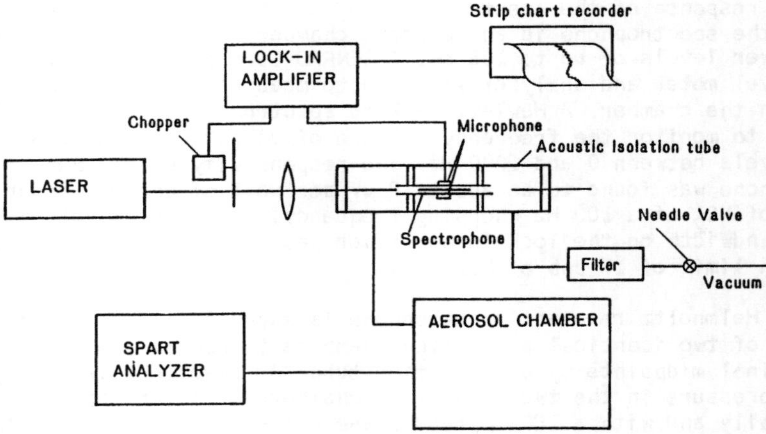


Figure 4. Experimental arrangement for measuring light absorption by aerosols.

using this spectrophone. The responsivity of the longitudinally resonant spectrophone was about 120 mV ppm$^{-1}$ W$^{-1}$. The resulting sensitivity for $NO_2$ detection was about 0.7 ppb with 1 watt of chopped laser power and optimum noise conditions. This corresponds to an absorption coefficient of about $3 \times 10^{-7}$ m$^{-1}$.

The Burleigh spectrophone was found to have a response to sinusoidal vibrations of 1.3 mV/g where g is the acceleration due to gravity. Broadband vibration was applied with an rms acceleration of 0.02 g/(Hz)$^{1/2}$ at 400 Hz. The noise was reduced by nearly 40 dB with the summing circuit. These results were obtained with the valves closed, with $N_2$ in the spectrophone and with 250 mW of chopped laser power passing through it. Figure 5 shows an absorption measurement for

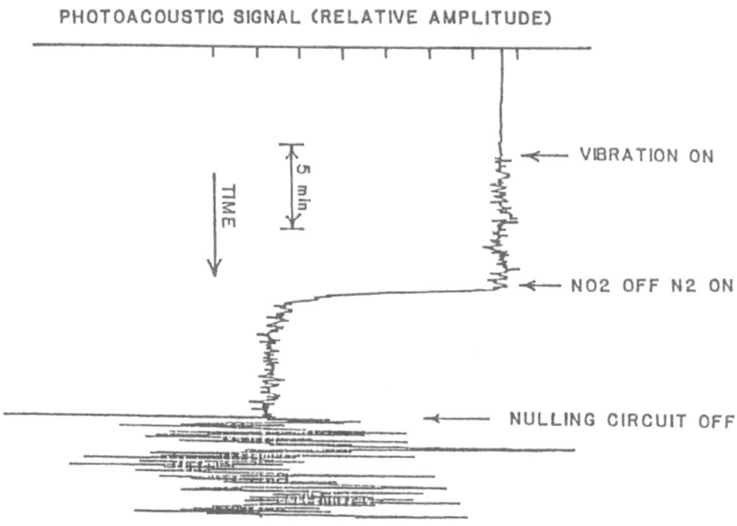


Figure 5. Photoacoustic measurements under vibrationally noisy conditions. The lock-in amplifier output shows the Burleigh spectrophone response to light absorption by $NO_2$ when the instrument is subjected to broadband vibrations with and without nulling circuitry.

7.6 ppm $NO_2$ in air while nulling vibrations with the Knowles microphone circuit. The gas was flowing through the spectrophone at 1.1 $cm^3/s$.

Photoacoustic measurements were made on aerosols of cigarette smoke, sodium chloride, and polystyrene latex spheres (PLS). Simultaneous measurements of the particle size distribution and the approximate number concentration were made using a Single Particle Aerodynamic Relaxation Time (SPART) analyzer.[8] Figure 6 shows a strip chart recording of the photoacoustic signal for cigarette smoke using the longitudinally resonant spectrophone. The count median aerodynamic diameter was about 0.6 $\mu m$.

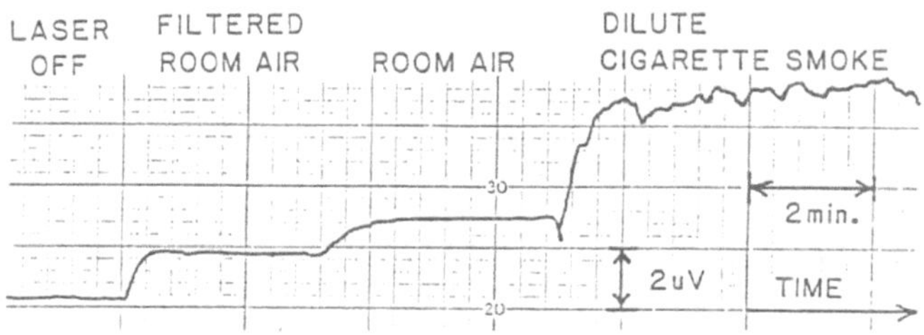


Figure 6. Strip chart record of measurements of the absorption of light by aerosols using a longitudinally resonant spectrophone.

Extrapolation of the signal down to the background noise level gave a detection limit of about 3 particles/$cm^3$ for one watt of average laser power. However, this corresponds to less than one particle in the illuminated volume at a time. The count median aerodynamic diameter for the NaCl particle was about 0.9 μm. The photoacoustic signal for the NaCl particles was found to be about three orders of magnitude smaller than for the cigarette smoke. A dependence on relative humidity was also observed. The detection limit for the PLS (diameter = 1.09 μm) was estimated at about 500 particles/$cm^3$. Some of this signal may have resulted from the high scattering intensity of the particles.

CONCLUSIONS

In the laboratory environment, the two-microphone vibration nulling system reduced the sensitivity of the non-resonant spectrophone to vibration by 40 dB. Acoustic sensitivity of the Burleigh spectrophone to broadband noise sound power levels of 100 dB was negligible. Single frequency acoustic noise at the same sound power levels in the frequency range between 50 and 2000 Hz was detected by the Burleigh cell at 20 dB above background. Both the Burleigh spectrophone and the Helmholtz resonant spectrophone suffered from window noise arising from window absorption and scattering of the incident light beam. The longitudinally resonant spectrophone did not have optical windows and provided the greatest sensitivity ($3\times10^{-7}$ $m^{-1}$) of the three spectrophones tested.

A detection limit for 1 um carbon particles was calculated to be approximately 0.39 μg/$m^3$. Measurements on cigarette smoke gave an extrapolated detection limit of 0.39 μg/$m^3$. For NaCl and PLS the mass detection limit is 2.7 mg/$m^3$ and 0.35 mg/$m^3$, respectively.

Photoacoustic detection techniques can be employed in a wide range of applications, including monitoring trace contaminants in clean room environments. This paper summarizes the experimental results in which photoacoustic signal enhancement was obtained by: (1) decreasing ambient acoustic noise by employing subtraction techniques in electronic signal processing, and (2) improving vibration isolation. Thus, photoacoustic detection capabilities were improved.

ACKNOWLEDGEMENTS

This work was supported in part by: 1) a NASA Grant #NAG 2-147, and 2) a NASA Training Grant #NGT-70013. The authors are grateful to D. Belk and M. Dugan for assisting in manuscript preparation, and to D. Watson and P. Roberson for editing the manuscript.

REFERENCES

1. A. Rosencwaig, "Photoacoustics and Photoacoustic Spectroscopy," Wiley & Sons, New York, 1980.
2. R.C. Weast, editor, "CRC Handbook of Chemistry and Physics," 61st edition, p. 260, CRC Press, Inc., Boca Raton, FL (1980-1981).
3. M.J.D. Low and G.A. Parodi, Dispersive photoacoustic spectroscopy of solids in the infrared range, J. Photoacoustics, 1(1), 131-144 (1984).
4. A.C. Szkarlat and S.M. Japar, Light absorption by airborne aerosols: Comparison of integrating plate and spectrophone techniques, Appl. Optics, 20, 1151-1155 (1981).
5. R.W. Terhune and J.E. Anderson, Spectrophone measurements of the absorption of visible light by aerosols in the atmosphere, Opt. Lett., 1, 70-72 (1977).

6. C.W. Bruce and R.G. Pinnick, In-situ measurements of absorption with a resonant CW laser spectrophone, Appl. Optics, 16, 1762-1765 (1977).
7. E. Saltzman, Colorimetric microdetermination of nitrogen dioxide in the atmosphere, Anal. Chem., 92(12), 1949-1955 (December 1954).
8. R.G. Renninger, M.K. Mazumder and M.K. Testerman, Particle sizing by electrical single particle aerodynamic relaxation time analyzer, Rev. Sci. Instrum., 52(2), 242-246 (February 1981).

# TRANSPORT OF CHARGED PARTICLES IN GAS STREAMS

C. P. Wu and M. K. Mazumder

University of Arkansas at Little Rock
Dept. of Electronics and Instrumentation
2801 S. University, Little Rock, AR 72204

This paper presents a brief description of electrodynamic screens and their applications for: (1) minimizing deposition of charged particles inside sampling tubes and (2) transporting charged particles away from nonconducting substrates for shielding and removal of contaminants. Ring electrodes attached to the inner wall of a sampling tube were used to produce a radial confinement force on the charged particles when an ac field was applied. Effects of varying the applied voltage and the frequency of the ac field are discussed. The higher the voltage, the stronger the confinement force on the charged particles, but the upper limit of the voltage is set by the breakdown of the dielectric medium between the rings. The distribution of the electrical field was calculated using a single-phase electrodynamic field, and the motion of particles inside the tube was computed using a theoretical model. Experiments were performed to design an optimal configuration of the electrodynamic screen. A computer model of the electrodynamic sampling tube for several ring-type electrodes with different wire and ring diameters, and for different spacings between rings, has been developed. Experimental data are compared with the values predicted from the model. Applications of the electrodynamic screen for sampling and measuring charge-to-mass ratio distributions of particles are discussed.

## INTRODUCTION

The transport of charged particles encounter significant particle deposition on the inner surface of the tube caused by electrostatic forces. While enhanced deposition of electrostatically charged particles in a gas stream flowing through conducting or nonconducting tubes has been reported[1], quantitative measurement of the deposition of

*Particles in Gases and Liquids 2*, Edited by K.L. Mittal
Plenum Press, New York, 1990

charged aerosol particles often has been hindered by deposition losses within the sampling tubes of the measuring instruments.

The primary goal of this work was the development of a sampling system for particle charge analyzers that minimizes sampling losses of charged particles. For example, the Electrical Single Particle Aerodynamic Relaxation Time (E-SPART) analyzer[2] can make rapid real-time measurements of the aerodynamic diameter and electrostatic charge of single aerosol particles, but particles having an electrostatic charge greater than 20 μC/g in magnitude encounter significant sampling losses. The primary motivation of this study was the design and application of an electrodynamic sampling tube to minimize deposition of charged particles with charge-to-mass ratio up to 50 μC/g and diameters from 0.3 μm to 70 μm.

In order to minimize particle deposition caused by electrostatic forces, it is necessary to counteract depositional forces with a confinement force. Investigation into the transport and removal of airborne particles using electrodynamic screens was first reported by Masuda.[3,4,5] He used electrical curtains for the confinement and transportation of charged aerosols. Applications of the "Masuda Curtain" or "Masuda Screen" in different particle technology areas have been reported.[5,6]

Basically, two operational modes of the screen are possible for minimizing particle deposition due to electrostatic forces.[3,4] In one mode, a single phase driving power supply is used, and in the other, a polyphase supply is used. While both modes of operation can provide particle confinement, the polyphase supply provides both confinement and transportation of charged particles because of the traveling electric field produced in this mode. However, only a single phase, ring-type, electrical screen was studied in our application since (1) particle transport is provided by the aerosol flow stream (by the sampling pump) and (2) our objective was the confinement of aerosol particles in the flow stream avoiding particle deposition on the walls of the sampling tube.

## PRINCIPLES OF ELECTRODYNAMIC TRANSPORT

A Masuda Curtain is formed by a set of ring electrodes insulated from each other and connected to an ac power supply as shown in Figure 1. Charged particles falling through the screen are attracted to the wall by both space charge repulsion and image charges in the wall; they are repelled by the electrical field externally applied to the rings. To minimize particle deposition on walls, the total radial depositional forces acting on the charged particles (mutual repulsion force and image force) must be smaller than the radial confinement force. The confinment force then provides a barrier against the deposition of charged particles on the inner surface of the screen.

The dynamic behavior of a charged particle in an alternating electric field is described by Mathieu's equation[3-4] as follows:

$$m d^2R/dt^2 + 6\pi\eta a dR/dt = qE \cos(\omega t) + F_e, \quad (1)$$

where q is the charge on particle (C),
m is the mass of the particle (kg),
$\omega$ is the angular frequency of electric field (Radians/s),

$\eta$ is the coefficient of viscosity of air (kg/ms),
a is the radius of the particle (m),
R is the position vector of the particle (m),
E is the peak value of the field strength (V/m),
$F_e$ is the external force (N).

The field intensity E can be obtained by solving the Laplace equation,

$$\nabla^2 P = 0 \tag{2}$$

where P is the peak value of the potential function, and

$$E = -\text{grad } P. \tag{3}$$

Since the analytical solution of the Laplace equation is complicated, a numerical method, the Charge Simulation Method,[7] is used for determining the field distribution. Using a linear approximation, the motion of the particle in the radial direction can be written as

$$m d^2 r/dt^2 + 6\pi\eta a\, dr/dt = qE_r \cos\omega t + F_{er}, \tag{4}$$

where r is the radial distance of the particle from the axis of the sampling tube. If the amplitude of charge oscillation is small, r can be expressed as

$$r = r_o + \Delta r, \tag{5}$$

where $\Delta r$ is the oscillation component, and $r_o$ is the coordinate of the particle at the center of its oscillation. Substituting equation (5) into (4), and using Taylor's expansion for E, we can separate equation (4) into two equations.

$$m\frac{d^2(\Delta r)}{dt^2} + 6\pi\eta a\frac{d(\Delta r)}{dt} = qE_{r_o} \cos\omega t \tag{6}$$

and

$$m\frac{d^2 r_o}{dt^2} + 6\pi\eta a\frac{dr_o}{dt} = q\left(\Delta r\frac{\partial E_r}{\partial r}\Big|_o + \Delta z\frac{\partial E_r}{\partial z}\Big|_o + \ldots\right) \cos\omega t + F_{er} \tag{7}$$

The steady state solution of equation (6) can be expressed as

$$\Delta r = C\, E_{r_o} \cos(\omega t - \theta) \tag{8}$$

where $E_{r_o}$ is the electric field intensity at the center of oscillation in the radial direction,

$$C = q/[m\omega^2\{1 + (6\pi\eta a/m\omega)^2\}^{1/2}] \tag{9}$$

$$\theta = \pi - \tan^{-1}(6\pi\eta q/m\omega). \tag{10}$$

The oscillation component of the particle in the axial direction z is given by

$$\Delta z = C\, E_{z_o} \cos(\omega t - \theta) \tag{11}$$

where $E_{z_o}$ is the electric field intensity at the center of oscillation in the axial direction.

Substituting Equations (8) and (11) into Equation (7) and neglecting higher terms in the Taylor expansion, we have

$$m\frac{d^2r}{dt^2} + 6\pi\eta a\frac{dr}{dt} = qC\cos(\omega t - \theta)\cos\omega t[E_{r_o}\frac{\partial E_r}{\partial r}|_o + E_{z_o}\frac{\partial E_r}{\partial z}|_o] + F_{er} \tag{12}$$

However, the field intensity E can be obtained from the potential P.

$$E_r = -\frac{\partial P}{\partial r}\ ;\ E_z = -\frac{\partial P}{\partial z} \tag{13}$$

$$E_{r_o}\frac{\partial E_r}{\partial r}|_o + E_{z_o}\frac{\partial E_r}{\partial z}|_o = \frac{1}{2}\frac{\partial E^2}{\partial r}|_o \tag{14}$$

where

$$E^2 = E_r^{\,2} + E_z^{\,2}.$$

Substituting Equations (14) and (9) into (12) and noting that the average of $\cos(2\omega t - \theta)$ is zero, we have

$$m\frac{d^2r}{dt^2} + 6\pi\eta a\frac{dr}{dt} = -\frac{1}{4}m\omega^2C^2\frac{\partial E_o^{\,2}}{\partial r} + F_{er}. \tag{15}$$

The first term on the right side of Equation (15) represents the electrodynamic repulsion force $F_{cr}$ in the radial direction:

$$F_{cr} = -\text{grad}(\phi_r), \tag{16}$$

where the scalar potential

$$(\phi_r) = q^2E^2/4m\omega^2[1 + (6\pi\eta a/m\omega)^2]. \tag{17}$$

A simplification is made by considering a normalized coordinate system. The field strength E is normalized by the average field strength V/b between the rings to obtain the electrical field utilization factor f, where V is the peak voltage applied and b is the minimal spacing between rings (m),

$$f = E/(V/b), \tag{18}$$

Let $(\phi_r) = b\,K\,f^2$, then

$$F_r = -K\,(\text{grad})_b\,(f)^2, \tag{19}$$

where $(\text{grad})_b$ denotes the gradient operator in the normalized coordinate system. K is a physical constant which includes the effect of all physical parameters except those of electrode configuration.

$$K = q^2V^2/4m\omega^2\{1 + (6\pi\eta a/m\omega)^2\}b^3. \tag{20}$$

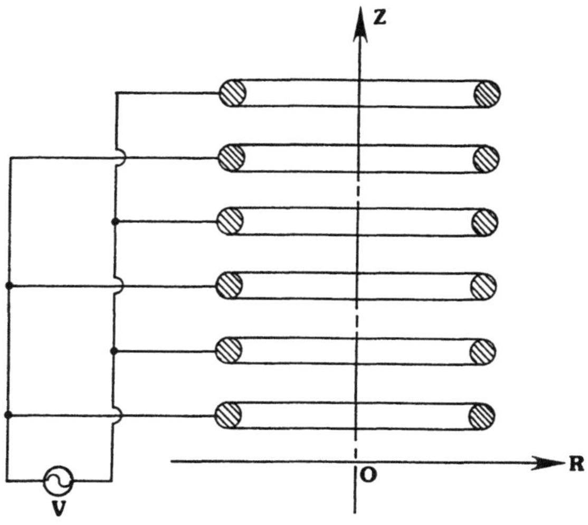


Figure 1. Configuration of a single phase, ring-type electrode screen.

In Figure 1, charged particles approaching the screen will experience a confinement force in two directions: the axial confinement force $F_{cz}$ and the radial confinement force $F_{cr}$.

$$F_{cz} = K\ P_z, \qquad F_{cr} = K\ P_r. \tag{21}$$

$P_z$ and $P_r$ are normalized confinement forces at the point where a particle experiences the smallest electrodynamic forces. Their approximate values are shown in Table I.

Table I. Approximate values of $P_r$ and $P_z$.

| d/D* | b/d** | $P_r$ | $P_r/P_z$ |
|---|---|---|---|
| 0.10 | 1.0 | 0.80 | >1000 |
| 0.10 | 1.5 | 0.85 | >1000 |
| 0.10 | 2.0 | 0.80 | 1000 |
| 0.15 | 1.0 | 0.78 | 1000 |
| 0.15 | 1.5 | 0.79 | 90 |
| 0.15 | 2.0 | 0.78 | 40 |

*d/D is the ratio of electrode diameter to the center ring diameter,
**b/d is the ratio of spacing between electrodes to electrode diameter.

Table I shows that the normalized confinement force in the axial direction $P_z$ is negligible compared to that in the radial direction if the diameter of the tube is more than 10 times that of the electrodes and the ratio of the spacing between the rings to the electrode diameter is not greater than 1.5. We designed the electrodes to obtain maximum radial confinement force; consequently, the normalized repulsion force in the axial direction was negligible.

In addition to the confinement forces, a charged particle falling through the screen will experience four additional forces: (1) image force, (2) mutual repulsion force due to the space charge of the particles, (3) drag force due to the motion of the particles relative to the gas, and (4) gravitational force.

For a conducting wall, the image force on a particle near the wall is

$$F_{ic} = q^2/4\pi\varepsilon_o(2S_w)^2, \tag{22}$$

where $\varepsilon_o$ is the permittivity of air, and $S_w$ is the distance between the particle and the wall.

For a dielectric wall, the image force is

$$F_{id} = F_{ic}(\varepsilon_r - 1)^2/(\varepsilon_r + 1)^2, \tag{23}$$

where $\varepsilon_r$ is the relative permittivity of the dielectric material.

The mutual repulsion of the unipolar charged particles causes precipitation of particles on the wall of the sampling tube. The electrical field caused by this space charge is time dependent and can be simulated by a series of ring charges. To estimate the mutual repulsion force, we assume a cylinder of radius $r_c$ containing uniform charges. This cylinder is assumed coaxial to the sampling tube. According to the Gauss law, the electrical field at the surface of a cylinder is

$$E_c = r_c\rho/2\varepsilon_o, \tag{24}$$

where $\rho$ is the particle space charge density. The mutual repulsion force on a particle can be approximated by

$$F_{mr} = qE_c = \rho r_c\, q/2\varepsilon_o. \tag{25}$$

As charged particles deposit on the insulated inner wall, they generate a time dependent electric field that affects further precipitation.[8] However, this effect is significant only when deposited particles nearly coat the inner surface of the tube. Since our objective is the minimal deposition of charged particles, this effect is not considered for the case of an electrodynamic screen. The total electrostatic depositional force, $F_t$, is then

$$F_t = F_{id} + F_{mr}. \tag{26}$$

The major translational forces that carry particles in and out of the tube are the drag force and the gravitational force. As an approximation, a constant gas velocity U is considered downward in a vertical sampling tube. The drag force on the particle in the air flow at speed U is then

$$F_d = 6\pi\eta a\, U. \tag{27}$$

In the axial direction, the total force is

$$F_{ez} = mg + F_d \ . \tag{28}$$

In the radial direction, the total external force is

$$F_e = F_{cr} - F_{er} \ , \tag{29}$$

Thus, the condition for transport without wall deposition, neglecting any radial component of the fluid flow, is

$$F_{cr} \geq F_{er} \ . \tag{30}$$

Let us consider the following electrode configuration:

$d$ = 0.72 mm
$b$ = 1.28 mm
$D$ = 15.82 mm
$n_o$ = 1,000 particles/cc
$q$ = $3.2 \times 10^{-15}$ C
$U$ = 0.08 m/s
$V$ = 3,000 Volts

The ratio of the confinement forces to the depositional forces in the radial direction is calculated for different frequencies of the applied field. The result is plotted in Figure 2. When the ratio is greater than 1.0, there is no deposition of the particles with charge equal to or greater than the given charge.

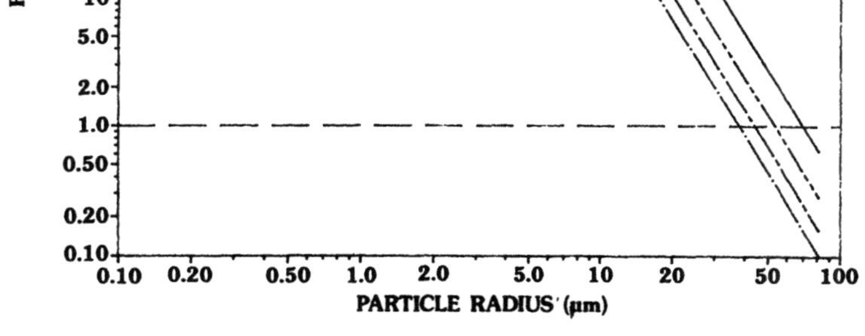


Figure 2. Ratio of confinement to deposition forces plotted as a function of particle size. V=3000, Q=0.32E-14 C, U=0.080 m/s. (1) 200 Hz, (2) 300 Hz, (3) 400 Hz, (4) 500 Hz.

## EXPERIMENTAL STUDIES

Experiments were performed using electrode screens of both plane and cylindrical geometrical configurations. Repulsion of the charged particles from the screen was observed for both types of screens and for an optimal configuration and operational mode.

A flat screen, as shown in Figure 3, was made using a computer communication cable. The cable contained 20 insulated wires in the form of a ribbon. An ac field was applied using a 300 Hz, 2 kV power supply. With the aid of a telescope of magnification 24, movement of particles near the surface of the screen was observed with and without the ac field. The particles were illuminated by laser light.

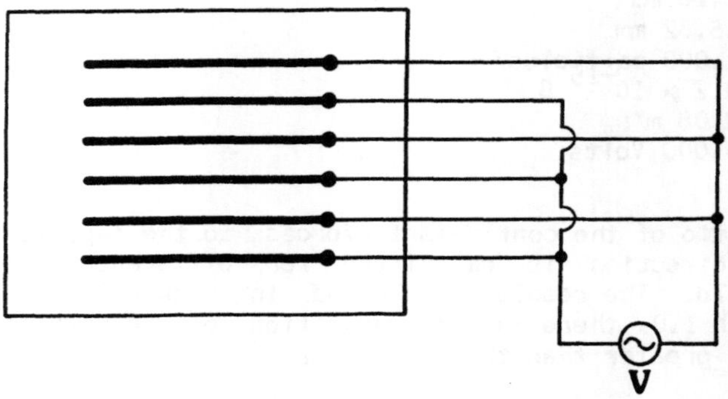


Figure 3. Schematic of a plane-type electrode screen.

Triboelectrically charged toner particles were deposited on the flat screen with the power supply turned off. Charged toner particles were seen to levitate and move along the electrical field flux lines as soon as the field was applied. The levitation of the particles was photographed. Figure 4 is a reproduction of the typical electrodynamic levitation achieved. Figure 5 shows the experimental setup used for measuring performance of the electrodynamic sampling system. A cylindrical ring-type electrode screen was constructed using insulated, equally-spaced rings, to increase the breakdown voltage. Epoxy cement was used as the insulating material. A powder dispenser was used to feed triboelectrically charged toners to the screen electrode. A humidity-controlled chamber was used in the experiment to maintain a relative humidity of 50 percent or less.

Evidence of repulsion of charged particles from the cylindrical screen was observed to be similar to the levitation experienced with the flat screen. When the supply voltage to the cylindrical electrodes increased to 2 kV, a large amount of toners, previously deposited due to electrostatic forces, was dislodged back into the flow field and was collected on the Faraday cage.

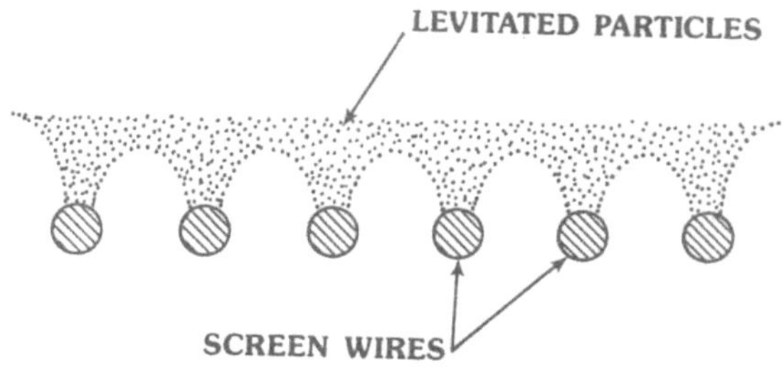


Figure 4. Levitation of toners on plane-type electrode screen.

To measure the charge-to-mass ratio on the toner particles, a Faraday cage with an electrometer was used as shown in Figure 5. Inside the Faraday cage an aluminum cup was connected to the electrometer. The cup was used to hold particles falling through the screen. By measuring the total charge and mass of particles collected in the cup the average charge-to-mass ratio was calculated. In these experiments, ionization was detected when the applied voltage exceeded 2 kV.

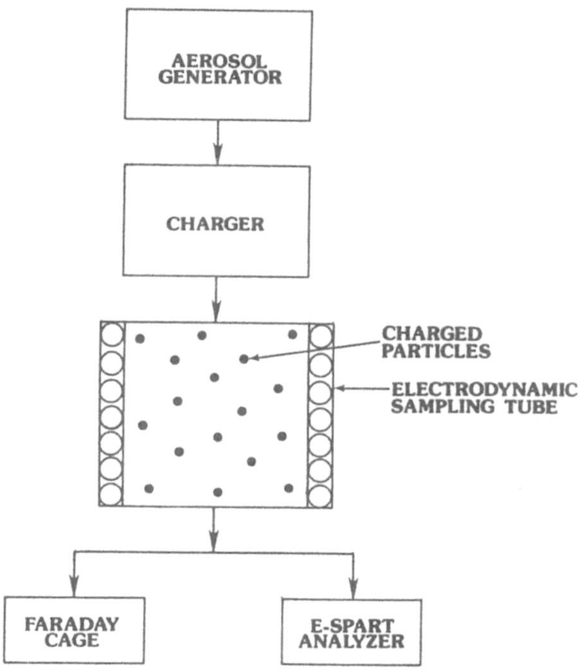


Figure 5. Experimental setup for measuring performance of electrode screen.

Evaluation of the performance of a cylindrical screen was made using an E-SPART analyzer.[2] The sampling tube used in the analyzer was replaced by the electrodynamic screen (Fig. 5). Toners were dispensed by using a magnetic brush device where the negatively charged toners were separated from carriers (iron particles) with an air jet. The mechanism of separating a charged toner particle is similar to the process used in electrophotography.

Results from the E-SPART are shown in Figure 6. The results show that for particles with an aerodynamic diameter larger than 10 μm, the charge/particle was higher when the electrodynamic screen was used.

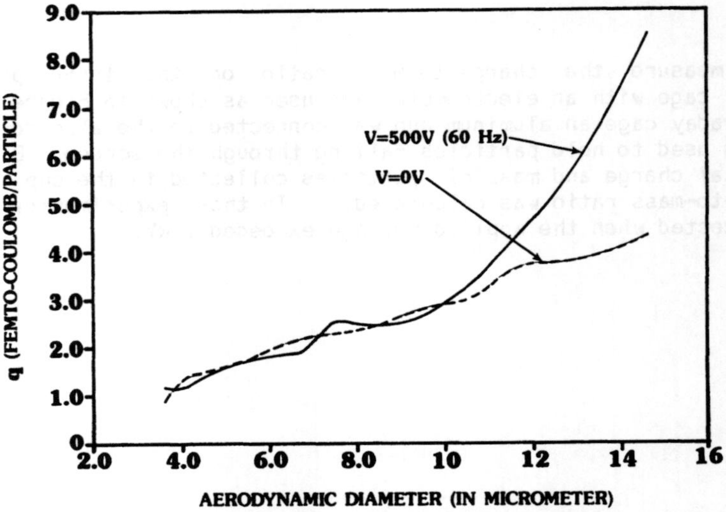


Figure 6. Electrostatic Charge (q/particle) distributions of toner particles as measured by the E-SPART analyzer with and without electric field applied to the electrodynamic sampling tube.

## SUMMARY


A theoretical analysis of the electrodynamic screen and electrostatic effects of charged particles on transport properties was conducted. A primary model has been established for building the screen. Preliminary experiments showed the expected confinement of particles within the screen. Experiments with the screen showed that deposition can be minimized but quantitative measurements of the sampling efficiency require generating and feeding uniformly charged particles to the screen.


## ACKNOWLEDGEMENTS


This work was supported, in part, by a research contract from Hosokawa Micron Corporation, Osaka, Japan. The authors are grateful to R. Cole, K. Tennal, P. Roberson, and D. Watson for review, and to D. Belk and M. Dugan for the preparation of the manuscript.

## REFERENCES


1. D. S. Ensor, A. C. Clayton, T. Yamamoto, and R. P. Donovan, Particle deposition velocity studies in silicon technology, in "Particles in Gases and Liquids I: Detection, Characterization and Control," K. L. Mittal, editor, p. 195, Plenum Press, New York (1989).

2. M. K. Mazumder, R. E. Ware, T. Yokoyama, B. Rubin, and D. Kamp, Measurement of particle size and electrostatic charge distributions on toners using E-SPART analyzers, IEEE ISA Conf. Proceedings 87CH2499-2, p. 1606 (1987).

3. S. Masuda, Electric curtain for confinement and transport of charged aerosol particles, Proceedings of the Albany Conference on Electrostatics (1971).

4. S. Masuda, K. Fujibayashi, K. Ishida, and H. Inaba, Confinement and transportation of charged aerosol clouds via electric curtain, Electrical Eng. in Japan, 92(1) 43 (1972).

5. S. Masuda and Y. Matsumoto, Theoretical characteristics of standing-wave electric curtains, Electrical Eng. in Japan, 93(1) 71 (1973).

6. J. R. Melcher, E. P. Warren, and R. H. Kotwal, Traveling-wave delivery of single component developer, IEEE ISA Conf. Proceedings 87CH2499-2, 1585 (1987).

7. H. Singer, H. Steinbigler, and P. Weiss, A charge simulation method for the calculation of high voltage fields, IEEE PAS 93 1660 (1974).

8. B. J. Savilonis and J. S. Lee, Particle deposition in a charged aerosol flowing through a conducting tube, J. Fluids Eng., 109, 449 (December 1978).

# RESOLUTION, SENSITIVITY, COUNTING EFFICIENCY, AND COINCIDENCE LIMIT OF OPTICAL AEROSOL PARTICLE COUNTERS

Holger T. Sommer

Hiac/Royco, Division of Pacific Scientific
2431 Linden Lane
Silver Spring, MD 20910

Aerosol contamination monitoring in the semiconductor industry is accepted as the key measure to improve microprocessor performance and reduce the number of failing microcircuits due to particulate contamination. This paper addresses the most important performance parameters and specifications of optical aerosol contamination particle counters (OACPC) and makes an effort to uncover their relationships. The need for definitions of these quantities will be documented. It is the intention of the paper to start the discussion of these quantities and to illustrate some of the obvious interconnections.

## INTRODUCTION

The importance of aerosol contamination monitoring in the semiconductor industry is well accepted and essential to achieve high yields in Integrated Circuit (IC) production. The trend in this industry is to smaller dimensions and it is projected that the linewidth of conductors will soon reach 0.5 µm. Since a 0.05 µm contaminating particle can have a damaging effect on these conductors, electronic chip manufacturers will need to be able to measure contaminating particles of this size, and through this measurement control clean room environments. Optical particle counters are routinely used to monitor particulate contamination levels. These sensors have the capability of efficiently and accurately detecting and sizing particles down to 0.05 µm if the sensor is designed properly. There are four parameters that characterize the performance of optical aerosol contamination particle counters (OACPCs): resolution, sensitivity, counting efficiency, and coincidence limit. These quantities will be defined and their interrelationships described.

## OPTICAL AEROSOL CONTAMINATION PARTICLE COUNTERS

The development of optical single particle counters accelerated during the last 30 years contributing to the " electronic revolution". Instruments as described in the literature[1-7] are based on the effect of small particles scattering the light of an incident beam in all directions. An OACPC consists in general of an optical sensor,

*Particles in Gases and Liquids 2,* Edited by K.L. Mittal
Plenum Press, New York, 1990

signal processing and counting electronics, and a sample handling system (pump) which supplies a controlled air stream to the optical sampling volume. Each of these components has an effect on the resolution, sensitivity, and counting efficiency. Fig. 1 illustrates these components. Each of the components, as well as the interaction between the components, affects the performance of the OACPC. These effects originate not only in the overall instrument design, but also in variations within different instruments of the same design.

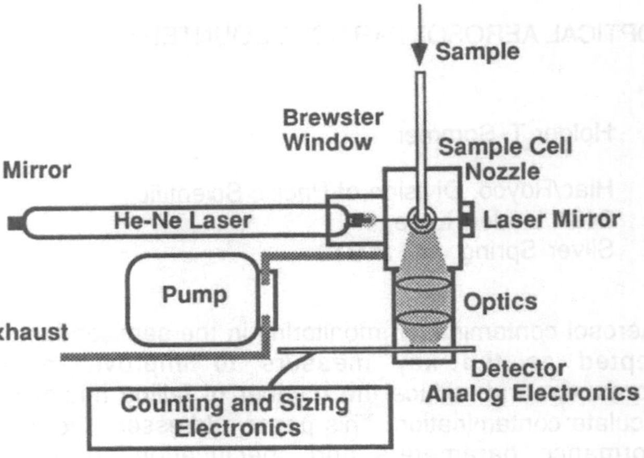


Figure 1 - Major Components of Optical Aerosol Contamination Particle Counter. This is an example of an OACPC that detects 90 ° scattered light.

RESOLUTION

Resolution is the ability of an OACPC to discriminate between two particles of different sizes. The problem of resolution or cross-sensitivity was discussed in earlier works[8-11]. The measure of resolution is the smallest difference in the characteristic length dimension of two particles that will allow the instrument to distinguish them as different sizes. For this to occur, the optical sensor must provide voltage signals different enough to register as two different particle sizes with the counting electronics. If we assume that the electronics can be made ideal and can distinguish between very small pulse heights, then the resolution is governed by the optical part of the OACPC. The variance resulting from the optical subsystem is much larger than the variance from the electronics, and therefore in this section the electronics can be considered "ideal". In this case, physics, the sensor design and the sensor geometry will determine the resolution of the instrument. The geometry of the illuminating laser beam and the focused air jet is critical. The design of most commercial high resolution OACPCs places the sensing volume within the cavity of a He-Ne laser. Fig. 2 shows a typical arrangement where the jet intercepts the laser beam in the external laser cavity between the Brewster window and the external mirror. Fig. 3 shows the intensity distribution of a typical multimode laser beam.

From the geometry of the laser beam and the air jet formed by the nozzle, it can be seen that particles of the same size might generate different scattered light intensities depending on the path that they take through the beam. If particles of the same size cross the beam at locations with different light intensities, they generate a range of signals, and will appear to the counter electronics as different sized particles. If particles pass through the laser beam where the intensity is falling off, the signal for those particles would be reduced from a path through the center of the beam. It would be impossible to distinguish between a particle in the center of

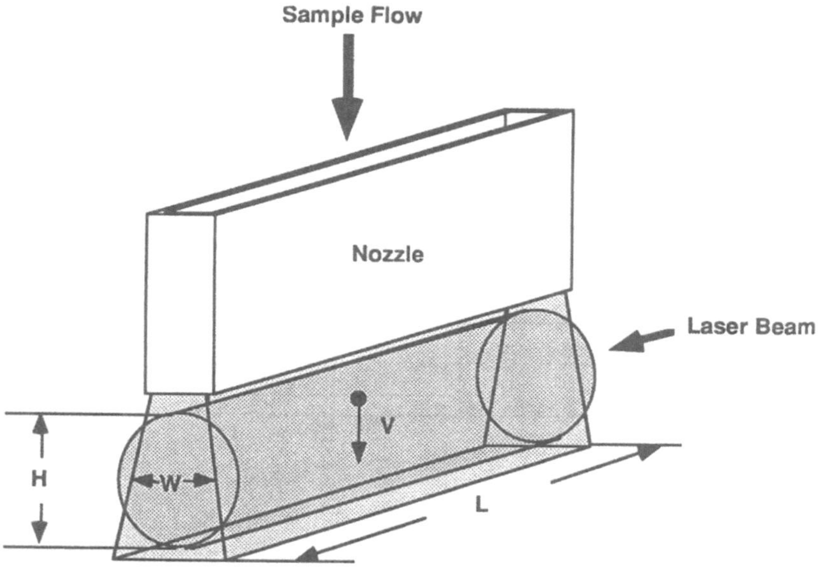


Figure 2 - Aerosol Jet and Laser Beam Geometry. The intersection of the laser beam and the sample determines the sensing volume. It has dimensions L x W x H. L and W will vary slightly over the sensing volume.

the beam and a larger particle at the edge of the beam.This degrades the instrument resolution. Since the exact laser beam profile and air jet geometry determine the magnitude of this effect, the resolution of the instrument is a function of both the instrument design and each individual instrument's characteristics. These effects combine to give each instrument a specific instrumental broadening function. The measurement of monodisperse particles with a very narrow size distribution will result in a measured distribution representing the convolution of the instrument function with the true size distribution. The non-uniformity of the illumination of the sensing volume is the most significant contribution to the broadening of the instrument function. However, an asymmetric flow through the air jet nozzle, e.g. due to an obstruction in the flow path, can also cause broadening. It could result in a distribution of particle paths that is not passing through the central most evenly illuminated region of the laser beam. The resolution can also be degraded by the presence of more than one particle in the sensing volume due to a large particulate concentration of the sample. This is described below under "Coincidence Limit", along with methods of reducing the effect. If the instrument function is a wide distribution, the OACPC reports a particle size distribution which is excessively broadened by the characteristics of the sensor.

## SENSITIVITY

The sensitivity, or sensitivity limit, of an OACPC is defined as the smallest particle size the counter will detect with a "reasonably" small error in number of counts. ("Reasonably" is not defined here. The section on "Counting Efficiency" below discusses this further.) The sensitivity of an instrument is limited by the inability to distinguish between instrument noise and the signal produced by a small particle.The distribution of the noise sets a lower limit on the particle size that can be detected. Therefore, the signal-to-noise ratio at the sensitivity limit characterizes the sensitivity and determines how much noise will be counted as particles and how many particles will not be counted because they are lost in the noise. Fig. 4a and 4b illustrate schematically the voltage distribution at the sensitivity limit of an OACPC.

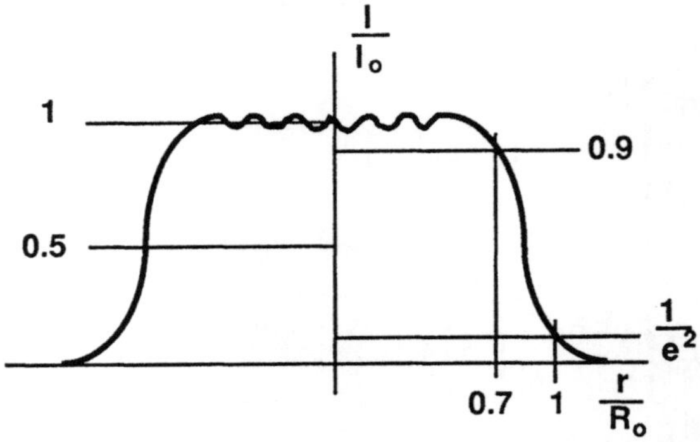


Figure 3 - Normalized Intensity Profile of a Multimode Laser Beam.
A multimode laser has an intensity profile that has a nearly constant intensity over a wider area than a single mode laser.

Fig. 4a shows the pulse height analyzer plot of a sensor with poor resolution. It is impossible to determine whether a pulse is due to a particle or noise in the region where the distributions overlap. If the noise rejection threshold is set in the region where the probabilities overlap, a number of noise pulses will be counted, shifting the peak of the combined distribution of noise and particles to slightly smaller values than the peak of the true particle distribution. The total number of particles counted can be smaller or larger than the true number of particles supplied to the sensor. The "particle counts" at small sizes will depend on the setting of the noise rejection threshold.

Fig. 4b shows the sensitivity limit of an instrument with good resolution. The noise is well separated from the particle size defined as the sensitivity limit. At this point the instrument noise rejection threshold can be set at a level where very few noise counts will be recorded as particle counts. All of the particles are accounted for because all of the particle signals are above the noise rejection threshold.

COUNTING EFFICIENCTY

Counting efficiency, $\eta$ , characterizes the fraction of particles of a particular size that a sensor is able to detect. It is calculated as:

$$\eta = \frac{\text{Measured Number of Particles}}{\text{Actual Number of Particles}}$$

To determine the counting efficiency, the true particle concentration of a sample must be known, or a sample of known concentration must be produced.

With a sample of a known concentration of particles, the number of particle counts measured by the OACPC can be compared with the true number of particles passing through the sensor. Fig. 5 illustrates the dependency of the counting efficiency on the size of the particles. It is apparent that at the lowest sizes, near the sensitivity limit of the sensor, the counting efficiency decreases with particle size. An ideal instrument would have 100% counting efficiency down to the noise rejection threshold and 0% counting efficiency below the threshold. The slopes in Fig. 5 are due to the convolution of the instrumental function described in "Resolution"with the ideal curve. A sensor characterized with "poor" resolution (Fig. 4a) will show a shallow drop of counting efficiency as the particle size approaches the sensitivity limit.

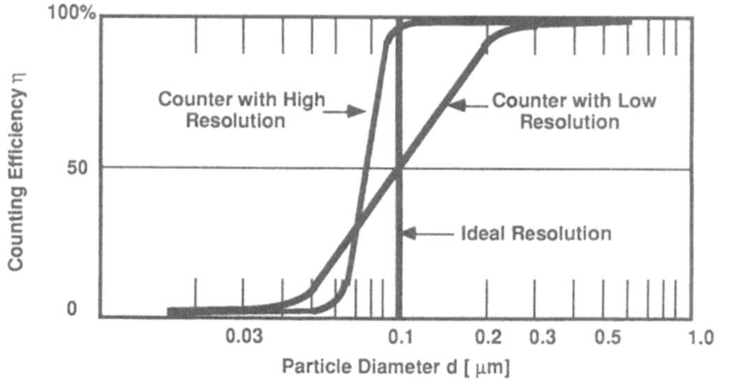


Fig. 4 Sensitivity Limit of OACPC

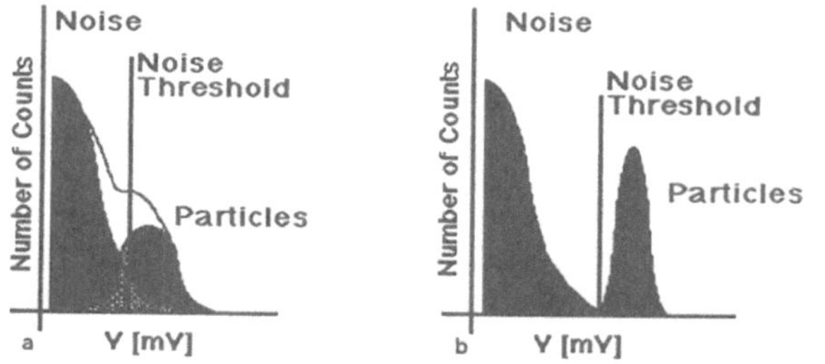


Figure 5 - Counting Efficiency as Function of Particle Size. Sensors with good and poor resolution. By using the different definitions these two instruments could both be defined with 0.1 μm sensitivity.

As the semiconductor industry demands more accurate counting of small particles, a counting efficiency curve with a steep drop of counting efficiency at the sensitivity limit of the instrument is desired. This is also characteristic of high resolution. Instruments with poor resolution show low efficiency in the smallest size ranges. If the sensitivity limit of the instrument is specified as the smallest particles at which near 100% counting efficiency is maintained, the true particle counts will be measured down to the specified detection limit.

## COINCIDENCE LIMIT

The coincidence limit is defined as the maximum error that will occur due to more than one particle being present in the sensing volume at the same time. Coincidence of particles in the measuring volume has been the subject of many investigations[12-18]. A coincidence limit of 10% indicates that in a sample with a concentration of particles less than the specified maximum, there is at most a 10% probability of having 2 particles simultaneously in the sensing volume. The coincidence limit is defined by two related parameters: the maximum probability of having more than one particle in the sensing volume concurrently and the concentration of particles at which this probability will occur. The significant aspect of these parameters to the user of an OACPC is that they define the maximum concentration of particles that the sample can contain without requiring dilution before measurement.

Most instruments are specified at 10% probability, and can therefore be compared directly. However, as the probability increases, the allowable concentration increases; and as the probability decreases, the concentration decreases (although coincidence limits less than 10% are not likely to be encountered).

As an example, for a 10% coincidence limit , the maximum concentration is 0.105 particles per sensing volume. A typical sensing volume, which is defined by the laser beam and the air jet cross-section, is

$$V = 1 \text{ mm} \times 0.8 \text{ mm} \times 11 \text{ mm} = 8.8 \text{ mm}^3$$

The maximum concentration is therefore:

$$0.105 \text{ particles} / 8.8 \text{ mm}^3 = 338{,}700 \text{ Particles / cu.ft.}$$

The maximum concentration for the coincidence limit is inversely related to the size of the sensing volume. An OACPC with a small sensing volume can accept a higher concentration of particles than a sensor with a larger volume. However, there are limitations on the dimensions that are set by the laser beam profile, the electronics processing time, the required sample flow rate, and the requirement of laminar flow through the sensing volume. The width is defined by the air jet (or nozzle) and the height of the volume by the laser beam. The height and width of the volume can also be defined optically by a mask in the sensor.

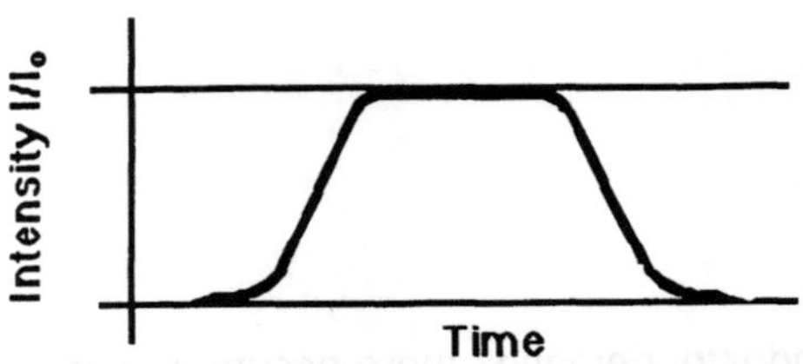


Fig. 6 Signal Pulse : Intensity as Function of Time

The height of the volume and the flow rate defines the residence time of a particle in the beam and thus the signal pulse width, as illustrated in Fig. 6. Decreasing the sensing volume by decreasing the height requires a smaller flow rate of air through the sensing volume since the electronics processing time is finite. Currently the semiconductor industry requires a flow rate of 1 CFM. The width of the volume is restricted by the profile of the laser beam. To maximize resolution, the width is limited to the central region of constant intensity in the laser beam. After the width is defined, the length of the nozzle must be adjusted such that at the given flow rate of 1 CFM the air flow is laminar through the laser beam. These considerations all affect the design of the sensing volume, which will then define the maximum concentration of particles that the sensor can process.

## SUMMARY

The user of an OACPC should be aware of the above definitions and interrelations. Specifications provided by instrument manufacturers should be carefully evaluated and instruments should be selected to meet application needs. The interrelationship between the above defined quantities is complex. However, in order to get true technical performance information for an instrument, the quantities Resolution, Sensitivity, Counting Efficiency and Coincidence Limit should be clearly stated and disclosed since these specifications define the performance of the instrument and the boundary of the measurable size range.

## ACKNOWLEDGEMENT

The author would like to thank Charles Harrison for stimulating discussions

## REFERENCES

1. F.T. Gucker, Proc. Ind. Instr. Conf., Lafayette, IN., pp. 284-312,(1956)
2. F.T. Gucker and C.T. O'Konski, *J. Colloid Sci.* 4,541,(1949)
3. F.T.Gucker, C.T. O'Konski,H.B. Pickard, and J.H.Pitts, *J. Am.Chem.Soc.* 69,2422, (1947)
4. F.T.Gucker and D.G. Rose, *J. Appl. Phys.Suppl.*3,5138, (1954)
5. C.T. O'Konski, M.D. Biton and W.I. Higuchi, *ASTM Spec.Tech.Publ.* 250,(1958)
6. C.T. O'Konski, M.D. Biton and W.I. Higuchi, *ASTM Spec.Tech.Publ.* 234,(1958)
7. M.A Fisher, S. Katz, A.Lieberman and N.E. Alexander, *Proc.Third.Nat. Air Pollution Symp.*, Pasadena, CA, April 18-22, (1955)
8. T.E. Greadle, *Aerosol Sci.*, 5,125-131,(1974)
9. K.T. Whitby and B.Y.H. Lui, *J. Colloid Sci.* 25,537-546,(1967)
10. O. Preining, *Staub-Reinhalt.Luft* 28,22-24,(1968)
11. K.T. Whitby, *Environ.Sci.Technol.* 1,801-814,(1967)
12. R. Jaenicke, *Aerosol Sci.* 30,.95-111,(1972)
13. W. Bennert and G. Hilbig, *Staub* 27,187, (1967)
14. Y.S. Sedunov, *Izv.Akad.Nauk. SSSR*, 8, 810-817,(1969)
15. R.Gold, *Rev.Sci.Instr.* 36,No.6, (1965)
16. B. Soucek, *Rev.Sci.Instr.* 36,No.11,(1965)
17. J.H. Williamson, *Rev.Sci.Instr.* 37,No.11, (1966)
18. A.V.Pazman, *Pribori i tekhnika experimenta*, 3,112-114,(1969)

# PART IV. PARTICLES IN LIQUIDS: DETECTION CHARACTERIZATION AND CONTROL

# CHARACTERIZATION OF THE PARTICLE LOADING IN DEIONIZED WATER SYSTEMS BY AUTOMATED SEM ANALYSIS

T.B. Vander Wood and L.D. Detter

McCrone Associates-Atlanta
1412 Oakbrook Drive, Suite 100
Norcross, Georgia 30093

Characterization of the particles found in ultrapure water systems should be a standard complement to routine monitoring of particle concentration. Scanning electron microscopy combined with energy dispersive x-ray spectrometry is the technique of choice for particle characterization, but can be very time consuming. Automation of the system for particle detection and elemental analysis largely overcomes this disadvantage, and allows for detection, sizing and characterization of hundreds to thousands of particles isolated from a pure water system (e.g., by a point-of-use filter), all without human intervention after initial setup. Automated systems commonly in use provide accurate sizing and compositional information, but are limited to particles larger than approximately 0.5 μm, and rely on the operator to anticipate possible particle types (e.g., stainless steel, silica). Improvements anticipated for the near future include a smaller minimum size, ultimately approximately 0.05 μm by use of transmission electron microscopy, and _a posteriori_ particle classification by statistical techniques.

## INTRODUCTION

Ultrapure water systems are in use in many applications, ranging from silicon wafer rinse systems to contact lens packaging. Great attention has been paid to the design of and materials used in these systems;[1,2] however, particulate contaminants can be found in the best of them. The standard solution to this problem is to employ point-of-use filtration systems to remove particles just at the dispense nozzle, before they can have an effect on the product.[3] This system is usually effective, but can be overwhelmed and/or clogged by high particle loading. Point-of-use filters may also serve to disguise particulate contamination problems upstream and may serve as a medium for bacterial growth, further compounding the problem.

In view of these difficulties, and the fact that particles do penetrate even the most carefully designed and implemented filtration systems, monitoring of the particle loading of pure water systems is called for. Some controversy exists over the most effective technique. Csikai and Barnard[4] have pointed out the growing acceptance of laser-based light scattering techniques for particle counting in water while discussing some of the problems in interpretation of the data produced. Dillenbeck[5] contends that light scattering techniques have "proved to be most successful for counting particles". On the other hand, Yabe, et al.,[6] indicate that the "most reliable method of analyzing particles in ultrapure water" involves scanning electron microscope (SEM) examination of filtered samples, but they note the significant time and effort required. Balazs[7] has compared the performance of SEM and in-line particle counting methods for ultrapure water and found that the SEM shows superior detectability for particles smaller than one micrometer diameter. It appears that these differing opinions can be resolved and the strengths of both methods captured by a combination of techniques. The ease of use and real-time capabilities of light-scattering techniques are most applicable for day-to-day system monitoring. This should be supplemented by a periodic check using more accurate and informative SEM techniques.

This disagreement over the most useful technique for monitoring the particulate loading of pure water systems has largely ignored one capability of most modern SEM's that greatly enhances their utility--the ability to determine the elemental composition of a particle via energy dispersive x-ray spectrometry (EDS), a technique that has been applied to the analysis of particulate contaminants in a wide variety of products and systems.[8] In an EDS analysis, x-ray photons generated by elements in the sample under the influence of high energy electrons from the instrument are collected and sorted on the basis of energy. For each photon detected, a count is placed in the appropriate bin of a multichannel analyzer, and counts are accumulated until a discernible spectrum (Figure 1) is produced. The spectrum consists of a broad background on which are superimposed peaks at energies corresponding to the characteristic x-ray energies of the elements present in the sample. Most EDS systems are sensitive only to elements heavier than neon (that

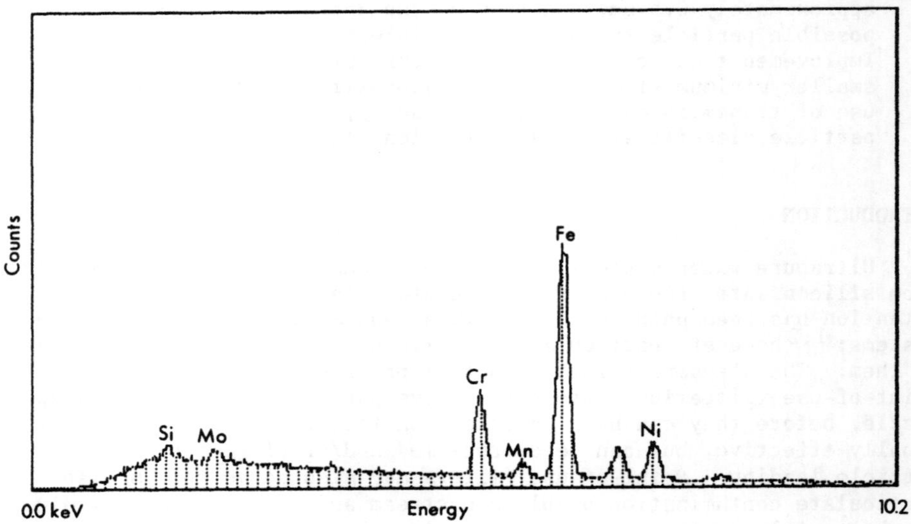


Figure 1. EDS spectrum in the range 0-10 kV. Peaks correspond to the characteristic x-rays detected from elements in the particle. Vertical full scale is 2048 counts. This spectrum is typical of 316 stainless steel.

is, sodium through uranium), but newer type detectors, sensitive to all elements from carbon up through the periodic table, are available and are becoming increasingly reliable. Under ideal circumstances, the detection limit for most elements is approximately 0.25 percent by weight.[9]

This capability for elemental analysis greatly enhances the utility of the SEM in monitoring ultrapure water systems for particulate contaminants: not only can the SEM provide superior reliability in the detection of particles but also offers the ability to determine what the particulate contaminants are. This material identification can and should play a critical role in the determination of particle origin in efforts to improve the cleanliness of water delivery systems and in trouble shooting system breakdowns.

**AUTOMATED SEM ANALYSIS**

Not unexpectedly, this valuable information, like the greater reliability in sizing and detection, comes at a price in time and effort. To manually detect, size, and analyze the hundreds to thousands of particles which make up a typical sample would require tens to hundreds of hours of both SEM and operator time. The expense associated with the effort involved just in counting has been the main drawback to use of the SEM as a system monitoring tool, and the time associated with elemental analysis only exacerbates the problem. Fortunately, there is a solution: automation of the SEM, which allows the detection, sizing, and elemental analysis of as many as a thousand particles in as little as eight hours, all without operator intervention after initial setup. This method has proven its applicability in the monitoring and identification of cleanroom contaminants[10] and high purity gases,[11] and offers the same advantages of high information content and efficiency in application to ultrapure water systems.

## Sample Preparation

Successful automated SEM analysis depends critically on the sample preparation. The substrate containing the sample must be smooth and as free of defects as possible, since the automated SEM procedure will identify defects as particles, thus skewing the results. Experience has shown that polished carbon planchets and small pore size (<0.5 μm) membrane filters made of polycarbonate are acceptable substrate materials if properly prepared and handled. Point-of-use filters are commonly polycarbonate or teflon membrane filters, and these filters can be analyzed directly by the automated SEM, with a minimum of further

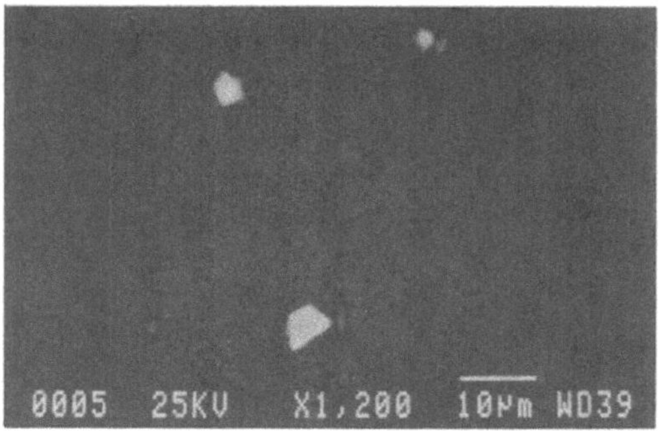


Figure 2. Backscattered electron image of particles dispersed on a substrate. Scale bar is 10 μm.

preparation, so long as other criteria of an acceptable sample are met. Particle loading of the substrate is a crucial aspect of the sample. The distribution of particles on the filter must be uniform, in order that valid generalizations concerning the particle population can be made from only a portion of the filter. A balance must be struck between a high density loading, which allows most efficient use of the instrument, and a low density loading, which helps minimize agglomeration and overlapping of particles, which confuse the analytical results. In practice, we have found that a particle loading of approximately five percent by area is acceptable (Figure 2).

Mounting of the filter for analysis involves selection and cutting of a representative portion (for 2.5 cm diameter filters, the whole filter is often mounted), fastening of the filter to a support, for example, a polished carbon or aluminum planchet, and coating with a thin layer of carbon in order to minimize sample charging under the influence of the electron beam. Singular care must be taken throughout this operation to avoid the introduction of extraneous particles to the sample. To ensure that the sample represents the particle loading of the ultrapure water supply, sample preparation should be done in a Class 100 environment, and a process blank should go through the procedure in parallel to any sample.

## Instrumentation

In order to carry out automated SEM analyses, the microscope must be interfaced to a computer capable of controlling the scan of the electron beam, automatically changing the sample position (i.e., moving the sample stage), acquiring digitized electron images of the sample for processing, and acquiring and storing EDS information. In practice, we find that for a variety of reasons, secondary electron imaging of the sample is not suitable in this application, and backscattered electron imaging capability is required. The computer capabilities and software required are generally associated with the EDS system, rather than the SEM itself. While some SEM/EDS systems are available as an integrated package from a single manufacturer, usually they are from two separate suppliers and the degree of interaction of the SEM and EDS systems required for automation necessitates close coordination between them. Fortunately, a variety of sufficiently integrated systems are available, along with commercial software packages capable of automated analysis.

The instrumentation now in use in our laboratory consists of JEOL electron microprobes (both JXA 733 and JXA 8600) interfaced with Tracor Northern 5500 EDS systems and 5600 stage automation. We have also successfully interfaced Tracor Northern systems to JEOL SEM's (800 series) and an A.R.L. SEM-Q electron microprobe. We are currently using a slightly modified version of Tracor Northern's program "Particle Recognition and Characterization" (PRC) as automation software, and are preparing more advanced software which should significantly improve the chemical characterization capabilities of the system by using off-line data reduction on a mini-computer. A rough schematic of the instrumentation necessary is given in Figure 3.

## Particle Location and Sizing

Particle location for subsequent sizing and analysis is based on recognition of intensity difference between the backscattered electron signal originating from the particles on the filter and the filter itself. These intensity differences arise from both topographic effects (i.e., particle shape) and from differences between the chemical composition of the particle and the substrate,[12] and cause the particle

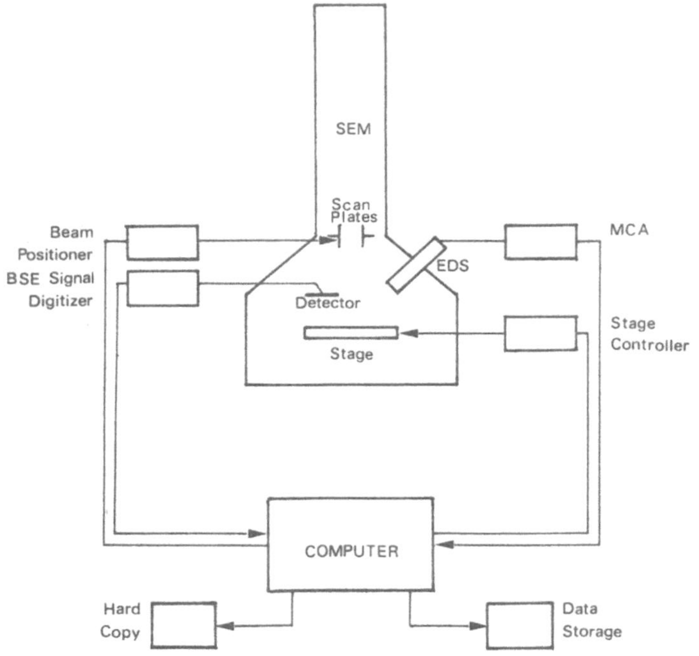


Figure 3. Schematic diagram of the SEM and interfaces necessary for automation.

to appear "bright" against the dark background of the substrate. Particle location is accomplished by computer controlled stepping of the electron beam over the field-of-view while simultaneously digitizing and monitoring the intensity of the backscattered electron signal. When this intensity falls between preset upper and lower threshold values, a particle is detected and the position recorded. The next step in particle sizing is determination of the particle centroid, found from a series of perpendicular diameters scanned through the particle. Once the centroid is located, a series of eight rotated diameters through the centroid is drawn (Figure 4), the perimeter and area are estimated, and this information along with

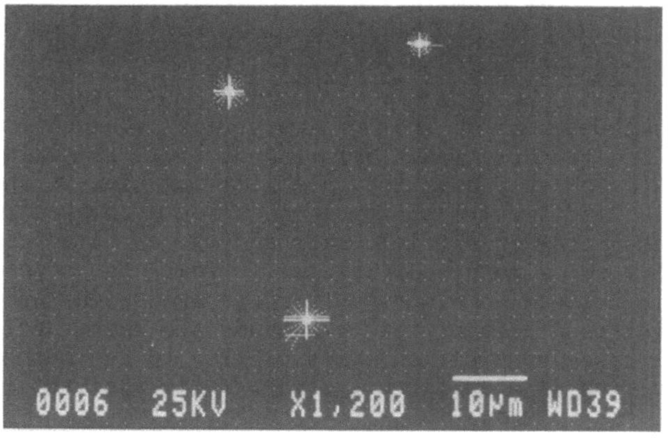


Figure 4. Digital beam search and sizing of the same field of view as shown in Figure 2.

average diameter, maximum diameter, minimum diameter, and centroid position (relative to the field-of-view and stage position) are stored. If no chemical analysis is requested, the computer then continues stepping the electron beam until the next particle is found or the entire field has been searched. When the search of each field is completed, the computer moves the stage to the next field and the process is repeated until a preset number of particles have been found or a preset area of the filter has been searched.

The final results of the sizing of a sample of particles can be presented as histograms for average diameter, minimum diameter, area or shape factor (which is related to aspect ratio, i.e., maximum diameter divided by minimum diameter). Figure 5 shows the histogram of average diameters of latex spheres whose diameters and distribution are 3.009μm ± 0.027 μm, traceable to NBS standards (Duke Scientific, cat. #4203). The average diameter determined by automated sizing is 3.11 ± 0.11 μm.

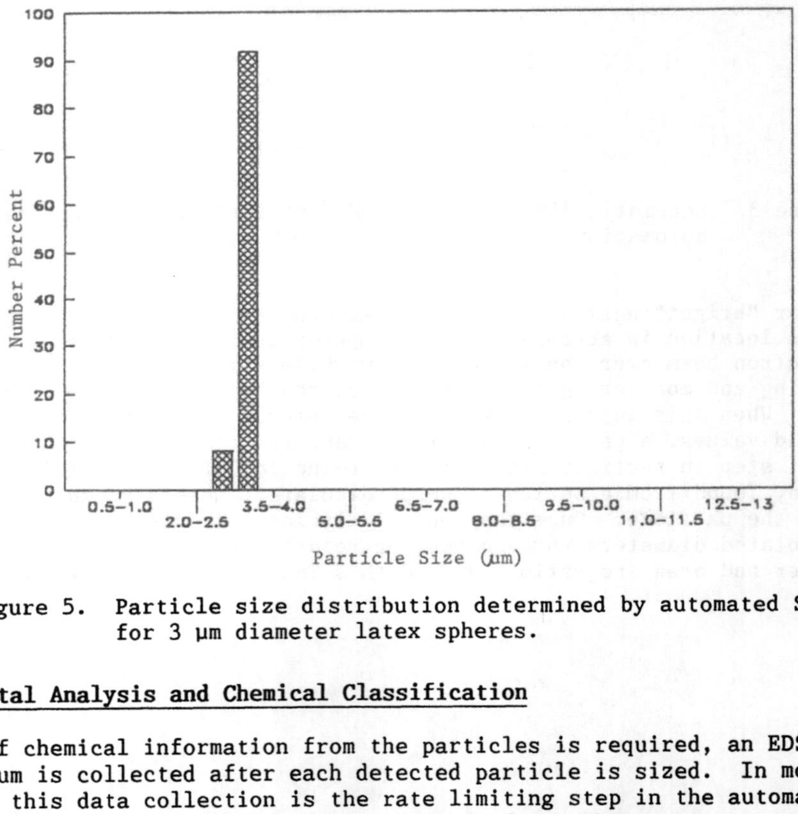


Figure 5. Particle size distribution determined by automated SEM for 3 μm diameter latex spheres.

## Elemental Analysis and Chemical Classification

If chemical information from the particles is required, an EDS spectrum is collected after each detected particle is sized. In most cases, this data collection is the rate limiting step in the automated procedure, requiring five to thirty seconds for each spectrum acquisition, depending on particle size and the expected concentration in the particles for elements of interest. Thus, for a thousand particles, data collection alone may require over eight hours if spectrum acquisition times of thirty seconds are specified. However, in the more usual case, acquisition times of five to ten seconds are adequate, resulting in significantly lower total analysis time.

In the current generation of automated analysis routines, it is not possible to store the entire EDS spectrum collected for each particle. Rather, the number of x-ray photons falling in a selection of "regions-of-interest" (ROIs) is calculated from each spectrum and stored along with the particle location and size data. ROIs are defined by the operator to encompass the EDS peak for the elements of interest, and are set-up for each element anticipated to be of importance in the sample. Chemical typing of particles is then based on the number of x-ray photons detected in the established regions-of-interest.

In the program we use, two broad types of chemical categories may be established. The first is "element-rich", designed to include particles which are dominantly composed of a single element (or more accurately, particles in which a single element for which EDS is sensitive predominates). For example, to characterize a sample known to be composed of alumina and silica, two particle types would be setup: Si90 and Al90. With this setup, and ROIs defined for silicon and aluminum, all particles with 90% or more of the recorded photons falling in the silicon ROI would be classed as Si90; that is, silica. All those with 90% or more in the aluminum ROI would be classed as Al90, or alumina. Those with intermediate proportions (e.g., 50:50, such as an agglomerated alumina/silica pair) would be classified as "unknown". The second broad type of chemical categories is designed to sort particles composed of mixtures of elements. For each particle type desired for this category, the elements and their associated proportions of ROI counts are assigned. Other elements for which ROIs have been defined but ranges not assigned may be listed as present, absent, or possible. Thus a stainless steel may be defined as Fe-50, Cr-25, Ni-present. To distinguish between 304 and 316 stainless steels, two types would be required: 1) Fe-50, Cr-25, Ni-present, Mo-absent, and 2) Fe-50, Cr-25, Ni-present, Mo-present.[13]

Utilizing preassigned classifications, a wide variety of chemical types can be identified, their proportion in the sample determined, and size and shape statistics for each type generated. With this information in hand, the particle loading of an ultrapure water system is characterized. Probable particle sources can be identified based on chemical types, and each source's contribution to the loading can be determined from the relative abundance of the type and the threat to the product assessed from this information combined with the size and shape measurements.

## Limitations

The automated sizing and analysis methods in use have broad applicability at present, but will be severely challenged by the need to characterize smaller and more diverse materials in the future. In the detection and sizing of particles, we are currently limited to particles with average diameters greater than or equal to 0.5 μm, so that smaller particles pass undetected. Particle sizing is also limited to only two dimensions: what is actually measured is the particle's projection onto a plane perpendicular to the electron beam. While knowledge of two dimensions is a great improvement over single "diameters" derived from light scattering or fluid dynamic measurements commonly in use, some measure of the third dimension would provide more complete characterization of the particle's true shape. The current scheme also has difficulty with particles with very high aspect ratios (such as fibers) or complex particles with large re-entrant angles. Size and shape measurements of these kinds of particles are easily handled by modern image processing software, but these programs run off-line on a digitized image of a whole field and have not been integrated in our laboratory into an on-line package capable of detection and sizing as well as directing a chemical analysis of each particle detected.

The elemental analysis of detected particles is limited by the limitations inherent in EDS: for most systems, only elements heavier than neon can be detected, and then only at levels above a few parts per thousand. Another difficulty in EDS measurement stems from the width of the spectral peaks, which can often result in the unresolvable overlap of peaks due to x-rays arising from different elements. Thus it may be difficult or impossible to discriminate sulfur in the presence of lead, or calcium in the presence of barium.

The storage of only ROI data precludes the application of software for quantitative analysis of the particle elemental data, so that all that is stored is raw x-ray counts. Therefore, chemical typing is done not on the basis of element concentration in the particle but rather on proportions of raw counts recorded. This is not an insurmountable obstacle but does dictate that considerable care be taken in the definitions of chemical types. We should also note that no ROI information is recorded for elements not in a predefined list, so that elements that are present but unanticipated will not be detected as such.

For particulate samples where the possible sources are known, (e.g., ultrapure water systems), the limitations of predefined element lists and chemical types are not serious; these restrictions can however be severe for completely unknown samples. In these cases, elements not usually considered can provide key clues to origin, and the development of type classifications requires a priori assumptions about the materials anticipated.

**APPLICATION**

In a recent application, automated SEM analysis was applied to a sample of purportedly ultrapure water used in packaging a prosthetic device intended for use in humans. The laboratory received a final filter that was in use on the water line in question. The filter was clogged, requiring resuspension of the particulate matter and filtration through 0.2 μm pore size polycarbonate membrane, followed by carbon coating of the membrane for automated analysis. Prior to the automated analysis, a brief manual survey of the collected particulate matter was carried out in order to determine which elements were present so that ROIs could be set up. The elements chosen based on this brief survey are shown in Table I. This survey also indicated the particle types anticipated, and the types chosen (supplemented by a post-analysis survey of data taken from individual particles by the automated analysis) are shown in Table II, along with a more descriptive name indicating the probable material identification of the particle type.

Other parameters requiring setup prior to analysis involve instrument control. SEM parameters used were a 20 kV, 10 nA electron beam with a raster equivalent to a magnification of 540X. The minimum particle size used as a criterion for particle recognition was 0.5 μm and a maximum particle diameter of 100 μm was set. EDS spectra acquisition times were thirty seconds. Particles were recognized when their brightness exceeded a level of approximately 30 on a gray scale of 256 levels. Backscattered electrons were used for imaging, and brightness and contrast of the detector/amplifier were set so that essentially all particles had gray levels of $\geq$30, and the substrate had a brightness of approximately 20. Analysis of 1000 particles was requested (due to an anomaly in the software, 1001 particles were analyzed), and a matrix of fields-of-view of sufficient size was established for particle searching. After this initial setup, the

**Table I. Element and Region of Interest (ROI) List for Ultrapure Water Particulate Analysis.**

| ROI | Element | Energy (keV) Low | High |
|---|---|---|---|
| 1 | Al | 1.400 | 1.580 |
| 2 | Si | 1.640 | 1.880 |
| 3 | Ti | 4.400 | 4.660 |
| 4 | Ca | 3.580 | 3.840 |
| 5 | Fe | 6.320 | 6.540 |
| 6 | Mg | 1.180 | 1.360 |
| 7 | K | 3.200 | 3.440 |
| 8 | P | 1.920 | 2.120 |
| 9 | Cl | 2.540 | 2.760 |
| 10 | Cr | 5.300 | 5.560 |
| 11 | Ni | 7.360 | 7.620 |
| 12 | S | 2.200 | 2.440 |
| 13 | Cu | 7.940 | 8.180 |
| 14 | Zn | 8.480 | 8.780 |
| 15 | Na | 0.960 | 1.160 |
| 16 | Pb | 10.360 | 10.740 |

**Table II. Particle Type List for Ultrapure Water Analysis.**

| Type | Definition | Description |
|---|---|---|
| 0 | --- | all particles detected |
| 1 | Si≥85 | quartz, silica |
| 2 | Fe≥70 | steel, iron oxide |
| 3 | Al≥85 | aluminum, aluminum oxide |
| 4 | Ti≥55 | white pigment ($TiO_2$) |
| 5 | Fe=50; Cr=25 | stainless steel |
| 6 | Al=40; Si=40; Fe=20 | soil derived dust |
| 7 | Al=20; Si=20; Fe=60 | " " " |
| 8 | Al=45; Si=45 | " " " |
| 9 | Al=25; Si=25; Fe=25 | " " " |
| 10 | Si=70 | " " " |
| 11 | Fe=50; Si=50 | " " " |
| 12 | Non-integrable | organic, polymers |
| 13 | Unknown | all other types |

**Note: All elements not specifically listed for a type are considered "possible".**

instrument ran in automated mode, searching each field-of-view in turn, locating, sizing, and analyzing particles until the desired number of particles were recorded.

Table III summarizes the results of the analyses of 1001 particles from the ultrapure water system. Included are the total number of particles, percentage of the total number, percentage of the area searched occupied by each particle type, average diameters, average maximum diameter, and average aspect ratio ($d_{max}/d_{min}$). From these data it can be seen that the average particle size for all types (based on number) is 1.5 µm, with an average aspect ratio of approximately 2. The single most abundant identified particle type is type 9, a mixture of

Table III. Data Summary for Ultrapure Water Analysis.

| Type | Number of Particles | Number Percent | Area Percent | $d \pm 1\sigma$ (µm) | $d_{max}$ (µm) | $d_{max}/d_{min}$ |
|---|---|---|---|---|---|---|
| 0 | 1001 | 100 | 0.30 | 1.52±1.48 | 2.26 | 2.10 |
| 1 | 51 | 5.1 | 0.017 | 1.85±1.42 | 2.59 | 1.92 |
| 2 | 71 | 7.1 | 0.008 | 1.25±0.60 | 1.54 | 1.47 |
| 3 | 2 | 0.2 | 0.000 | 2.46±0.70 | 2.82 | 1.29 |
| 4 | 14 | 1.4 | 0.001 | 1.12±0.34 | 1.34 | 1.40 |
| 5 | 92 | 9.2 | 0.017 | 1.37±0.86 | 1.88 | 1.74 |
| 6 | 55 | 5.5 | 0.046 | 2.37±2.71 | 3.82 | 2.51 |
| 7 | 86 | 8.6 | 0.008 | 1.09±0.57 | 1.54 | 1.81 |
| 8 | 93 | 9.3 | 0.043 | 2.09±1.68 | 2.91 | 1.91 |
| 9 | 113 | 11.3 | 0.062 | 1.87±2.10 | 3.07 | 2.56 |
| 10 | 42 | 4.2 | 0.018 | 1.85±1.78 | 3.10 | 2.61 |
| 11 | 81 | 8.1 | 0.039 | 1.74±2.17 | 2.62 | 2.13 |
| 12 | 140 | 14.0 | 0.018 | 1.18±0.73 | 1.86 | 2.24 |
| 13 | 161 | 16.1 | 0.019 | 1.13±0.69 | 1.60 | 1.90 |

aluminum, silicon, and iron in proportions indicative of soil dust. All of the particle types that indicate soil-derived material (6-11) make up a combined total of 47% of the particles detected, and reflect a combination of aluminosilicates and iron oxide typical of southeastern soils. Type 13 (unknown) was the single most common particle type reported. Classification as "unknown" indicates that these particles did not fall into any of the predefined types. However, ROI data from these particles were stored, and average proportions of counts in each ROI were computed for the type and are reproduced in Table IV. In this table a comparison is made between the average composition of type 9 (Al=25; Si=25; Fe=25) and "unknowns". The average composition and general similarities between type 9 and "unknown" suggest that both classes represent mixtures of iron oxides, aluminosilicates, and other trace components with "unknown" having a relatively higher proportion of iron oxide. If this conclusion is valid, the type "unknown" can be included with other soil derived components. This leads to the conclusion that over 60% of the particulate matter found in this water system originates from outside the system, rather than from system components. The most likely source is the feedwater loading, which implies a breakdown of the particulate filtering system. Other possibilities include introduction of contamination during filter maintenance and breakdown of the air filtration system on the pure water storage tank.

Aside from the soil derived component, there is evidence that system components may be contributing significantly to the particle loading of the system. The second most abundant category of particles reported (14%) is type "12", "non-integrable". This category contains all particles which were found but from which no elements were detected, that is no elements heavier than neon were present above the detection limit (defined as twice the standard deviation of the background, determined by counting statistics in the data channels immediately adjacent to the ROI in the EDS spectrum). Inspection of the elemental composition recorded for these particles confirms that no elements were detected. Therefore, these particles must be composed principally of hydrogen, helium, lithium, beryllium, boron, carbon, nitrogen, oxygen, fluorine, neon, and/or scandium, (scandium is not included in the table of elements monitored and is the only element for which our EDS is sensitive, that is both not included in the element list and which does

Table IV. Average Proportion of X-Ray Counts in Each ROI for Particle Type "Unknown" and Particle Type "Al=25; Si=25; Fe=25".

| Element | Percent Counts in Type "Unknown" | Percent Counts in Type "Al=25; Si=25; Fe=25" |
|---|---|---|
| Al | 13 | 22 |
| Si | 17 | 25 |
| Ti | 4 | 4 |
| Ca | 2 | 1 |
| Fe | 35 | 26 |
| Mg | 2 | 2 |
| K | 3 | 2 |
| P | 3 | 2 |
| Cl | 3 | 3 |
| Cr | 6 | 6 |
| Ni | 1 | 1 |
| S | 4 | 4 |
| Cu | 1 | 1 |
| Zn | 1 | 1 |
| Na | 1 | 1 |
| Pb | 4 | 0 |

not display significant overlap into a defined ROI). Clearly the most abundant materials composed solely of these elements in most environments are organic, and in pure water systems these are most likely bacteria or polymeric components of the delivery system.

A third major category of particles detected is stainless steel (type 5), and makes up 9.2% of the sample by number. Non-stainless steels or their corrosion products (type 2) make up 5% of the sample. The most probable origins of steels in the system are fittings, valves and positive displacement pump blades, which wear and shed during operation. Quartz (or silica), possibly from silica glass system components (e.g., silica glass filter fibers) and $TiO_2$ (a white pigment) also contribute to the particle loading in the system.

Knowledge of the chemistry of the particulate loading of an ultrapure water systems is of use not only in determining its source, but also in assessing the threat to the product or process in which it is used. Another important aspect of this assessment is particle size. The data summary shown in Table III gives the mean diameter of the particles in each type as well as the variation in size. This information indicates that the particles are small (generally less than 2 μm average diameter), but the variation in size is large. A more complete characterization of the particulate size distribution can be obtained in the form of a histogram of particle size. Figure 6 shows the particle size distribution for all particles measured in the sample, with bin sizes of 0.5 μm. The distribution shows that 43.6% of the particles fall in the smallest bin (0.5 - 1.0 μm), and 81.7% are smaller than 2 μm diameter. However, particles as large as 13 μm diameter were present. Inspection of the histograms for individual particle types shows that the largest particles are in the types derived from soil (all particles ≥10 μm diameter fall into these types; 35 of 36 particles ≥5 μm diameter are soil derived). Therefore, it is easy to conclude that exclusion of soil from the water system will remove 60% of the particles ≥0.5 μm diameter and 97% of particles larger than 5 μm diameter.

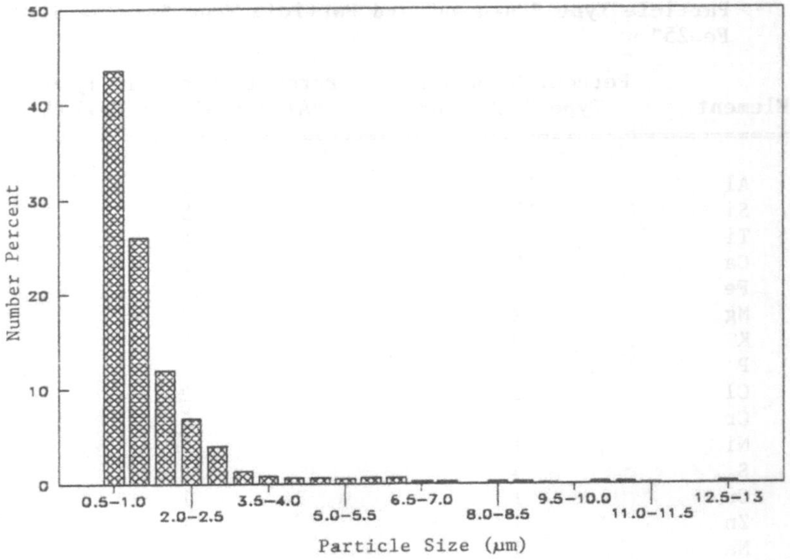


Figure 6. Measured particle size distribution in a sample from an ultrapure water system.

**IMPROVEMENTS FOR THE FUTURE**

Several improvements in the current method of automated analysis are underway. Perhaps the most important is the pushing of the lower boundary for particle size below 0.5 µm. Ultimately, as the critical particle size for deleterious effects continues to shrink, this will involve use of the transmission electron microscope and development of new procedures for sample preparation. A second improvement under development is a removal of the necessity for a priori operator-defined particle types, with the concomitant risk of skewing the detected types by operator prejudice concerning what materials are expected. The approach being taken is cluster analysis[14] in N dimensions (N=number of elements) of a data set for the definition of types represented in the sample. Carrying this procedure one step forward, principal component analysis may be applied to discern the original source materials giving rise to detected clusters.

Improvements in the quantitative capabilities are also being sought. The limits of ROI count ratio used in defining materials can be severe, and can be removed by application of standard electron microprobe data reduction techniques to the collected EDS spectra to generate element concentrations. These techniques are computer intensive, and dictate storage of the collected EDS spectra for off-line data reduction. The inherent limitations of EDS due to spectral peak overlaps may be overcome by incorporation of wavelength dispersive x-ray spectrometric (WDS) measurements into automated analysis.

The inability of the on-line particle sizing routines to handle high aspect ratio materials like fibers may be resolved through integration of currently existing image processing hardware and software with EDS data collection, a requirement that also places strong demands on computation capabilities. Investigation of the use of backscattered electron signal intensity as a measure of particle thickness is underway.

**SUMMARY**

Automated particle analysis by scanning electron microscopy provides a means of directly sizing and chemically characterizing the particulate loading of ultrapure water systems. The information generated is invaluable in monitoring the performance of the system, trouble shooting system breakdowns, and assessing particulate threats to the product or process which the systems serve. Future improvements in the method will be computer intensive but will result in enhanced abilities to monitor smaller and more complex particles and determine their chemical composition.


**REFERENCES**


1. K. Yabe, Y. Motomura, H. Ishikawa, T. Mizuniwa and T. Ohmi, Responding to the future quality demands of ultrapure water, Microcontamination, 7(2), 35-46, (1989).
2. P.A. McConnelee, S.J. Poirer and R. Hanselka, Water quality improvements and VLSI defect density, Semiconductor International, 9(8), 82-85, (1986).
3. K. Skidmore, Filter chemicals at point of use for ultrapurity, Semiconductor International, 11(10), 66-71, (1988).
4. N.J. Csikai and D.J. Barnard, Jr., Counting particles in mobile liquid chemicals for semiconductor processing, Microcontamination, 4(11), 45-50, (1986).
5. K. Dillenbeck, Advances in particle-counting techniques for semiconductor process chemicals, Microcontamination, 5(2), 30-38, (1987).
6. K. Yabe, T. Kumagai, H. Ishikawa, S. Akigama, T. Mizuniwa and T. Ohmi, Evaluating equipment technologies for future monitoring demands of ultrapure water, Microcontamination, 7(3), 25-30, (1989).
7. M.K. Balazs, Measuring and identifying particles in ultrapure water, Microcontamination, 6(5), 35-39, (1988).
8. T.B. Vander Wood, Identification of particulate contaminants in I.C. manufacture, Solid State Technology, 28(8), 177-182, (1985).
9. V.D. Scott and G. Love, "Quantitative Electron-Probe Microanalysis," John Wiley & Sons, New York, 1983.
10. J.M. Rebstock and T.B. Vander Wood, Automated analysis of cleanroom particulate contaminants, paper presented at the American Association for Aerosol Research Annual Meeting, (October, 1988).
11. W.R. Gerristead, Automated microprobe characterization of particles in VLSI gases, paper presented at the American Association for Aerosol Research Annual Meeting, (October, 1988).
12. J.I. Goldstein and A. Yakowitz, eds., "Practical Scanning Electron Microscopy," Plenum Press, New York, 1975.
13. American Iron and Steel Institute, "AISI Standard Steels," 1977.
14. D.L. Massart and L. Kaufman, "The Interpretation of Analytical Chemical Data by the Use of Cluster Analysis," John Wiley & Sons, New York, 1983.

# A SENSITIVE MONITOR FOR PARTICLES IN LIQUIDS

John Gregory

Department of Civil Engineering
University College London
Gower St., London WC1E 6BT, UK

A new technique for monitoring low concentrations of particulate impurities in flowing liquids is introduced. The method is based on measurements of fluctuations in the intensity of light transmitted through the flowing sample. These fluctuations arise from variations in the number of particles in the probe volume and follow Poisson statistics, so that the root mean square value of the fluctuating signal depends on the square root of the particle concentration. Simple analysis indicates that, for particle sizes of around 1 μm or greater, the fluctuating signal provides a much more sensitive measure of particulate contamination than turbidity measurements (either by light scattering or transmission). A practical device based on the new technique is described, which provides a very simple means for continuous, flow-through monitoring of relatively clean liquids.

## INTRODUCTION

When liquids are treated to remove suspended particles, often by some form of filtration process, it is important to test the effectiveness of the treatment. Depending on the application and the tolerable level of particulate contaminants, monitoring may take the form of turbidity measurement or particle counting, both of which can be carried out in an on-line, flow-through manner.

Turbidity monitoring is very often applied in water treatment plants to provide a check on the quality of the treated water and to indicate when "breakthrough" of particles from a filter occurs, so that a backwashing procedure can be initiated. Recently, turbidity standards for drinking waters have been subject to increased scrutiny and more stringent standards may be introduced[1] . Although originally, turbidity standards were set to ensure that treated waters were free from visible cloudiness, current concerns are with very low levels of particulate impurities, especially micro-organisms, which may be harmful. There have been several incidents where the parasitic organisms *Giardia lamblia*[2] and *Cryptosporidium*[3] have been found in water supplies. These organisms form spores a few μm in size and such particles are not easily detectable by conventional turbidity measurements, which are more sensitive to smaller

*Particles in Gases and Liquids 2,* Edited by K.L. Mittal
Plenum Press, New York, 1990

particles. So, a very low turbidity may not imply that a treated water is free of such organisms.

A more satisfactory alternative to turbidity monitoring for these purposes would be to employ on-line particle counting and sizing[4], but such an approach requires expensive instrumentation and generates more data than can easily be utilized for control purposes. An alternative approach is to derive information from fluctuations in the intensity of light transmitted through a flowing sample[5]. The application of this method to the detection of low levels of particulate impurities is described below.

## TURBIDITY FLUCTUATIONS

When a flowing suspension is illuminated by a narrow light beam (Figure 1), the transmitted light intensity shows random fluctuations about some mean value. The mean (DC) value is simply a measure of the average transmitted light intensity and depends on the concentration and nature of the suspended particles, through the Lambert-Beer law:

$$I/I_0 = V/V_0 = \exp(-NCL) \tag{1}$$

where N is the number concentration of particles, with light scattering cross-section C and L is the optical path length. I is the transmitted light intensity and $I_0$ is the incident light intensity. V and $V_0$ are the corresponding voltage outputs from a suitable photodetector, which depend linearly on the light intensity. (For simplicity, only one type of particle is considered, i.e. a monodisperse suspension).

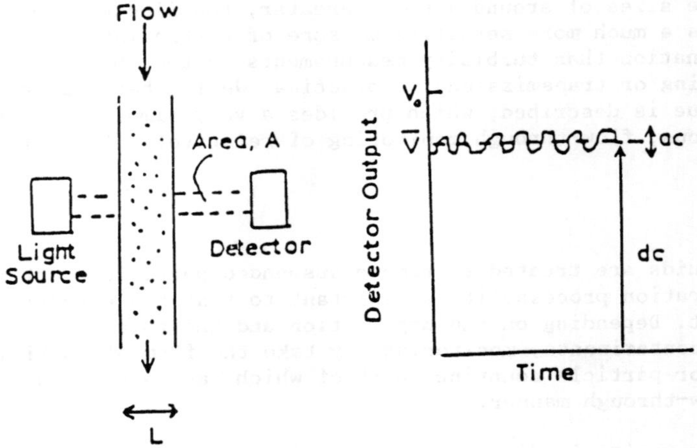


Figure 1. Schematic diagram showing the principle of the turbidity fluctuation technique. Flow of sample through the light beam causes fluctuations in the transmitted light intensity.

If we consider the volume of suspension illuminated by a beam with cross-sectional area A (assumed constant, i.e. a parallel beam) and length L, the number of particles, on average, contained within this volume is $\bar{n}$ = NAL, so that the Lambert-Beer law can be re-written:

$$V/V_0 = \exp(-\bar{n}C/A) \tag{2}$$

When the suspension is flowing, the actual number of particles in the light beam varies because of the fundamental non-uniformity of such systems[6] and the fact that the sample is continually being renewed by the flow. The variation in number follows the Poisson distribution, so that the variance is the same as the mean, n, and the standard deviation is $n^{1/2}$. It is important to realize that the Poisson distribution applies to cases where the average number of particles in the beam may be less than one, as well as to more concentrated suspensions where there may be hundreds or thousands of particles in the sample volume. In the latter case, the Poisson result is very like a Gaussian distribution with mean n and standard deviation $\bar{n}^{1/2}$.

It can be shown[5] that these variations in particle number lead to corresponding fluctuations in the transmitted light intensity and that the standard deviation of the intensity about the mean value, or the *root mean square (rms)* value of the fluctuating voltage signal is given by:

$$V_{RMS} = \bar{V}(\bar{n}^{1/2} C/A) \tag{3}$$

where V is the average transmitted light signal and $V_{RMS}$ is the rms value of the fluctuating component of the signal.

It is convenient to write this expression in terms of the particle number concentration, N, rather than the average number in the beam, n, and to introduce the dimensionless term R, which is simply the ratio of $V_{RMS}$ to the mean value V:

$$R = V_{RMS}/\bar{V} = (NL/A)^{1/2} C \tag{4}$$

This important result shows that the fluctuating signal depends on the square root of the particle concentration and on the first power of the scattering cross section (which depends on particle size - see below).

Eq (4) was derived for a monodisperse suspension. The corresponding result for a suspension having $N_1$ particles with scattering cross section $C_1$, $N_2$ with $C_2$ etc. is:

$$R = (L/A)^{1/2} (\sum N_i C_i^2)^{1/2} \tag{5}$$

where the sum is taken over particles of all types.

This shows that there is still a dependence on the square root of the particle concentration and that larger particles (with larger scattering cross sections) will make a greater contribution to the result.

Returning to the monodisperse case, it is possible to write eq. (4) in terms of the volume fraction of suspended particles, $\Phi$, rather than their number concentration, N, and to introduce the *scattering coefficient*, Q, instead of the scattering cross section, C.

The volume fraction (assuming spherical particles, radius a) is simply:

$$\Phi = 4\pi N a^3/3 \tag{6}$$

The scattering coefficient is the ratio of the scattering cross section to the geometric cross sectional area:

$$Q = C/\pi a^2 \tag{7}$$

Combining these with eq. (4) gives:

$$R = (3\pi L/4A)^{1/2} (\Phi a)^{1/2} Q \tag{8}$$

The first bracketted term on the right hand side is entirely dependent on the effective dimensions of the light beam and hence on the design of a particular instrument. A longer path length and a narrower beam will give greater values of R. The remaining terms show the expected square root dependence on the volume fraction and that the ratio should be proportional to the square root of the particle size and the first power of the scattering coefficient.

Fluctuations are observed only for flowing suspensions - if the flow is stopped the fluctuating signal is lost, since then there is a fixed number of particles in the beam. (There may be changes due to Brownian motion or sedimentation, but these are negligible compared with the number variations caused by flow). The frequencies of the observed fluctuations depend on the transit time of particles through the beam and hence on the flow rate. It is best to maintain a flow rate which gives particle velocities of a few cm/second through the light beam, which, for typical beam widths, corresponds to transit times of around 1 msec and frequencies of a few hundred Hz. Changing the flow rate causes the frequency of the fluctuating signal to change, but should have no effect on the amplitude, or rms value.

It will be shown that measurements of fluctuations in transmitted light ("turbidity fluctuations") can give a much more sensitive indication of low particle concentrations than straightforward extinction measurements, based on the Lambert-Beer law, eq. (1). For this purpose we need to define turbidity, $\tau$, as $\tau = NC$, which, when divided by the volume fraction and substituting the scattering coefficient, gives the *specific turbidity*:

$$\tau/\Phi = 3Q/4a \tag{9}$$

The Lambert-Beer law then becomes:

$$V/V_0 = \exp(-\ 3Q\Phi L/4a) \tag{10}$$

In order to proceed further, the dependence of the scattering coefficient, Q, on particle size has to be considered. For particle sizes of interest here (from sub-micron to several microns), the Mie theory of light scattering should be used (see Kerker[7]). This gives values of Q which rise sharply with size for small particles and then pass through a series of regularly-spaced maxima and minima, eventually reaching a constant value of Q = 2. The values also depend on the wavelength of light in the medium, $\lambda$ ($= \lambda_0/n_0$, where $\lambda_0$ is the wavelength *in vacuo* and $n_0$ is the refractive index of the medium) and the relative refractive index, m ($= n/n_0$, where n is the refractive index of the particles). An example is shown in Fig. 2, where the scattering coefficient for spherical particles, with m = 1.2, is shown as a function of the dimensionless size parameter, $\alpha = 2\pi a/\lambda$. The values are taken from the computations of Heller and Pangonis[8].

Also shown in Fig. 2 are results of calculations based on the "anomalous diffraction" approximation for the scattering coefficient[7]:

$$Q = 2 - (4/\rho)\sin\rho + (4/\rho^2)(1 - \cos\rho) \tag{11}$$

where $\rho = 2\alpha(m - 1)$.

Eq. (11) applies only when the size parameter, $\alpha$ is fairly large and the refractive index m is quite small. These conditions apply to many particles of interest. Even when m = 1.2 (i.e for a particle refractive index of about 1.6 in water) eq. (11) is not too bad an approximation to

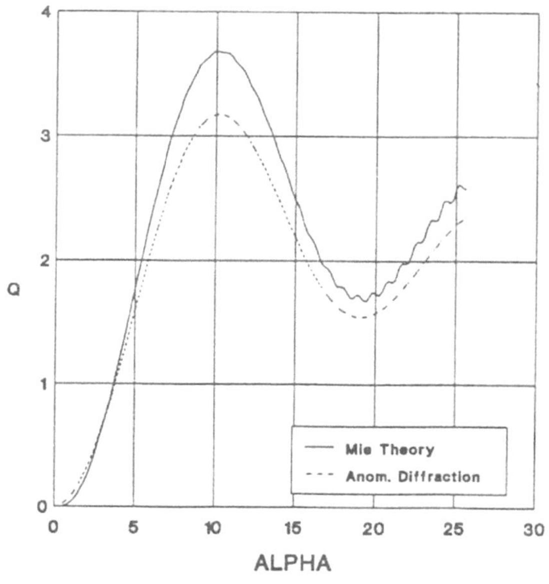


Figure 2. Scattering coefficients for spherical particles, with refractive index m = 1.20, as a function of the dimensionless size parameter $\alpha$. The values are from exact computations based on Mie theory and from the anomalous diffraction approximation, eq. (11).

the exact Mie results. At worst, the approximate value is about 14% too low. For lower values of m the anomalous diffraction approach gives better agreement. For this reason, and for ease of computation, the "theoretical" results in what follows will be calculated from eq (11). Since this slightly underestimates the "correct" scattering coefficients, the conclusions concerning detection limits will be on the conservative side.

## EFFECT OF PARTICLE SIZE AND REFRACTIVE INDEX

The specific turbidity, calculated from eqs. (9) and (11) is shown in fig. 3 for a range of refractive index values m = 1.05 to 1.20 and for particle diameters up to 10 µm. The wavelength chosen for these calculations is 850 nm, for comparison with predicted R values in fig. 4, which, in turn, are intended for comparison with experimental measurements made at $\lambda_0$ = 850 nm. The results show the expected form - a rise in specific turbidity with particle size up to a maximum value and then a decline, followed by further slight undulations and a general downward trend. At smaller values of refractive index, the maximum occurs at larger particle sizes and the turbidity value at the maximum is lower. For very large particles, the specific turbidity varies inversely with size, as expected from eq. (9) with Q = 2. Similar calculations for shorter wavelengths, in the visible region, show the turbidity maximum occurring at smaller particle sizes, with a higher peak value, but the form of the curves is similar to those in fig. 3.

It should be mentioned that many "turbidity" measurements carried out in practice, especially in the water treatment field, are not based on transmission (extinction) methods, but on scattering at a specific angle (often 90°) to the incident beam. These *nephelometric* methods can be more sensitive than transmission measurements, but they show a different

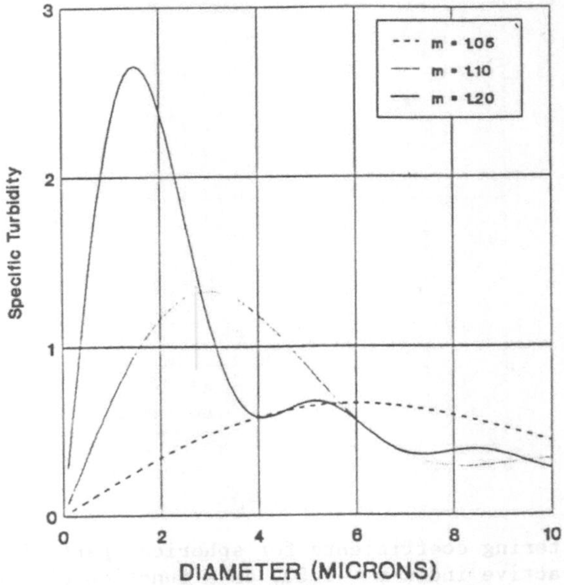


Figure 3. Specific turbidities ($\tau/\Phi$) for spherical particles, as a function of the particle diameter, for three values of the relative refractive index, m, and a wavelength, $\lambda_0$ = 850 nm. Scattering coefficients calculated from eq (11).

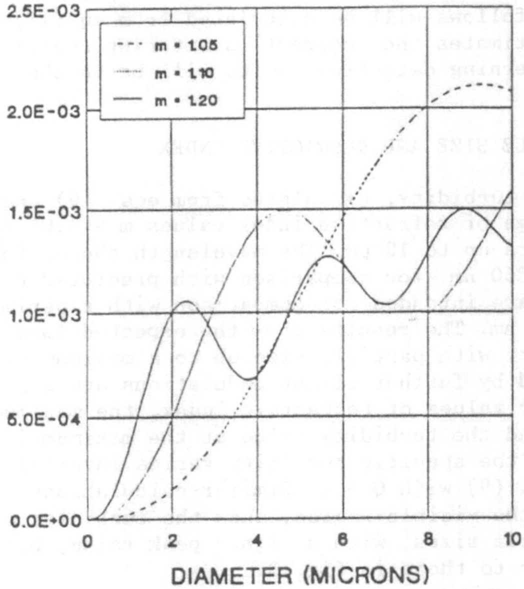


Figure 4. Ratio ($V_{RMS}/\bar{V}$) values as a function of particle diameter. Same conditions as Figure 3.

dependence on particle size - - the maximum in specific turbidity is found for much smaller particles, usually in the sub-micron range[9].

The corresponding calculations for the ratio, R, of the fluctuating signal to the mean value, are plotted in fig. 4, from eqs. (8) and (11). In this case certain other values have to be chosen, as well as the wavelength and refractive index (for which the same values have been assumed as in the specific turbidity examples). For comparison with experimental results, the following values have been assumed: L = 3 mm, A = 0.07 $mm^2$ (corresponding to a cylindrical light beam about 0.3 mm in diameter) and $\Phi = 10^{-6}$ (1 ppm by volume). The results show very different behavior to the specific turbidity trends in fig. 3. There is a rapid initial rise in R with increasing particle size, especially for the higher refractive indices, where the initial rise is followed by a series of maxima and minima superimposed on a gradual upward trend. For very large particles, where Q = 2, eq (8) predicts that R should increase as the square root of particle size. For the low refractive index case (m = 1.05) the initial rise is less pronounced, but at diameters of a few microns, the Ratio value can be higher than those for particles with greater m values. Ultimately, at particle sizes much greater than those shown, all of the curves coincide.

## DETECTION LIMITS

Normal transmission measurements and the fluctuation method are both subject to limitations when very low particle concentrations are involved. In the case of transmission, there are problems when the transmitted light intensity is only slightly less than the incident intensity. Usually, the incident intensity is determined with pure, particle-free liquid in the light beam and compared with subsequent values when sample suspensions are present. For almost "clean" liquids, the difference can be very small and a minute change in the signal, caused by, for instance, a temperature change and consequent drift in electronic and electro-optic components, can have a major effect on the measured turbidity. For an incident value $V_0$ = 10.000 volts and $\overline{V}$ = 9.999 volts, a change of only 0.5 mV in the measured signal (50 ppm) would cause a 50% error in the turbidity. It is virtually impossible, except under rigorously controlled conditions, to avoid drift of this order, so that transmission measurements become unreliable at such levels. In flow-through cells, another source of error is the coating ("fouling") of optical surfaces by attached particles, which causes a reduction in the effective incident value. Frequent checks with pure liquid are necessary and this can be very inconvenient. For the present computations, a limit for $\overline{V}$ of $0.9998V_0$ will be assumed. Under practical conditions, values closer to $V_0$ could not confidently be interpreted in terms of particulate contamination. Inserting this value in eq. (10) and re-arranging gives a value for the minimum volume fraction of particles detectable by extinction measurements:

$$\Phi_{MIN} = 2 \times 10^{-4} (4a/3QL) \qquad (12)$$

In the case of the fluctuating signal, long-term drift is not a serious problem, and, in any case, division of $V_{RMS}$ by V to give the Ratio value effectively compensates for any change that does occur. The only serious limitation is then the inherent noise in the equipment, especially in the frequency range of interest (50 - 2000 Hz). By using good quality electronic and electro-optic components and following proper grounding techniques, it is possible to restrict the rms noise level to less than 1 part in $10^5$ of the mean (DC) signal, V. This noise level contributes to the measured total rms value, together with the signal due to particle number fluctuations. If the noise level is X and the signal S, the total rms value is given by:

$$V_{RMS} = \sqrt{(X^2 + S^2)} \tag{13}$$

So, if the signal level exceeds the noise by a factor of 2, the noise contributes only about 11% to the result.

In calculating the minimum detectable concentration, it has been assumed that a ratio value of 2 x $10^{-}$ due to turbidity fluctuations (i.e. at least twice the typical noise level) is the lower limit. In fact, since the noise level often remains nearly constant over long periods, signals only slightly greater than the noise could be measured with some confidence. Nevertheless, the limit chosen represents a value which will be virtually unaffected by noise and is a convenient boundary. Putting R = 2 x $10^{-5}$ and re-arranging eq. (8) gives an expression for the minimum detectable volume fraction:

$$\Phi_{MIN} = 4 \times 10^{-10} (4A/3\pi L)(1/aQ^2) \tag{14}$$

Results from eqs. (12) and (14) are plotted in figs. 5 and 6 for two values of refractive index, m = 1.20 and 1.05, and $\lambda_0$ = 850 nm. The turbidity and Ratio limits are shown on the same figure for ease of comparison. These results show very different particle size behavior, as might be expected from the corresponding data in figs. 3 and 4. In the case of turbidity measurements, sensitivity increases with increasing particle size, up to some critical size, above which the sensitivity shows a decreasing trend. For m = 1.2, the critical size is about 1.5 µm (as expected from the location of the appropriate peak in fig. 3) and the minimum detectable concentration would be about 3 x $10^{-8}$ (30 ppb by volume). For m = 1.05 (fig 6) the greatest sensitivity would be found for a particle size of about 6 µm, at which a concentration of about 100 ppb would be detectable.

The fluctuation method shows a very dramatic improvement in sensitivity as particle size increases from sub-micron levels up to a few microns and then a more gradual change. For m = 1.2, the Ratio value gives a more sensitive measure for all particle sizes above about 0.6 µm and for particles larger than about 1.5 µm the limiting concentration is always below 1 x $10^{-9}$ (1 ppb) and at least 100 times lower than the turbidity limit. If the refractive index is 1.05, the cross-over with the turbidity limit occurs at a particle diameter of about 1.2 µm and for larger particles the two techniques have sensitivities which can differ by a factor of about a thousand.

Another way of expressing concentration limits is in terms of the minimum *number* concentration, which is easily derived from the corresponding volume fraction using eq. (6). The volume fraction limits for the fluctuation method in figs. 5 and 6 have been converted to number concentrations and are shown in fig 7. This shows that, for 2 µm particles with m = 1.2, a concentration of about 100 particles per mL would be detectable. For larger particles, the limit drops to 1 particle per mL or less, although, in practice, such low levels may not be attainable for a number of reasons. With the assumed path length and effective beam area (L = 3 mm, A = 0.07 $mm^2$), the illuminated sample volume is about 0.2 $mm^3$, so that, for a particle concentration of 5000 per mL, there should be just one particle, on average, in the light beam. At lower concentrations, the fluctuating signal would be due mainly to single particles passing through the beam. With higher number concentrations, there would be fluctuations about some mean value. These two conditions cannot be distinguished from

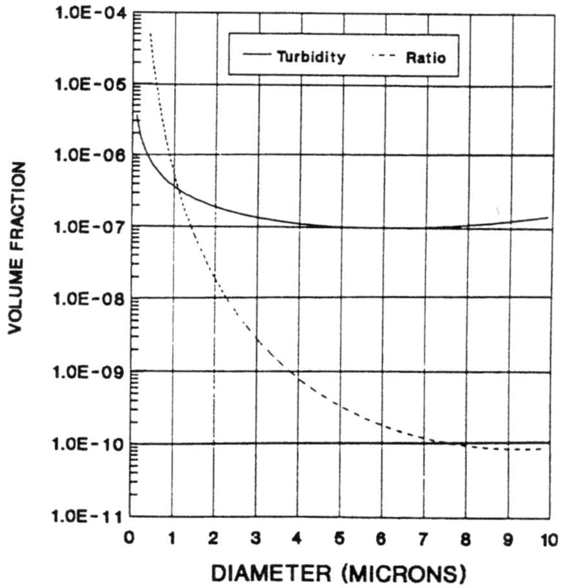


Figure 5. Lowest particle concentrations (volume fractions) detectable by turbidity and fluctuation (ratio) techniques. Refractive index m = 1.20.

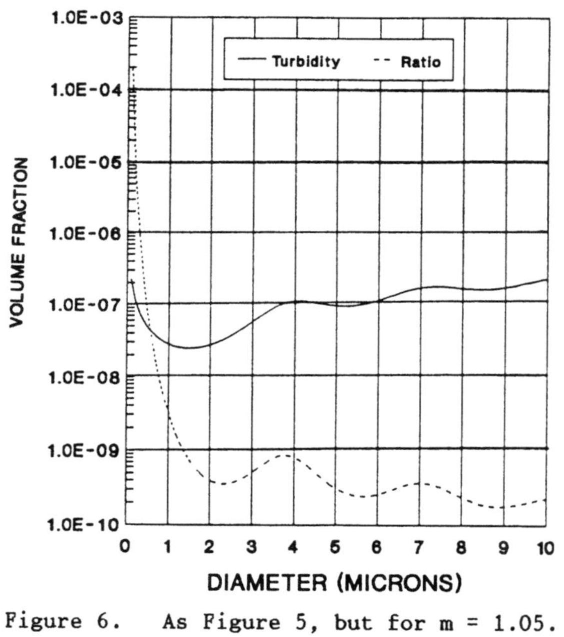


Figure 6. As Figure 5, but for m = 1.05.

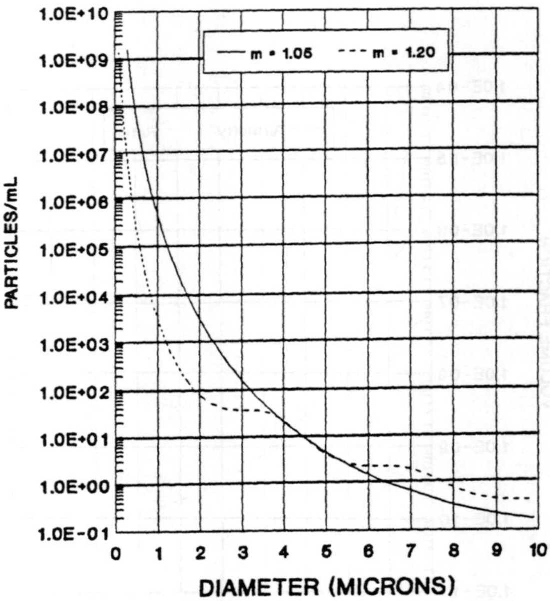


Figure 7. Lowest detectable particle number concentrations by the fluctuation technique, for two values of refractive index, m.

the Ratio value alone. A given result may be due to a high concentration of small particles or a much lower concentration of larger particles.

These computations were based on arbitrary, but realistic, assumptions and the general conclusions would not be altered by choosing a different set of assumptions. Apart from the properties of the particles (size and scattering coefficient), the dimensions of the light beam are the only other factors which affect the sensitivity. An increased path length would increase the sensitivity of both techniques in the same way. A narrower beam would improve the sensitivity of the fluctuation technique, but would have no effect on turbidity. However, it is difficult to increase path length significantly while maintaining a narrow beam diameter, so that the sensitivity of the turbidity technique could be more easily improved by choosing a longer light path. Nevertheless, increasing the path length for turbidity measurement from the assumed value of 3 mm to a few cm would still leave the fluctuation technique (with L = 3 mm) far more sensitive for most particles.

Nephelometric turbidity measurements on latex suspensions, using a commercial instrument, show a decreasing sensitivity as the particle size increases above about 0.3 µm. The estimated detection limits (0.1 NTU) for micron-sized particles and larger were of the same order as those shown in fig 5 (50 - 200 ppb)[9].

## EXPERIMENTAL

A particle monitor based on the principles outlined above has been developed and is commercially available (Rank Brothers Ltd. Bottisham, Cambridge, England). A narrow light beam illuminates the flowing sample in a transparent plastic tube, of 3 mm internal diameter. This tube is clamped in the sensing head and is slightly compressed, giving nearly flat surfaces where the light beam passes through the tube walls. The compression of the tube also gives considerable immunity to vibration, which can be a problem in this type of measurement when rigid (e.g. glass) flow cells are used. The light source is a high-intensity light-emitting diode with a wavelength of 850 nm and the transmitted radiation is monitored by a PIN photodiode with a closely matched spectral sensitivity. The effective beam width in the sample is about 0.3 mm, determined by precisely-aligned pinholes. (The parameters in the above computations were chosen to match these conditions).

The fluctuating component of the photodiode output is separated from the DC component and its RMS value is derived. Since, for "clean" fluid applications, changes in the DC value provide little or no information, the DC level is kept constant, by a feedback technique. This avoids the need to derive the ratio value (RMS/DC), since the RMS value is always referenced to the same DC level. Electronic drift and contamination of the tube walls are automatically compensated. For the same reason, a new tube can be fitted with no need for re-calibration (provided the internal diameter and hence the optical path length are not changed).

Sample flows at a rate of 60 mL/min or more and the RMS output can be monitored continuously. The noise level in the instrument is usually very low (1 part in $10^5$ or less) and can be checked simply by stopping the flow, since the fluctuations due to particles then cease. The noise level can be backed off by a discriminator, so that a truly particle-free liquid will give an RMS reading of zero. However, because the noise contributes to the signal in a non-additive manner, such a procedure is not strictly justified.

The response of the monitor to suspensions of polystyrene latex particles ranging in diameter from about 0.5 to 6 µm has been established[9]. The RMS value always varies as the square root of particle concentration, as expected. From the slopes of such plots it is possible to derive data for comparison with theoretical predictions. The Ratio value for a volume fraction $\Phi = 10^{-6}$ has been determined in each case and can be compared with calculated values for the same concentration, such as those in fig 4. In fact, rather than comparing experimental values with computations based on the anomalous diffraction approximation, eq. (11), Ratio values based on exact (Mie) values of the scattering coefficient have been calculated. The theoretical results in fig 8 are calculated from eq. (8), using values of Q in fig 2, for m = 1.2 (a value close to that for polystyrene latex) and assuming the same values of the other parameters as before. Experimental points for 8 different latex particle sizes are shown.

For the smaller particles, the agreement between the experimental and theoretical results is quite good, bearing in mind that some of the parameters chosen, such as the refractive index, m, and the effective beam area, A, are subject to some uncertainty. For larger particles, especially around 6 µm, the discrepancy is rather large, with the measured value falling well below the theoretical prediction. The reason for this behavior has not been established, but may be a result of the greater degree of forward scattering as particle size increases. Light scattered at a narrow forward angle to the incident beam may be received at the detector and give an apparent extinction less than the expected value[10]. There may also be problems associated with deriving the true RMS value of a fluctuating signal which consists of fairly infrequent large pulses, as would be the case with a suspension containing a low concentration of large particles. Another factor might be that the larger latex particles used were not as uniform as the smaller ones, with quite a wide variation in size apparent from microscopic observation.

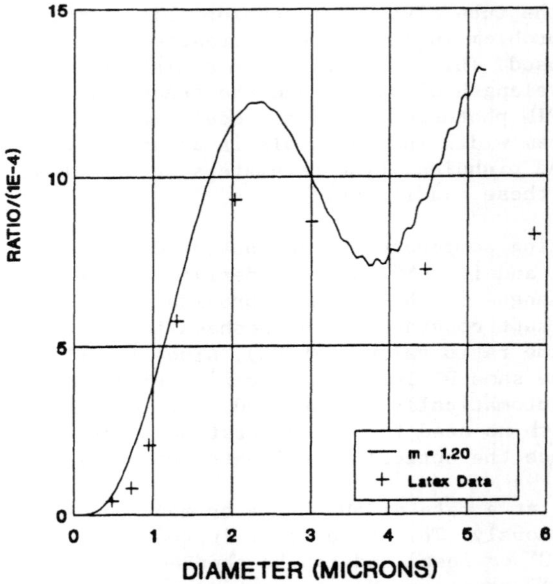


Figure 8. Full curve: computed Ratio values for m = 1.20, based on Mie scattering theory. Points: experimental results for polystyrene latex. $\Phi = 10^{-6}$, $\lambda_0$ = 850 nm.

Nevertheless, the conclusions drawn above concerning the sensitivity of the technique are only marginally affected and it has been established that sub-ppb concentrations of all latex particles larger than 2 µm give responses well above the noise level.

The new monitor has been found to give a sensitive indication of general particulate contamination in high quality water, for instance from a laboratory water purification unit. Day-to-day differences are found in the RMS reading which can only be explained by changes in particulate contaminants, possibly micro-organisms. Membrane filtration (0.22 µm) removes such particles and the monitor response then falls essentially to zero. These differences are not detectable by turbidity measurements, which show always the same, minimum reading for water from the purification unit.

## CONCLUSIONS

A simple monitoring technique has been described, which is based on measurements of fluctuations in the intensity of light transmitted through a flowing sample. The fluctuating signal gives a very sensitive indication of the presence of particulate impurities, especially those of a few microns in size. For such particles, the sensitivity is orders of magnitude higher than with turbidity measurements, either by transmission or nephelometry.

At very low concentrations, the response of the monitor arises essentially from the passage of single particles through the light beam and it may then appear similar to a particle counter based on the light blockage principle. However, in the new device, no attempt is made to count and size individual particles. An overall index (the RMS value or Ratio) is derived as a measure of the contaminant level. For a given type of particle, this depends on the square root of particle concentration, so that a wide range of concentrations can be covered without any change of scale. (With a 4-digit display, concentration changes over at least a million-fold range could, in principle, be followed).

A major advantage of the new method over particle counting is that coincidence problems do not arise. With all particle counters, the concentration must be low enough so that there is no more than one particle in the sensing zone at a time. Very often, this limits particle number concentrations to $10^6$ /mL or less.

Of course, particle counters give much more detailed information, such as particle size distributions, which may be essential in some applications. However, there are many cases where routine monitoring needs to be carried out in order to ensure that particulate contaminants are kept below some agreed limit. In these cases, a simple monitor, giving an empirical index, may be acceptable. The response could be compared with particle counting data and a threshold level established, below which the particulate contamination could be assumed to be at an acceptable level. If the monitor response rises above this threshold, then appropriate action could be taken.

## REFERENCES


1) E.R. Baumann, Water, its quality and YOU!, *Filtration & Separation*, 26(3), 204-211 (1989).

2) G.F. Craun, Surface water supplies and health, *J. Amer. Water Works Assoc.*, 80(2), 40-52 (1988).

3) J.B. Rose, Occurrence and significance of Cryptosporidium in water, *J. Amer. Water Works Assoc.*, 80(2), 53-58 (1988).

4) G.P. Treweek, Optimization of flocculation time prior to direct filtration, *J. Amer. Water Works Assoc.*, 71(2), 96-101 (1979)

5) J. Gregory, Turbidity fluctuations in flowing suspensions, *J. Colloid Interface Sci.*, 105(2), 357-371 (1985).

6) N.G. Van Kampen, "Stochastic Processes in Physics and Chemistry", North Holland, Amsterdam, 1981.

7) M. Kerker, "The Scattering of Light and Other Electromagnetic Radiation", Academic Press, New York, 1969.

8) W. Heller and W.J. Pangonis, Theoretical investigations on the light scattering of colloidal spheres. I. The specific turbidity. *J. Chem. Phys.*, 26(3), 498-506 (1957).

9) J. Gregory, A simple particle monitor for low-turbidity waters, Proc. Amer. Water Works Assoc. Water Quality Technology Conference, St. Louis, pp. 563-575, November 1988.

10) P. Latimer and F. Wamble, Light scattering by aggregates of large colloidal particles, *Appl. Optics*, 21(13), 2447-2455 (1982).

# NEW TECHNIQUES FOR SAMPLING SUBMICRON PARTICLE CONTAMINATION IN WATER

Paul Melanson and Clifford F. Frith

ANATEL CORPORATION
2200 Central Avenue
Boulder, CO 80301

Pure water specifications have been defined in many publications, depending on the industry and the application for the fluid. Very small particles are a significant parameter in water purity. The current on-line particle counters have limitations in measuring particles below 0.1μm. A new method for sampling ultrapure water containing very small particles is described in this paper. This automatic Scanning Electron Microscope (SEM) sampling instrument also provides a unique method for evaluating the flow decay or silting index of the sample, using the same microporous analytical filter employed for the SEM procedure.

## INTRODUCTION

The need for improved accuracy and sensitivity in the monitoring of contamination in ultrapure water environments has been recognized in technology-driven industries. Nowhere is the need more evident than in the measurement of particles, a discipline where specifications challenge current instrument technology. In many applications, inspection of analytical filters with a scanning electron microscope (SEM) has been adopted[1], to allow the accurate measurement of very small particles in liquids. The SEM technique, utilizing the inspection of a statistically significant portion of a membrane filter at extremely high magnification, has been widely accepted[2]. SEM analysis not only allows a direct measurement of particle concentration, it extends the limits of sensitivity. In addition, the actual particles are analyzed for composition which may assist in source identification. Information on sizing, type, and chemistry is invaluable in troubleshooting, failure analysis and material evaluation.

Control of all aspects of the SEM sampling procedure is necessary to minimize analytical errors. The numerous variables introduced by manual collection methods made some of the current data suspect. A new microprocessor-based instrument has been developed to significantly reduce potential sampling problems.Continuous on-line monitoring, necessary for such control, also provides data which can be used to evaluate changes in the sampling filter performance. This is referred to as flow decay[3], a qualitative indication of particle contamination. This paper outlines a sampling instrument and the new analytical methods which have been developed to further its utilization.

*Particles in Gases and Liquids 2,* Edited by K.L. Mittal
Plenum Press, New York, 1990

Traditionally, sampling for SEM analysis has been conducted using a membrane filter, mounted in an analytical filter holder. This is connected into the fluid stream in the process under study. Small surface area filters and large fluid volumes are used to maximize the concentration effects, and minimize the number of fields which must be counted to evaluate the entire population of particles. The volume filtered is recorded manually by intermittent flow measurements. A suitable flow totalizing monitor for ultrapure water at these low sampling rates is currently unavailable. Filter loading, sampling, and particle counting often are performed by multiple personnel at each site.

Engineers dedicated to improving the accuracy and repeatability of SEM analysis soon perceived the need for automating their sampling technique. Accuracy of SEM analysis in ultrapure water, where a small number of particles are collected from large volumes over extended time, requires the sampling periods discipline and standardization only made possible by automation. Not only will this approach increase control over the collection process, an automatic technique removes operator variables and provides quantitative information in an efficient manner. Additionally, an on-line automatic sampling apparatus provides information that indicates filtering parameters changes during actual sampling. Higher particle contamination and subsequent retention results in a marked decrease in filter performance. This "flow decay" phenomenon[4], observed in the data taken to estimate total volume for manual collection, has been a matter of interest for some time.

The degradation of flow through filters has been used for many years to evaluate liquids with high particle concentration or for testing filter efficiency. The simple monitoring of flow rate and pressure differential across the filter, along with temperature reading to compensate for viscosity, can provide continuous data over the duration of the analysis. The prospect of not only an automated SEM sampler, but a method that offers real-time quality indication, led to the concept that evolved into the A-300 Submicron Analyzer.

Anatel Corporation developed the special analyzer in cooperation with a major semiconductor manufacturer as a way of providing strict control of automated SEM sampling by on-line, real-time pressure and volume monitoring. The analyzer is microprocessor-based and utilizes accurate transducers for data collection. The design minimizes the possibility of the instrument contributing to the particle inventory, while furnishing a very accurate measurement of the filter's performance. Field trials substantiate the reproducible sampling. Several major semiconductor manufacturers report that this automated sampling procedure eliminates the variables which had produced confusing results. The design, operation, and results of the Submicron Analyzer and its testing program are presented.

## THE SEM ANALYTICAL TECHNIQUE

Detection of low concentrations of very small particles in ultrapure water is a critical requirement. In most instances, the characterization of these particles by number, size, type, and composition is essential. The limitations of liquid particle counters for very small particles (below 0.2 µm) in low concentrations is documented[5]. In addition, the analysis by size is limited, while information on particle type is nonexistent. For these reasons, the routine measurement of particles on a microporous membrane filter using the SEM technique has been standard procedure for many years. The Nuclepore® polycarbonate membrane is commonly used for SEM analysis due to its uniform surface morphology and low background readings.

Recently, specifications have been established for advanced manufacturing techniques which require detection of very low levels of 0.1 µm and smaller particles. The SEM procedure has been established as the most reliable method of obtaining meaningful results. A standardized sampling procedure is critical to the accuracy of collected data especially as process water receives more treatment and larger sample volumes (and longer times) are required.

The standard SEM procedure employs a 25 mm (diameter) membrane filter holder. The operator periodically measures the volume per unit time and integrates these measurements over the total sampling time, to estimate the total volume filtered. Non-uniform changes in flow, especially those which are transient or caused by varying feed pressures, increase the inaccuracy of this measurement. Changes in differential pressures relate directly to the rate of permeate flow for an ideal liquid across an ideal filter. In most industrial applications, this pressure differential varies widely due to the motive pressures in the process and the constant changes in normal operating facilities.

Pressures maintained during day shift operations, when flow measurements are usually taken, differ dramatically from those recorded during less active periods of plant operations. These variables are almost impossible to quantify due to the dynamic nature of water usage, and certainly cannot be documented by the intermittent collection of data. Upstream pressure control is not desirable due to its propensity to add contamination to the sample under study. Suitable instrumentation for both pressure and flow recording, as well as automatic monitoring and control, simply do not exist for ultrapure water. The uncertainty in volume measurement propagated by existing techniques transfer into the calculation of particles per unit volume, limiting the accuracy of this technique.

## THE NEW SUBMICRON PARTICLE ANALYZER

The Submicron Analyzer records continuous readings of differential pressure, flow, time, and temperature, using state-of-the-art components with high accuracy and a wide dynamic range. Not only does this provide detailed and extremely accurate measurement of flow, and pressure, the instrument can also control sampling based on selected parameters. Measurements and calculations by the analyzer also provide an indication of filter performance and its changes.

The flow decay, silt density (or silting) index, and filter plugging tests have been used in liquid analysis for a long time[6], albeit in fluids with considerably higher particle concentrations than ultrapure water. The observance of a flow decrease with constant pressure or an increase in pressure differential over the sampling period are standard methods[7]. Continuous measurement of all parameters is obviously superior in industrial situations where pressures and flows vary. In heavily contaminated liquids, this does not cause significant problems, as the sample period is short and the conditions are easy to control during that time. However, ultrapure water, as well as today's liquids for critical applications, are purified with high efficiency filtering media and this produces very low "counts" per unit volume.

The Submicron Analyzer was designed first and foremost to be an effective SEM sampling device. This is reflected in the care which was taken to avoid contamination from the instrument itself or the procedures necessary to utilize it. The filter holder is designed to accommodate a standard 25 mm filter membrane and all positioning and sealing is done with a single fluoroplastic encapsulated O-ring. The holder is machined from 316L stainless steel, and electropolished and passivated, which experience has shown to be an effective treatment for this material in ultrapure water.

The two components of the filter holder are assembled easily using a plastic flange clamp (Figure 1) which can be removed with minimal particle generation when compared with the standard metal or plastic screw threads. Also, as the surfaces of the holder have no threads or crevasses, it is easily cleaned before the actual filter is exposed for installation. The filter assembly fits into the instrument with two face mounted 0-rings as seals, avoiding particle generation by threads or compression fittings. This filter mounting system assures a very high level of cleanliness.

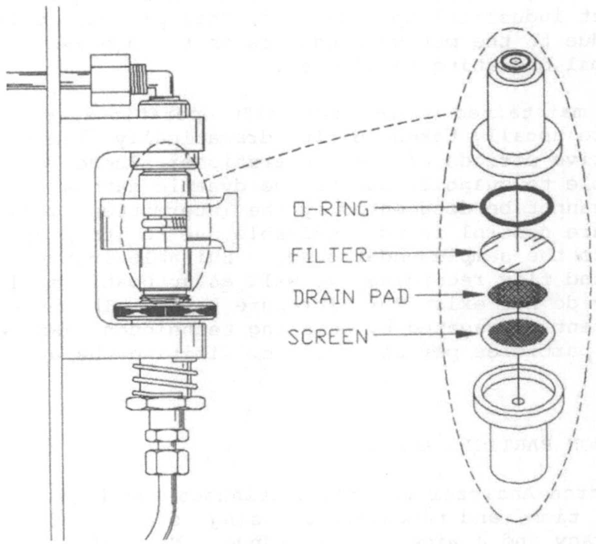


Figure 1. A-300 Filter Holder Assembly (Exploded View).

The internal components (Figure 2) of the instrument were designed with ultraclean sampling in mind. The components include a valve for turning the sample flow on and off, a differential pressure transducer across the filter holder, a flow monitor which measures the entire flow through the filter, and a temperature probe used to correct the pressure and flow transducers as well as to indirectly measure the viscosity of the water. Of these, only the valve and one leg of the differential pressure transducer need be "upstream" of the filter itself. Both of these could introduce significant contamination into the input of the filter holder: the valve through abrasion and erosion, the pressure transducer through the "dead leg" created. To avoid this, the valve is made from molded PFA (Perfluoroalkoxy resin), a material which has excellent chemical and erosion resistance, and is opened and shut at an extremely slow rate. A gear reduced stepper motor operating at a fraction of an RPM with positional feedback is used for this purpose. The gradual change in flow resulting from this actuation also causes a minimal disruption of the filter surface and the particles retained on it. The differential pressure transducer is equipped with a "purge" path to continually flush out the inlet pressure transducer chamber. This flush is positive enough to negate the possible effects of any particle accumulation or generation in the transducer assembly itself.

During operation, the data from the transducer are reported continuously and averaged over one minute increments. This is the minimum time base for subsequent analysis. Pressure differential, flow rate, temperature, and accumulated volume can be accessed through front panel display, or printer, at any point during analysis. Other values are calculated from these measurements, the most significant being "specific flow". As the pressure differential may vary with incoming pressure, the subsequent flow rate varies also. This relationship is expressed in Darcy's law, which relates flow to pressure, viscosity, number of pores, and filter restriction. The specific flow is calculated from this equation and can be thought of as either the scaled flow which would be noted at 1 psi differential pressure, 1 poise viscosity (near 23° C for water), or a direct measurement of filter restriction.

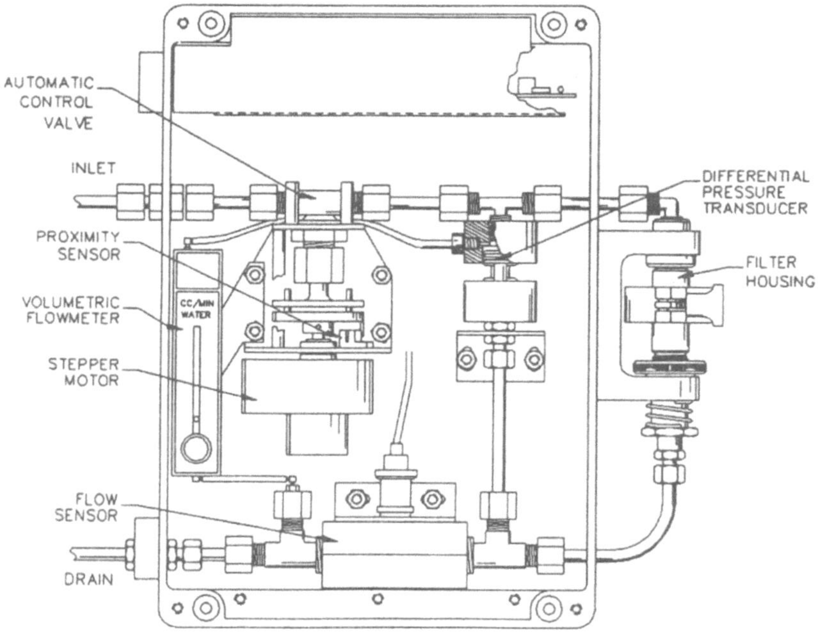


Figure 2. A-300 Sensor Assembly.

This specific flow provides consistent data no matter what the variations in system pressure. Any attenuation of the specific flow over time is considered "flow decay"[4], which is measured as a percentage of the initial flow. Most filters take some time to "condition" initially (through the wetting of pores and other physical chemical phenomena). The point at which the instrument reports the initial flow reading is programmable by the user to avoid the effect of this conditioning. This flow decay value can be used to compare readings at different sample locations, or different sample times, regardless of temperature and pressure of the system being tested. The instrument has three means of output: the front display, a scaled analog (4-20 mA) signal, and a compact printout (Figure 3).

The front display can be programmed to "roll through" the entire set of data (which also includes elapsed time) updated every minute. The analog output can be programmed to report either specific flow or percent flow decay.

The printout period is programmable from one minute to one hour, and consists of the time (real, with calendar date at the start of the test, and after each midnight), accumulated volume in liters, average flow decay over the printout period (blank until the calibration is done), flow in ml/min, time in minutes, pressure in psi, and water temperature in degrees centigrade. The printout header indicates the settings of the instrument parameters as well as providing an explanation of the printout, instrument type, serial number, and a reserved spot to record sample location and filter specifics. The instrument also has an optional RS232 serial port to provide detailed output to any data collection system which is capable of serial communications.

```
*************************
*         ANATEL        *
* Flow Decay Analysis *
********************1001
        09/14/87
        17:44:02
  Stabilization time =  60 min.
 Print-out Time Interval =  30 min.
   Continuous Analysis Mode

Filter Type =

Filter Lot # =

Filter Pore Rating =

Sample Location =
```

| Real Time (Hr:Mn) | Total Volume (L) | Flow Decay (%) | Raw Flow (mL/M) | Raw Press. (PSID) | Spec. Flow | Water Temp (C) |
|---|---|---|---|---|---|---|
| 17:45 | 0.0 | | 0.0 | 0.00 | 0.000 | 26.5 |
| 18:15 | 1.1 | | 39.7 | 6.46 | 5.100 | 27.4 |
| 18:45 | 2.5 | | 45.0 | 9.03 | 4.995 | 27.6 |

****** Start % Flow Decay Calculation ******

| | | | | | | |
|---|---|---|---|---|---|---|
| 18:45 | 2.5 | 0.0% | 41.9 | 9.19 | 4.559 | 27.6 |
| 19:15 | 3.7 | 7.1% | 39.4 | 9.33 | 4.233 | 27.6 |
| 19:45 | 4.7 | 18.9% | 35.1 | 9.51 | 3.695 | 27.6 |
| 20:15 | 5.7 | 28.0% | 31.5 | 9.61 | 3.280 | 27.4 |
| 20:45 | 6.6 | 31.2% | 30.3 | 9.68 | 3.136 | 27.4 |
| 21:15 | 7.5 | 33.3% | 29.6 | 9.76 | 3.038 | 27.3 |
| 21:45 | 8.4 | 35.1% | 29.0 | 9.81 | 2.956 | 27.2 |
| 22:15 | 9.2 | 37.0% | 28.2 | 9.82 | 2.872 | 27.1 |
| 22:45 | 10.0 | 38.4% | 27.5 | 9.80 | 2.807 | 27.0 |
| 23:15 | 10.8 | 39.6% | 27.0 | 9.83 | 2.753 | 27.0 |
| 23:45 | 11.6 | 41.0% | 26.5 | 9.86 | 2.688 | 26.9 |

*** 09/15/87 ***

| Real Time (Hr:Mn) | Total Volume (L) | Flow Decay (%) | Raw Flow (mL/M) | Raw Press. (PSID) | Spec. Flow | Water Temp (C) |
|---|---|---|---|---|---|---|
| 00:15 | 12.4 | 42.2% | 26.0 | 9.89 | 2.634 | 26.7 |
| 00:45 | 13.2 | 43.5% | 25.6 | 9.94 | 2.574 | 26.6 |
| 01:15 | 13.9 | 44.6% | 25.0 | 9.94 | 2.524 | 26.5 |
| 01:45 | 14.7 | 45.5% | 24.7 | 9.97 | 2.484 | 26.5 |

Figure 3. Typical A-300 Printout.

The operation of the instrument for automatic sampling initiation and termination is quite simple. After the instrument has undergone certain background tasks and transducer nulls, a green light on the front panel indicates that a test filter can be loaded. Once loaded and secured, the run is manually initiated and can be programmed to end at a preset volume, time, flow decay, or manually stopped. The A-300 operation can be stopped at any time, with a complete consolidation of its data as if a normal completion had taken place. This allows automatic sampling to defined end points and/or manual control.

TEST RESULTS

The graphical presentation of flow decay data obtained with the A-300 is similar to graphing data obtained using manual sampling procedures. The "X" axis can represent time or accumulated volumes. The "Y" axis is usually percent (%) flow decay. This axis could be pressure differential.

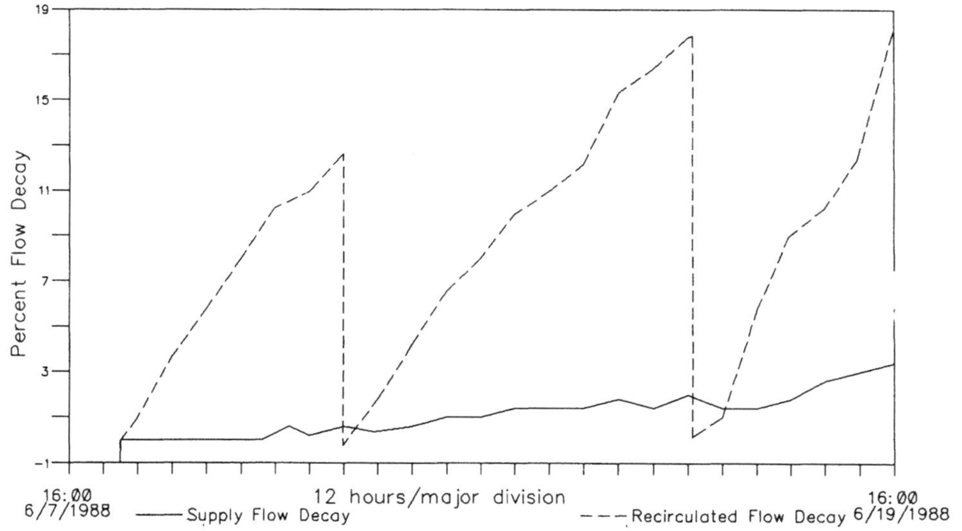


Figure 4. Illustrates Two Different Sampling Points (Supply and Return) in a Purewater System.

An example of flow decay in an ultrapure water system is illustrated in Figure 4. The initial purified water supply is very low in particles compared to the same water after recirculation. The dotted line is a sample point for the recirculated water prior to repurification. These data were collected over a period of eleven days, showing the importance of constant monitoring at multiple points in the pure water distribution system. The flow decay results for the recirculated water required two additional test membranes during this period. The original supply test membrane indicated no appreciable decay during the eleven days. Subsequent SEM analysis of the filters confirmed that the recirculated liquid was many orders of magnitude more contaminated with very fine particles than the ultrapure supply.

The relationship between flow decay and particle concentration, especially that recorded by SEM analysis, is currently under study. Data which have been collected, like the example above, indicate a possible correlation. At the least, flow decay produces a qualitative indication of contamination and provides significant information for improving purification technology. An increase in flow decay is an undisputable indication that further analysis of the test filter using chemical and SEM instrumentation will produce additional analytical information about the particles collected on the filter. The repeatability of this flow decay technique was verified utilizing test filters from the same manufacturing lot number, a standard volume of water through each filter and a 15 psi differential. Each filter tested was analyzed by SEM to confirm the results.

In our quality control testing, an Anatel instrument calibration station was used as a reference water source. Water of exceptional quality (in terms of dissolved ions, organics and particles) is produced and continuously recycled within the station. Conditions such as temperature and pressure can be controlled. Addition of fine contaminants was used in the evaluation process and the subsequent high particle counts were reported using a liquid particle counter and SEM analysis. The differences between levels of contamination at two sampling points can be observed in the flow decays (Figure 5). A small change in flow rate on the test system occurred at 10:00 PM, June 7th, which caused a change in particle distribution. This example indicates the sensitivity of the new analyzer.

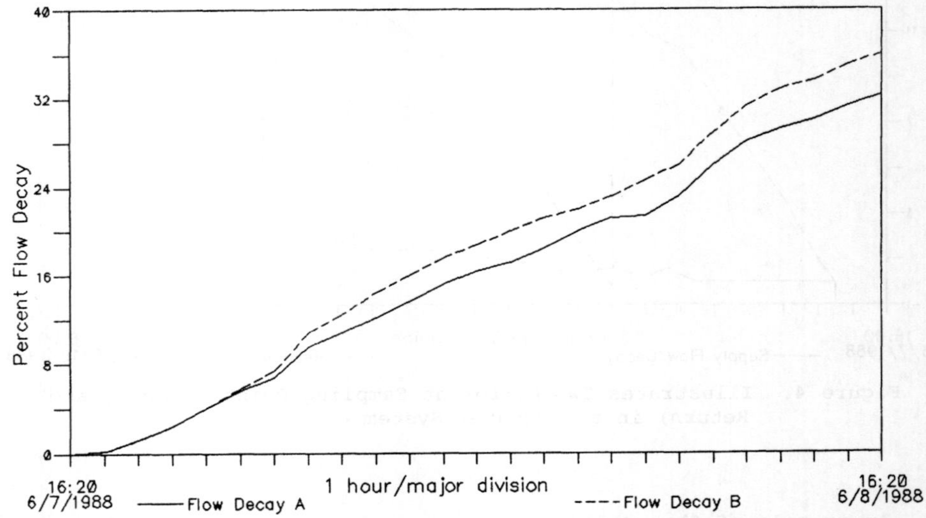


**Figure 5. Comparison at Two Separate Sampling Points on Test System.**

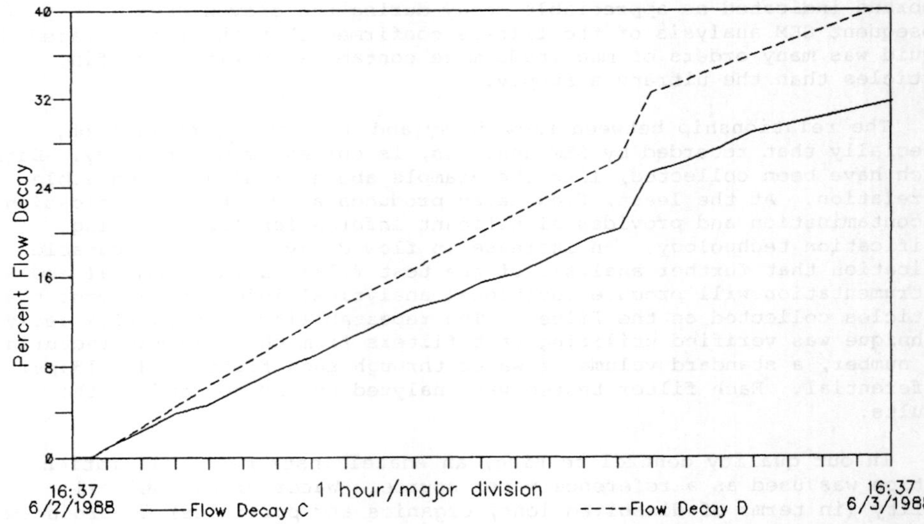


**Figure 6. Effect of Severe Flow Rate Change to Test System by Valve Actuation.**

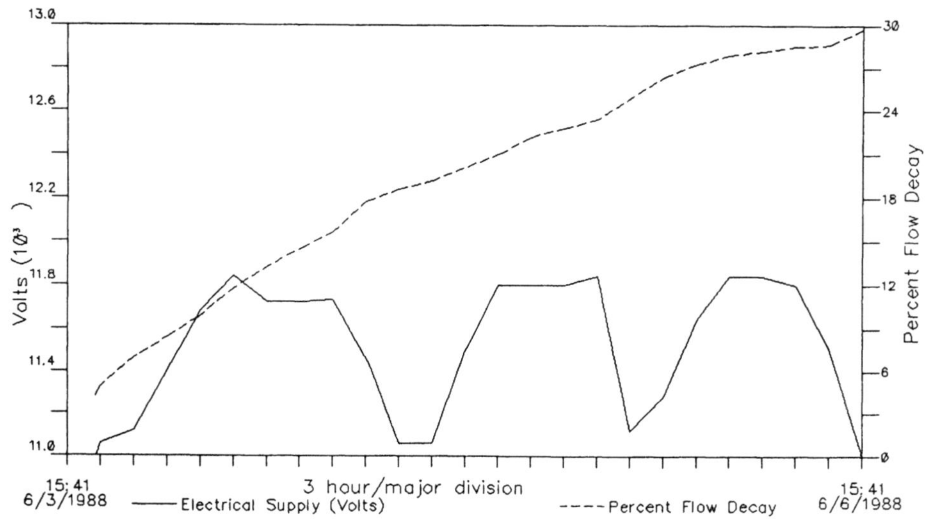


Figure 7. Effect on Flow Decay Due to Small Changes in Electrical Supply to Pump.

The decay information also demonstrates "upsets" in the calibration station caused by manipulation of a value (Figure 6) or variations of the pump performance due to electrical power "brown out" (Figure 7).

Data from two A-300 units, using filters from the same production batch, sampling a single point on the calibration station are illustrated in Figure 8.

This test was performed over a significant time period and the tracking flow decay shows that the increase in filter restriction was parallel in both units.

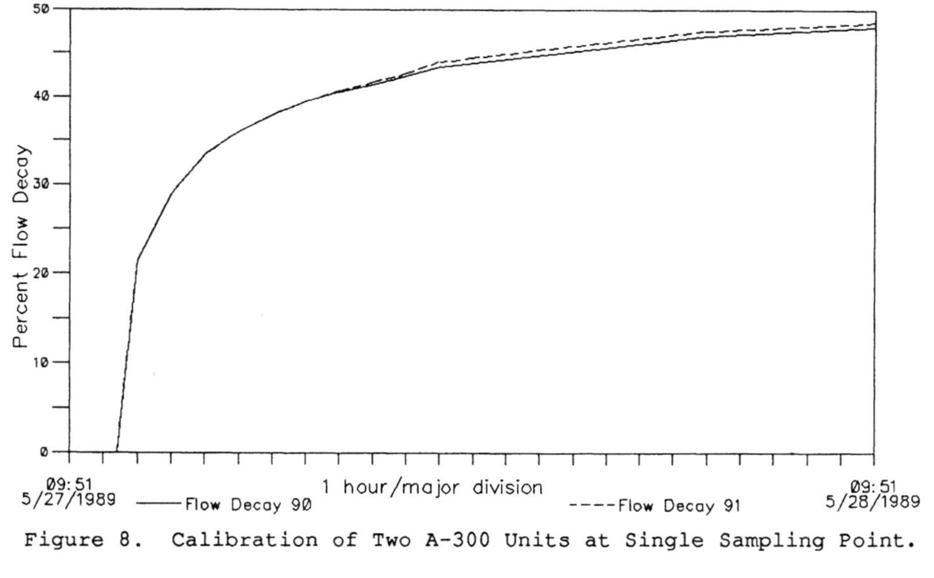


Figure 8. Calibration of Two A-300 Units at Single Sampling Point.

A more difficult task was the evaluation of actual SEM particle counts, using the A-300 Submicron Analyzer versus liquid laser particle counters (LPC). The initial results, as indicated in Table I, do not confirm this relationship.

Table I. Data From a Liquid Laser Particle Counter (LPC), Compared to an A-300's Flow Decay and SEM Analysis

| Accumulated Volume Liters | Particle Density LPC > 0.2 µm Counts/ml | Flow Decay % | SEM Total Counts >0.2 µm | SEM Total Counts >0.5 µm |
|---|---|---|---|---|
| 10 | 3 | 1.0 | – | – |
| 20 | 9 | 3.8 | – | – |
| 40 | 4 | 14.0 | – | – |
| 60 | 2 | 27.0 | – | – |
| 80 | 4 | 49.0 | – | – |
| 100 | 6 | 77.0 | 96000 | 54000 |
| Average | 4400/1 | – | 960/1 | 540/1 |

SEM analysis did not detect the large concentrations of very small particles observed by the particle counters. Microbubbles were suspect for the high numbers obtained by the LPC. These studies are continuing, with an emphasis on relating flow decay to actual SEM analysis. It is important to note that the filter characteristics are a critical component of the flow decay. It was realized early in the development program that the reproducibility of test membranes would be important when comparing data, and thus all of our initial users were supplied with filters from the same manufacturer batch. While certain data, such as parallel runs, suggest the statistical uniformity of these 0.2 µm Nuclepore filters, the amount of data collected at various ultrapure water quality levels is limited. Studies of variability in filter morphology and repetitive testing using different A-300 units are ongoing. Other filtration materials have been employed in the instrument even though in these cases subsequent SEM analysis cannot be performed due to the non-uniformity of the filtering surface. The flow decay observed across different rated pore sized filters should give significant information which will be helpful in evaluating ultrapure water systems with below >0.1 µm particle analysis.

Perhaps the most significant advancement found in using the A-300 Submicron Analyzer is the difference in decay reported on samples which produce continuous zero counts with particle counters in the 0.2 µm analysis mode. These data were collected from a system utilizing 0.2 µm cartridge membranes. Subsequent SEM analysis showed a low population of discernable particles although there was evidence of pore restriction due to absorption of even smaller particles and an organic film.

The ability to compare facilities and processes at very clean levels is a major advancement with the A-300. Our information is preliminary, but indications are that with improved methods of purification the flow decay is greatly reduced. In many cases, this is the only confirmation that treatment efficiency, compared to the previous treatment method, has been substantially improved.

In actual practice, the observance of sudden increases in flow decay are traceable to the exact period of process change, such as new filters installed or replaced, as well as DI columns regenerated. Unknown or unsuspected upsets can be detected in the same manner.

## SUMMARY

SEM sampling is an important technique for measuring particles in high purity liquids, and is destined to be more widely used for assuring quality in critical processes. Care and accuracy in sampling should be an objective for all SEM programs. The use of the A-300 Submicron Analyzer in ultrapure water facilities solves the problems associated with sampling and markedly increases reproducibility and consistency.

As the quality of liquids used in processes improves, the sampling volume necessary for useful SEM analysis (and thus sampling times) becomes greater. "Upsets" in the system can be lost in this time period if they are of short duration or periodic. Not only does the A-300 sample more accurately over the SEM collection period, the data accumulated will indicate sudden or unusual changes in the particle populations through flow decay. These upsets can be traced to changes in the system, or certain combinations of other factors which can not be resolved from only the SEM results. The combination of these techniques is a very powerful tool for the improvement of pure liquid specifications well into the future.

The advantages of real-time, on-line devices for monitoring all types of contamination in pure water is well documented. The use of membrane filtration is a source of such real-time data for particle contamination and offers exciting new possibilities for the improvement of water quality and for tracing facility deficiencies. As smaller particle control is required, filters with smaller pores can replace those used in the current procedure to sample more effectively and increase the particle retention performance. Even if the filter used is not suitable for the SEM procedure (which is true for many of the test membranes), the results of the flow decay provide clear insights into the condition and quality of the process. The application of this automatic monitor is limited only by the filter membranes available.

## REFERENCES

1. R.A. Hango, B. Eldred, and C.F. Frith, DI Water: A common sense approach, Semiconductor Purewater Conference, San Jose ,CA, January, 1984,pp.127-152.

2. M. Yang and D.L. Tolliver, Water particle monitoring for advanced semiconductor manufacturing, Semiconductor Purewater Conference, Santa Clara, CA, January 1989,pp. 100-119.

3. C.F. Frith, State-of-the-art for electronic grade ultrapure water manufacturing and distribution, proceedings of EMTAS Conference, Phoenix, AZ, 1983.

4. C.F. Frith, Membrane filtration: Application, techniques, and problems in "Membrane Filters in Industry", J. Dutka, Editor, Ch. 21, Marcel Dekker, New York, 1981.

5. B. Stranezyk and T. Chu, The value of SEM particle counting in monitoring DI water, Semiconductor Purewater Conference, San Francisco, CA, January 1987,pp.194-220

6. C.F. Frith, Reverse osmosis technology: Applications for high purity water production, in "Electronic Grade Water Production Using Reverse Osmosis Technology", S. Parekh, Editor, Ch. 6, Marcel Dekker, New York, 1988.

7. American Society for Testing and Materials, D4189-82, Test method for silt density index (SDI) of water, Vol. 11.01, 1986.

# SCALING LAWS FOR RAYLEIGH PARTICLE DETECTION IN LIQUIDS

J.S. Batchelder and M.A. Taubenblatt

IBM Research Division,
T.J. Watson Research Laboratory
Yorktown Heights, NY 10598

Different optical designs for particle detection in process fluids have generically different capabilities for scaling their performance to meet future semiconductor manufacturing requirements. We examine the ability of darkfield scattering intensity detection and brightfield amplitude detection to produce statistically significant measurements of liquid contamination levels for particles in the Rayleigh limit. We show that the flow rate dependences for the two techniques are particle size to the twelfth power and particle size to the sixth power, respectively. We find that both techniques require laser power levels to increase more than a factor of two every year if increased laser power is used to compensate for decreasing feature groundrules in the microelectronics industry. Finally we include data on the ability of brightfield amplitude detection to distinguish between particles and bubbles.

## INTRODUCTION

Groundrules for semiconductor devices have been shrinking roughly at a rate of a factor of two every five years. This, plus the reduction of film thickness used in these devices, has been the driver for improving the sensitivity of tools used to detect particulate contaminants. The quest for sensitivity in detecting particles in liquids is of special interest[1] due due to the dominance of process contamination as a yield detractor. In practice, sensitivity to a minimum particle size is not sufficient for process control and yield learning. It is necessary to count a sufficient number of particles such that the statistical significance of the data gives the desired accuracy of the contamination level measurement. The noise or false call rate must be

*Particles in Gases and Liquids 2*, Edited by K.L. Mittal
Plenum Press, New York, 1990

substantially lower than the detection rate, and the escape rate must at least be known and predictable. In this paper we will concentrate on how the design of an optical particle detector for liquids affects the rate at which statistics are acquired from the detector. We will derive the flow rate that is required for sensing particles of a given size for two geometries of optics: the standard darkfield light scatter technique, and forward scatter interferometric detection[2]. We restrict ourselves to non-absorbing particles on axis and in focus, and to single pass illumination optics.

## DARKFIELD DETECTION

We first examine the class of detectors that operate by collecting darkfield light scattered by a particle located at the focus of a gaussian (laser) light source. We assume no signal in the absence of a particle in the focus of the laser. This ignores molecular scatter, the effects of window scattering, and other background effects. Light is collected over as large a solid angle as possible to maximize the signal when a particle is present.

Let $\lambda$ be the vacuum wavelength of the light, $c$ be the speed of light, $\varepsilon_1$ and $\varepsilon_m$ be the complex dielectric constants of the particle and the liquid medium, respectively, $a$ be the radius of the spherical particle, and $\mu_m$ be the magnetic susceptibility of the liquid medium. The wavenumber of the light, $k$, in the liquid is

$$k = \frac{2\pi c}{\lambda}\sqrt{\varepsilon_m \mu_m} \qquad [1]$$

The scattering cross section for the particle[3] (effectively the size of the shadow cast by a non-absorbing Rayleigh particle) is $C_{\text{sca}}$ .

$$C_{\text{sca}} = \frac{8}{3}\pi k^4 a^6 \left| \frac{\varepsilon_1 - \varepsilon_m}{\varepsilon_1 + 2\varepsilon_m} \right|^2 \qquad [2]$$

If a gaussian beam of $I_0$ watts is focussed using optics with a numerical aperture of NA, the radius of the focussed spot at the $1/e^2$ point is[4]

$$\omega_0 = \frac{2c\sqrt{\varepsilon_m \mu_m}}{k\,\text{NA}} \qquad [3a]$$

and the depth of focus, or the distance from the plane of focus where the spot area has increased by a factor of two, is

$$z_0 = \frac{k\omega_0^2}{2} \qquad [3b]$$

Under these conditions, the peak irradiance, in power per area, on axis and in focus is

$$I_{\text{pk.irr.}} = \frac{2I_0}{\pi\omega_0^2} \qquad [3c]$$

and the volume of liquid enclosed by a cylinder $2z_0$ long and $2\omega_0$ in diameter is

$$V_0 = \pi k \omega_0^4 \qquad [3d]$$

The detector(s) will have a quantum efficiency $\eta$. We include the factor $\beta$ for the efficiency that scattered light is imaged onto the detector(s). This includes the numerical aperture of the collector, reflective losses, absorption, and aberrations. Combining Eqs.3a-d gives the observed signal for a Rayleigh particle in a darkfield detector:

$$S_{\text{dark sca}} = C_{\text{sca}} \frac{2\eta\beta I_0}{\pi\omega_0^2} \qquad [4a]$$

The dominant noise source in such a system is at best the Johnson noise from the load resistor $R$. In practice, background scattering gives a larger background signal than Johnson noise, causing the roll-off of flow rate with sensitivity to be faster than indicated in Eq 4e. If $k$ is Boltzmann constant, $B$ is the detection bandwidth, and T is the absolute temperature of the resistor, the background dark noise is

$$N_{\text{dark}} = \sqrt{\frac{4\,\text{kT}\,B}{R}} \qquad [4b]$$

For particles to be counted with negligible false calls, the ratio of the signal to the noise (SNR) must be at least some minimum value. Assuming that the flow rate of the liquid is set such that it matches the electronic bandwidth,

$$\text{Flow} = V_0 B \qquad [4c]$$

we can eliminate the bandwidth from Eqs.4b and 4c to obtain the maximum amount of liquid per unit time that a darkfield system can inspect:

$$\text{Flow}_{\text{dark scattering}} = C_{\text{sca}}^2 \frac{\beta^2 k R I_0^2 \eta^2}{\pi\,\text{kT SNR}^2} \qquad [4d]$$

$$\text{Flow}_{\text{dark scattering}} = \frac{64\pi R \beta^2 I_0^2 \eta^2 k^9 a^{12}}{9\,\text{kT SNR}^2} \left| \frac{\varepsilon_1 - \varepsilon_m}{\varepsilon_1 + 2\varepsilon_m} \right|^4 \qquad [4e]$$

There are some interesting features in this relation. The most important one is that the flow rate goes as the twelfth power of the particle size (decreasing the desired minimum particle size by two decreases the flow rate by 4096). Another result is that the flow rate is independent of the geometry of the illumination optics (NA). It is clear that the load resistor should be increased until the capacitance of the detector times the load resistance is roughly the system bandwidth. Equivalently, there is an upper limit for the allowable capacitance:

$$C_{\max} = \frac{k\omega_0^4}{R\,\text{Flow}_{\text{dark scattering}}} \qquad [4f]$$

This verifies the approach of using arrays of detectors, where each can have a lower capacitance and therefore a higher load resistor. There is also the possibility of cooling the detector and its load resistor to reduce the Johnson noise contribution.

## BRIGHTFIELD DETECTION

Next we derive the flow rate relationship for a brightfield interferometer type of particle detector. In this case the particle introduces a phase shift in one light beam, which is detected when the light from a reference beam is mixed with the first beam at the detectors. Figure 1 shows an example of a brightfield particle interferometer. In this example, helium neon laser light is expanded to fill the pupils

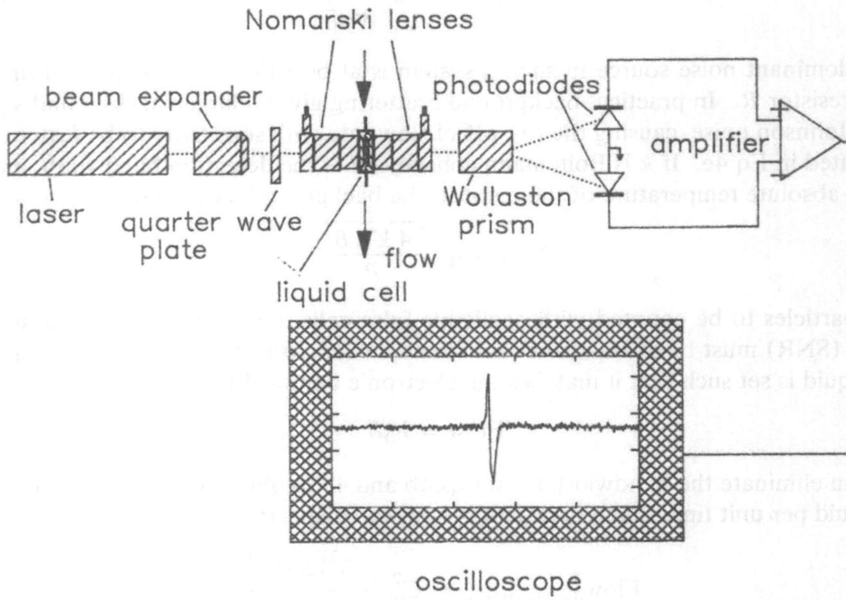


**Figure 1. Brightfield interferometer for particles in liquids.** Circularly polarized light fills the input pupil of the first of two standard Nomarski objectives. The final element of each objective forms a flow cell window. The emerging light is analyzed with a Wollaston prism oriented so that each detector received half of the energy from either spot produced by the first Nomarski wedge. Differential amplification minimizes common mode laser noise.

of 0.7 N.A. Nomarski immersion microscope objectives placed nose to nose on either side of a flow cell. Light emerging from the second objective is analyzed by a Wollaston prism oriented at 45 degrees with respect to the axis of the Nomarski wedges. The relative phase of the two beams produced in the flow cell is adjusted such that both detectors see the same irradiance in the absence of a particle. The detectors drive a differential amplifier. In the absence of a particle, all the light hits the two detectors equally. The particle's presence in the first beam on axis and in focus generates a phase shift in that beam:

$$\Delta\phi = \frac{k^2a^3}{z_0}\left(\frac{\varepsilon_1 - \varepsilon_m}{\varepsilon_1 + 2\varepsilon_m}\right) \quad [5a]$$

Taking the difference of the outputs of the two detectors gives the signal

$$S_{\text{bright phase}} = 2I_0\eta \sin(\Delta\phi)\,\frac{1 - 1/e}{1 - 1/e^2} \quad [5b]$$

$$S_{\text{bright phase}} = \frac{2I_0\eta k^2a^3}{z_0}\left(\frac{1 - 1/e}{1 - 1/e^2}\right)\left(\frac{\varepsilon_1 - \varepsilon_m}{\varepsilon_1 + 2\varepsilon_m}\right) \quad [5c]$$

If $\varepsilon_1 < \varepsilon_m$, the sign of $S_{\text{bright phase}}$ is negative. The fact that $S_{\text{bright phase}}$ is bipolar allows particles to be differentiated from bubbles. For example, using the apparatus in Figure 1, we have taken waveforms from two different events and have superim-

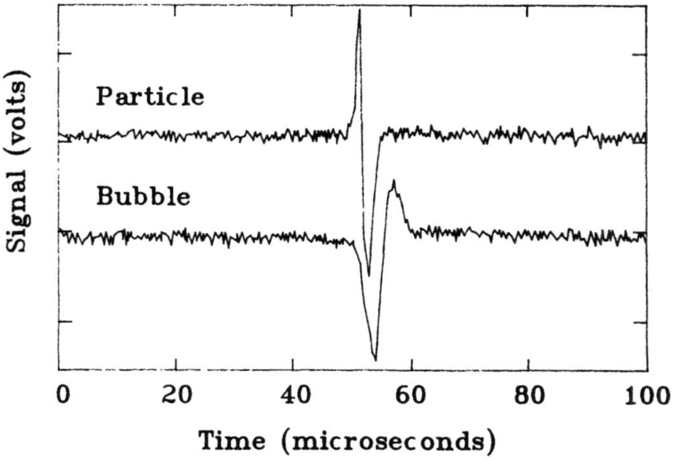


**Figure 2. Signal traces from two different samples.** The upper particle signal is typical for a 100 nanometer latex sphere in water. The lower trace is characteristic of a bubble, such as is produced when a Freon solvent is injected into the water up stream of the flow cell. The change in sign of the signals indicates that the particle has an index higher than water, and the bubble's index is lower than that of water.

posed them in Figure 2. The upper signal shown in Figure 2 is due to a 100 nanometer latex sphere injected into the flow cell. The sine wave shape occurs because the particle first is in the path of one beam, creating a positive phase delay with respect to the second beam, and then it crosses through the second beam, effectively giving the first beam a negative phase shift with respect to the second beam. The lower signal in Figure 2 occurred when Freon-TF (trichlorotrifluoroethane) was injected into the flow cell. It has the opposite sign of the particle signal, and is attributed to a small gas bubble passing through the inspection volume.

The dominant noise source in this case is optimally shot noise produced by random fluctuations in the optical field. With $q$ being the number of coulombs per electron, this noise is

$$N_{\text{bright}} = \sqrt{2qI_0\eta B} \qquad [5d]$$

Combining the above relations gives the amount of liquid per unit time that the brightfield interferometer can inspect:

$$\text{Flow}_{\text{bright phase}} = \frac{8\pi I_0 \eta k^3 a^6}{q\,\text{SNR}^2}\left(\frac{1 - 1/e}{1 - 1/e^2}\right)^2\left(\frac{\varepsilon_1 - \varepsilon_m}{\varepsilon_1 + 2\varepsilon_m}\right)^2 \qquad [5e]$$

In this case we note that the flow rate dependence goes as the sixth power of the particle size. The flow rate is again independent of the numerical aperture of the optics.

## COMPARISON OF DARKFIELD AND BRIGHTFIELD

To give a numerical example of the flow rates predicted by Eqs. 4e and 5e, we assume 100 nanometer diameter latex spheres ( $\varepsilon_1 = (1.59)^2$ ) in water ( $\varepsilon_m = (1.33)^2$ ) with 5 milliwatts of 780 nanometer light. We let $\beta = 0.6$, SNR = 5, $\eta = 0.45$ amps per watt, $R = 10^5$ ohms, and $kT = 4 \times 10^{-21}$ Joules at room temperature. Under these conditions the flow rates for the dark and bright field techniques are 0.27 and 0.26 milliliters per minute, respectively. The maximum darkfield detector capacitance is 0.9 picofarads. (We note that 100 nanometer sensitivity is roughly where the flow rate performance of the two techniques intersect.) We have plotted the change in inspectable flow rate with minimum detectable particle size in Figure 3 for the cases described above.

Comparing Eqs. 4e and 5e lets us draw the following conclusions about how well liquid inspection sensors based on these techniques will address future semiconductor applications:

- As long as the particle size exponent in the cumulative particle distribution is greater than minus six (typical values in practice are around minus two) the statistical validity of the measurement will be improved by increasing the size of the smallest detectable particle and inspecting at a higher flow rate.
- If the needed performance improvement required to satisfy shrinking groundrules is obtained by increasing laser power, with other parameters held constant, the power will have to increase by a factor of about 2.3 per year for either optical technique.
- If the needed performance improvement required to satisfy shrinking groundrules is obtained by reduced flow rates and longer sampling times, the flow rate inspected by the darkfield technique will degrade more quickly with particle size than the flow rate inspected by the brightfield technique.

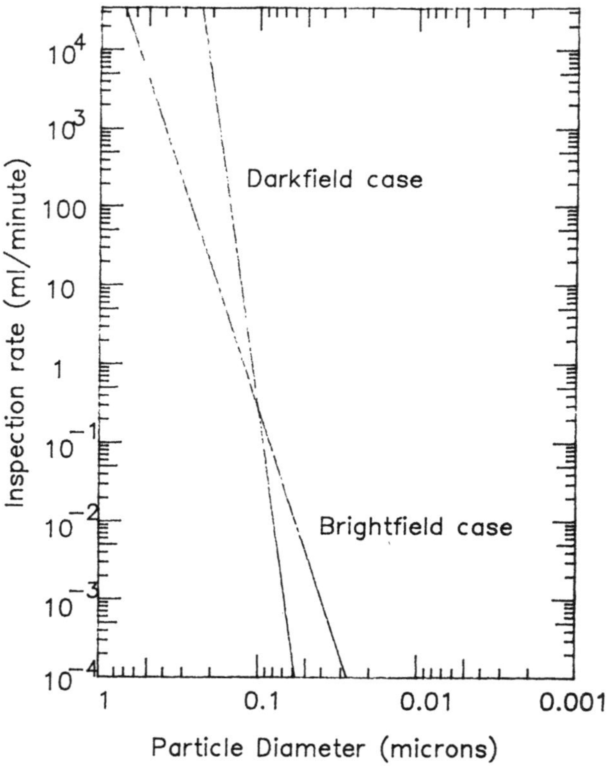


**Figure 3. Variation of inspectable flow rate with particle size** In this example, we have assumed a laser power of 5 milliwatts at 780 nanometers, that the particles are latex ( $\varepsilon_1 = (1.59)^2$) in water ( $\varepsilon_m = (1.33)^2$ ), that a minimum signal to noise ratio of 5 is required, that the detector conversion efficiency is $\eta = 0.45$ amps per watt, that the darkfield load resistance is $R = 10^5$ ohms and that the resistor is at room temperature. We have ignored molecular scatter and other background effects that degrade the performance of the darkfield case.

## REFERENCES


1. M.W.Mittelman, Contamination control activities in today's semiconductor market, ***Microcontamination*** , **5,** 18(1987).
2. J.S.Batchelder and M.A.Taubenblatt, Interferometric detection of forward scattered light from small particles, ***Appl.Phys. Lett.,*** , **55,** , 215(1989).
3. C.F.Bohren, "Absorption and Scattering of Light by Small Particles", John Wiley and Sons, New York, 1983.
4. A.E.Siegman, "Lasers", University Science Books, Mill Valley, CA., 1986.

# REMOVAL OF PARTICLES FROM DEIONIZED WATER IN A RECIRCULATING BATH BY ETCHANT FILTERS

John Robertson
Delco Electronics
2150 E. Lincoln Road
Kokomo, Indiana 46902

Robert Goozner
Pall Corporation
30 Sea Cliff Avenue
Glen Cove, New York 11542

The effectiveness of several etchant filters to remove particulates from a recirculating bath was evaluated in a semiconductor manufacturing cleanroom. The ability of the filters to remove a latex bead challenge from the bath as a function of time was measured using two methods:

1. Counting particles in the bath liquid using an automatic liquid particle counter equipped with a batch sampler.

2. Counting particles adhering to silicon wafers in the bath and counting the adhered particles using a surface particle scanner.

The data from the two analysis methods are correlated. The results are related to variables such as flow rate and filter rating. The optimal filter should have both the highest submicron removal rating, and a high flow rate.

The three filters evaluated showed appreciable differences in both bath cleanup and in the background level of particles in the bath. Of the three filters considered, Filter 1 showed the best performance characteristics in the recirculation system. Filter 1 should, therefore, be the filter specified for use in recirculating service.

*Particles in Gases and Liquids 2*, Edited by K.L. Mittal
Plenum Press, New York, 1990

## INTRODUCTION

Wet etching is a process frequently used during the manufacture of integrated circuits. It is performed by immersing silicon wafers in a chemical which dissolves the top layer of the wafer without dissolving the internal layers. Wet etching is one of the wet chemical processes which should be optimized using microfiltration.[1,2]

Oxide etching is the most frequently performed etch process. Oxide etchants are usually mixtures of HF with $NH_4F$ and $H_2O$. Pre-diffusion etch is a mixture of HF and $H_2O$. Deposited oxide etching or Silox etching is performed by a mixture of acetic acid and $NH_4F$.

The variables affecting removal by the etchant include etchant composition, immersion time and temperature.

Wet etching is an intrinsically dirty process which generates particles. Most contamination control in semiconductor liquids is passive, because the action performed by process steps such as applying a photoresist or rinsing with DI water does not generate particles as part of the process activity. Since the etch process does generate particles, contamination control in recirculating etch baths poses a vigorous challenge. Not only must ambient particles be removed from the fluid, but newly generated particles must also be removed as quickly as possible. The goal is to prevent a particle generated by the etch process from redepositing on the wafer to cause a defect.

The goal of this study was to demonstrate the effectiveness of different filters in removing particles from recirculating etch baths, taking into consideration filter pore size rating and material of construction. The study was performed using deionized water, not an actual etchant, in order to minimize the variables that could have unforeseen influence on the results of the study.

## EXPERIMENTAL

The experimental conditions were selected to best approximate the conditions actually existing in a recirculation bath during semiconductor manufacturing processes.

A Santa Clara Plastics recirculating bath was used for the evaluation. The recirculating bath was fitted with a paddle flow meter immediately downstream of the filter housing. The fluid used for the evaluation was 18 MΩ-cm DI water. The challenge was provided by 0.519 μm latex beads (Duke Scientific).

Analysis of the bath was carried out with a HIAC/ROYCO Model 4100 automatic liquid particle counter equipped with a Model 346-BCL sensor and using a HIAC/ROYCO Model 3200 batch sampler. The batch sampler was set to take 25 cc per min samples. The lower sensitivity limit of the liquid particle counter was to 0.5 μm.

Analysis of the bath was also performed by immersing Fluoroware boats containing 5 polished precounted 5 inch silicon wafers in the bath for 1 minute. Following the immersion, the wafers were spun dry for 120 seconds using a Semitool spin dryer. The wafers were then particle counted using an Aeronca WIS-150 wafer scanner having a resolution of 0.3 μm.

The protocol to evaluate each filter was as follows:

1. Install the filter in the bath.

2. Fill the bath with DI water (approximately 6 gallons).

3. Allow the bath to recirculate for 1 hour to obtain background level of counts using the liquid particle counter. The background level of counts was less than 10 counts per ml.

4. Shut off the recirculating bath. Add sufficient beads (via serial dilutions) to give a challenge of $1.0 \times 10^6$ (±5%) per 25 ml, as determined by the liquid particle counter.

5. Record challenge level prior to turning on pump.

6. Turn on the recirculating bath pump. Record liquid particle counts at 2-3 minute intervals. Boats of 5 inch silicon wafers were immersed in the bath for analysis. After several initial determinations, it was observed that the wafer immersion and the liquid particle counting procedures did not interact to change the absolute number of particles being counted. Therefore, the two different measurements were performed during the same run. But two separate runs were conducted for the same filter due to the difficulty of correctly performing the procedure and the potential for error when both measurements are performed simultaneously.

7. The flow rate is noted several times during the run.

The filters evaluated are listed in Table I. All the filters under evaluation had hydrophilic polyvinylidene difluoride membranes, polypropylene support and drainage layers, and polypropylene hardware. Filter 1, from manufacturer A, had a 0.1 μm removal rating in recirculating service.[3] Filter 2, also from manufacturer A, had a 0.1 μm absolute rating.[4] Filter 3, from manufacturer B, had a 0.1 μm rating. The filter ratings presented are those stated by the manufacturers. The rating of fine membrane filters has been reviewed by Krygier.[4]

Table I. Filters evaluated in the recirculating etch bath. All the filters had hydrophilic polyvinylidene difluoride membranes, polypropylene support and drainage layers, and polypropylene hardware.

| Filter | Manufacturer | Description |
|---|---|---|
| 1 | A | 0.1 μm removal rating in recirculating service |
| 2 | A | 0.1 μm absolute rating |
| 3 | B | 0.1 μm rating |

**RESULTS AND DISCUSSION**

Liquid Particle Counts

The reduction of particles measured in the liquid by the liquid particle counter for Filter 1 is shown in Figure 1. The reduction in

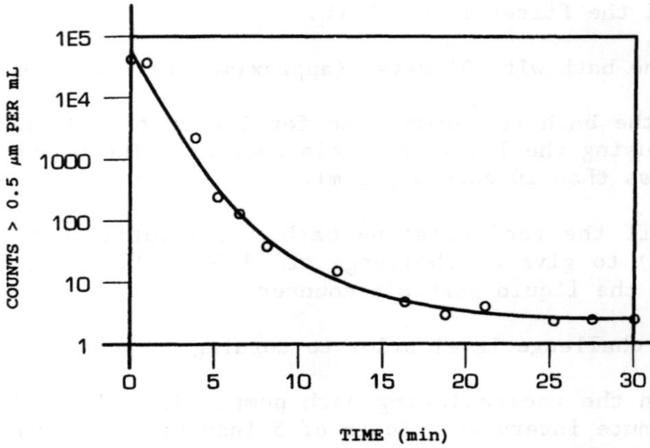


Figure 1. Filter 1 Liquid Particle Counts. Counts > 0.5 $\mu$m per mL Vs. Time (min).

particle counts from the challenge level was very rapid during the first several minutes, and a baseline level of less than 10 counts per ml was attained in approximately 15 minutes.

The reduction in particle counts from the challenge level performed by Filter 2 is shown in Figure 2. The level of particles measured by the liquid particle counter also falls off rapidly to reach the baseline level in approximately 15 minutes.

The reduction in particle counts from the challenge level for Filter 3, as determined by the liquid particle counter, is illustrated in Figure 3. Similar to the other two filters, there is a rapid drop in the particle levels to the baseline in a little more than 15 minutes.

A comparison of the liquid particle counts for Filter 1, Filter 2 and Filter 3 is shown in Figure 4. Figure 4 shows the liquid particle counts up to 15 minutes, which better illustrates the bath cleaning properties of the three filters. The rate of bath cleanup is the fastest for Filter 1, second fastest for Filter 2, and slowest for

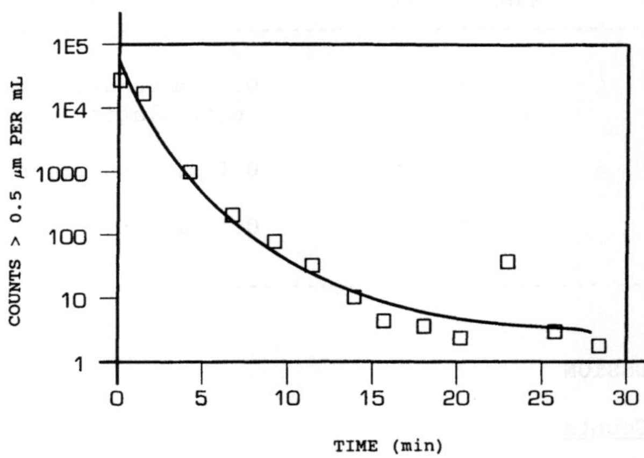


Figure 2. Filter 2 Liquid Particle Counts. Counts > 0.5 $\mu$m per mL Vs. Time (min).

Filter 3. The difference in cleanup is shown numerically in Table II which gives the time required for each filter to yield a liquid particle count of 50 per ml. Filter 1 attained this level the most rapidly at 6.5 minutes, Filter 2 was second fastest at 10 minutes, and Filter 3 was the slowest at 11.5 minutes.

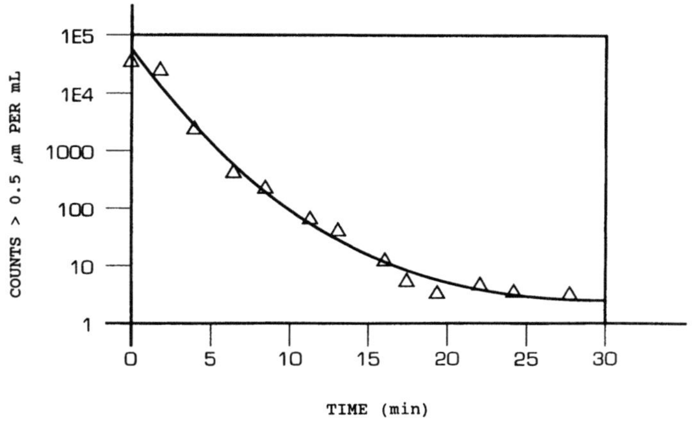


Figure 3. Filter 3 Liquid Particle Counts.
Counts > 0.5 μm per mL Vs. Time (min).

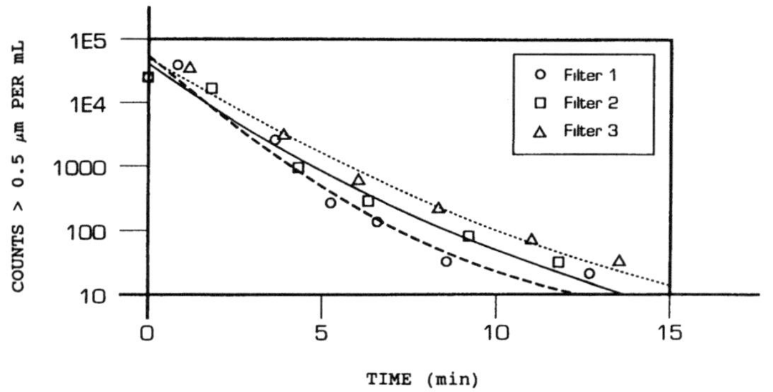


Figure 4. Filter 1, 2 and 3 Liquid Particle Counts.
Counts > 0.5 μm per mL Vs. Time (min).

Table II. Time required for liquid particle counts to reach 50 per ml.

| Filter | Time (min) |
|---|---|
| 1 | 6.5 |
| 2 | 10 |
| 3 | 11.5 |

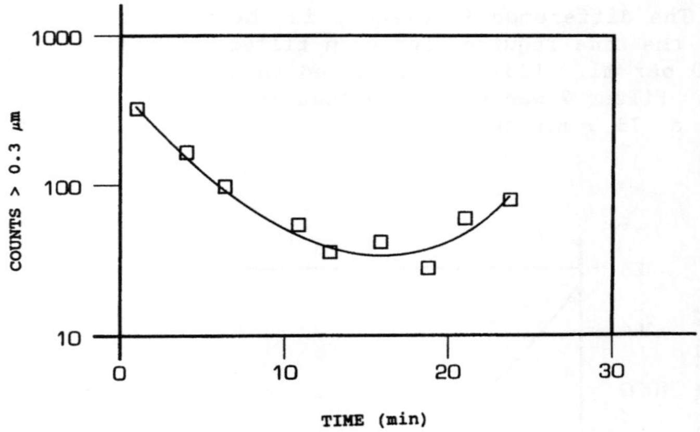


Figure 5. Filter 1 Wafer Particle Counts.
Counts Per Wafer 0.3-2 µm Vs. Time (min).
Challenge: 0.5 µm Latex Beads.

Wafer Particle Counts

The average of the 5 wafers used in each analysis boat was calculated. Figure 5 shows the counts per wafer between 0.3 µm and 2 µm as a function of time for Filter 1. After an initial reduction in particle counts a baseline of less than 100 particles per wafer was attained. There appears to have been some variability in the baseline counts, perhaps due to some type of particle generating mechanism in the bath itself.

Figure 6 shows the counts per wafer between 0.3 µm and 2 µm as a function of time for Filter 2. The initial drop in particles is not as steep as observed for Filter 1, but baseline level of less than 100 particles per wafer is attained in less than 10 minutes.

Figure 7 shows the counts per wafer between 0.3 µm and 2 µm as a function of time for Filter 3. The counts per wafer for Filter 3 drop off rapidly, but the baseline level of counts is higher than those ob-

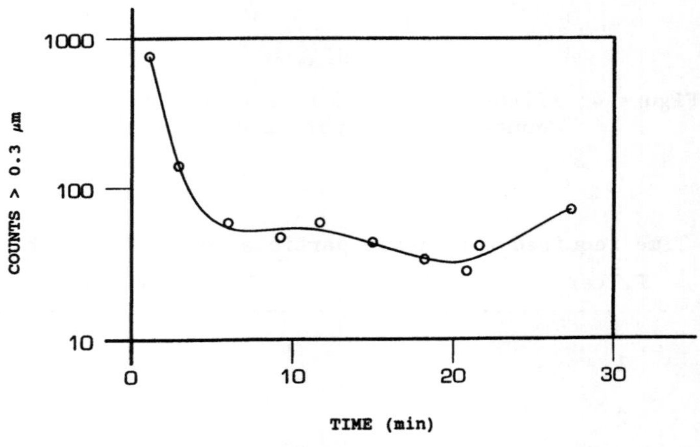


Figure 6. Filter 2 Wafer Particle Counts.
Counts Per Wafer 0.3-2 µm Vs. Time (min).
Challenge: 0.5 µm Latex Beads.

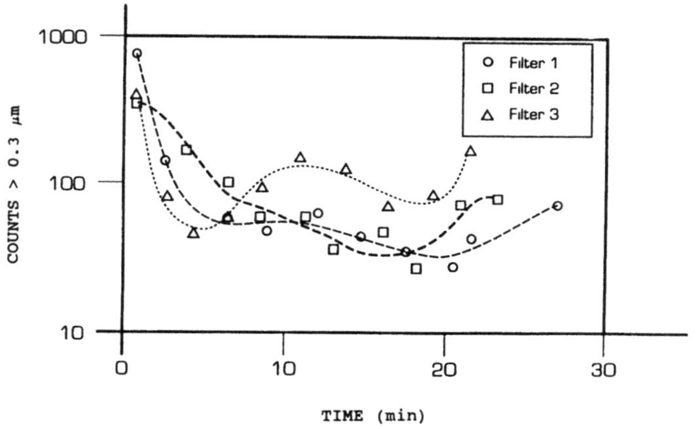


Figure 7. Filter 3 Wafer Particle Counts.
Counts Per Wafer 0.3-2 µm Vs. Time (min).
Challenge: 0.5 µm Latex Beads.

served for Filter 1 and Filter 2. Filter 3 also shows a variation in the baseline level of counts, which is probably due to some undetermined particle generating mechanism in the bath itself.

A comparison of the wafer particle counts for Filter 1, Filter 2 and Filter 3 can be seen in Figure 8, which shows the counts per wafer between 0.3 µm and 2 µm as a function of time for the 3 filters. Filter 3 clearly shows a higher level of baseline counts, while Filter 1 and Filter 2 consistently have a baseline count of less than 100 counts per wafer. The baseline counts for each filter after 10 minutes were averaged, and these values can be found in Table III. Filter 1 had an average baseline of 50 counts per wafer. Filter 2 had an average baseline of 52 counts per wafer. Filter 3 had the highest level of baseline counts, with an average of 124 particles per wafer being observed. It should be noted that most semiconductor etching processes are of relatively short duration, usually several minutes, and processes requiring extended immersion in an etchant are infrequently used.

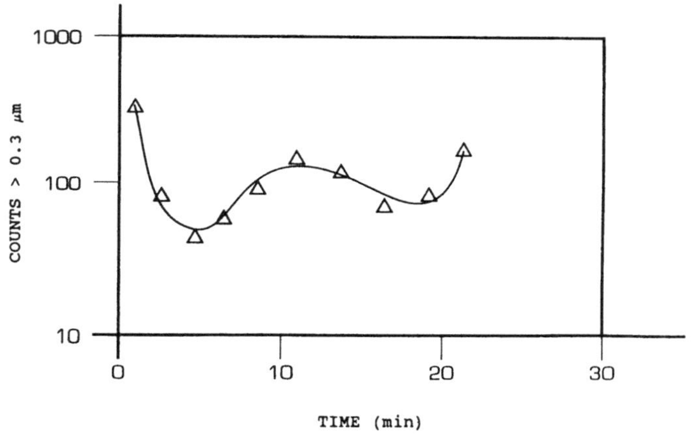


Figure 8. Filter 1, 2 and 3 Wafer Particle Counts.
Counts Per Wafer 0.3-2 µm Vs. Time (min).
Challenge: 0.5 µm Latex Beads.

Table III. Average particle counts per silicon wafer after 10 minutes.

| Filter | Counts 0.3 to 2 μm |
|---|---|
| 1 | 50 |
| 2 | 52 |
| 3 | 124 |

Table IV. Flow rates of the comparative filters.

| Filter | Flow (gpm) |
|---|---|
| 1 | 8.3 |
| 2 | 7.7 |
| 3 | 7.2 |

Flow Rate

The flow rates observed for the three filters when in service in the Santa Clara Plastics bath are given in Table IV. Filter 1 had the highest flow rate with 8.3 gpm. Filter 2 had an intermediate flow rate of 7.7 gpm. Filter 3 had the lowest flow rate, with 7.2 gpm being observed.

**CORRELATION OF THE RESULTS**

The dynamics of particle removal from a recirculating bath by filtration has been investigated.[3] A theoretical analysis of the variables of a recirculating bath fitted with a filter yielded the following first order equation:

$$C_t = C_o \exp(-K\{Qt/v\}^r)$$

where $C_t$ = concentration of particles at time t
$C_o$ = concentration of particles at the start of the test
Q = flow rate
t = time
v = volume of the system

r and K are constants whose values are dependent upon the geometry of the recirculation system.

The above equation will yield the graph of the theoretical relationship between particle concentration in the bath and time illustrated in Figure 9. Figure 9 shows a rapid exponential drop in the particle concentration to reach a steady baseline level of particles as

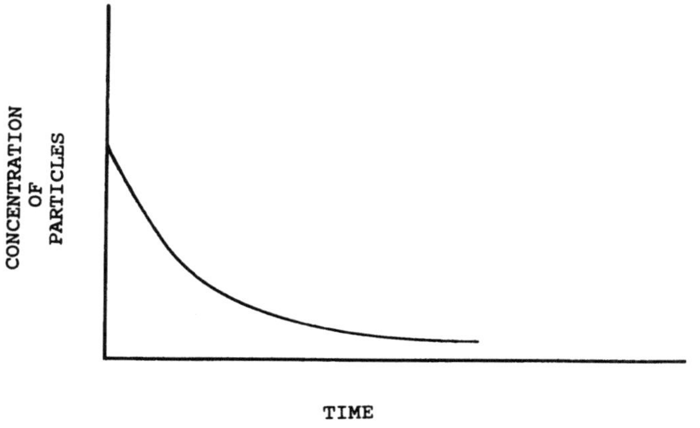


Figure 9. Theoretical Curve of the Reduction of Particles in a Recirculating Bath as a Function of Time.

time proceeds. This behavior is demonstrated by the liquid particle counts for the three comparative filters graphed in Figures 1-4.

The theoretical particle removal behavior shown in Figure 9 is also reflected in the wafer particle counts graphed in Figures 5-8. In this case, the stability of the baseline particle counts is less clear, and undetermined factors may have some effects. These factors can include:

1. The adhesion properties of the silicon wafers.

2. Particles being generated by the recirculating system which are not part of the latex bead challenge.

3. Particles being deposited on the wafers during handling, spin drying, etc.

A properly designed recirculating system will minimize $C_t$. In order to accomplish this, a number of variables must be optimized as follows:

1. $C_o$, the original concentration of particles, should be minimized. This is accomplished by minimizing all sources of particles which can be introduced to the bath. $C_o$ was optimized in this study by performing the experiment under ultraclean conditions in a Class 10 cleanroom.

2. The bath volume, v, should be minimized. This should be a factor when a recirculating bath is specified.

3. The recovery time, t, which follows a particle generating event, required for the particulate level in the bath to return to baseline levels should be minimized. This study has shown this recovery time to be dependent upon which recirculating filter is being used. Frequently there is insufficient time for the bath to clean up to baseline levels during semiconductor manufacture, and all the other factors must be optimized to assure that the bath is as clean as possible during the wet etch process.

4. The flow rate, Q, should be maximized. The different filters evaluated in this study exhibited differing flow rates under identical

conditions. The filter combining the highest flow rate with the smallest particle size removal rating should be selected, all other factors being equivalent.

The results for the three comparative filters are summarized in Table V, which for each filter gives the flow rate, time required for the liquid particle counts to reach 50 per ml, and the average of the wafer particle counts after 10 minutes. Filter 1 displays the highest flow rate, shortest time to reach 50 counts per ml, and the lowest average of baseline wafer particle counts. Filter 2 has the second highest flow rate, second fastest time required to reach 50 counts per ml, and the second lowest average of baseline particle counts. Filter 3 has the lowest flow rate, longest time required to reach 50 counts per ml, and the highest level of baseline wafer particle counts.

Table V. Summary of test results for the three comparative filters.

| Filter | Flow (gpm) | Time For Liquid Particle Counts to Reach 50 per ml (min) | Average Wafer Particle Counts After 10 Min (counts/wafer) |
|---|---|---|---|
| 1 | 8.3 | 6.5 | 50 |
| 2 | 7.7 | 10 | 52 |
| 3 | 7.2 | 11.5 | 124 |

## CONCLUSION

The optimization of many variables must be performed in order to gain the best performance of a recirculating etchant system, thereby minimizing yield losses stemming from particle deposition on the silicon wafers during the wet etching process. Among the variables to be optimized is the choice of filter to be used in the recirculating system. The optimal filter should have both the highest submicron removal rating and a high flow rate.

The three filters evaluated showed appreciable differences in both bath cleanup and in the background level of particles in the bath. Of the three filters considered, Filter 1 showed the best performance characteristics in the recirculation system. Filter 1 should, therefore, be the filter specified for use in recirculating service.

## REFERENCES

1. P. Burggraaf, Applied wet-chemical microfiltration, Semiconductor Intl., 86-89 (March 1985).

2. V. Krygier, M. Latham, R. Conway, Automatic particle measurements in liquids used by the semiconductor industry downstream of fine membrane filters, Proc. Fourth Annual Semiconductor Pure Water Conference, San Francisco (1985) pp. 71-87.

3. B. Gotlinsky, "Removal Rating of Pall Fluorodyne Super Etch Filter In Recirculating Service," Pall Corporation Publication STR REB100, East Hill, N.Y. (1987).

4. V. Krygier, Rating of fine membrane filter used in the semiconductor industry, Proc. Fifth Annual Semiconductor Pure Water Conference, San Francisco (1986) pp. 232-255.

# MODELLING OF PARTICLE REMOVAL FROM A CIRCULATING ETCH BATH

Joseph Zahka, Donald Grant and Cynthia Myhaver

Millipore Corporation
80 Ashby Road
Bedford, Massachusetts 01730

A model has been developed to assist in the design of recirculating etch baths. The model predicts particle concentrations in a bath as a function of the flow characteristics of the bath, the types of filter and pump used, and the pressure drop characteristics of the system. The model shows that high flow rate and high filter particle retention efficiency are desirable for rapid bath cleanup. However, since the particle retention efficiency and pressure drop of filters are inversely related, selection of the optimum filter for a specific bath depends upon the size of the smallest particle which must be efficiently removed.

## INTRODUCTION

The processes used to produce state-of-the-art microcircuits are very sensitive to particle contamination and are expected to become even more so in the future[1,2]. The chemical baths used to clean and/or etch wafer surfaces are a major potential source of particle contamination because the chemicals used in these baths are known to be particle laden and the etching/cleaning processes themselves often produce many particles[3].

The particle concentrations in the bath can be reduced by recirculating the bath fluid through a filter. Recirculation has been shown to decrease particle concentrations by several orders of magnitude within a few minutes[4,5,6]. The effectiveness of recirculation in reducing particle concentrations is dependent upon many variables including the recirculation rate, the flow pattern in the bath, the circulation pump used, and the type of filter used. This paper is an initial attempt to develop a model which quantifies the effects of some of these variables so that etch baths can be more efficiently designed.

*Particles in Gases and Liquids 2*, Edited by K.L. Mittal
Plenum Press, New York, 1990

## MODEL DESCRIPTION

The objective of a model describing a recirculating etch bath (REB) is to predict the concentration of particles in the bath as a function of time and system operating variables. This is accomplished by performing a mass balance on the bath.

A simplified flow schematic of a recirculating etch bath is shown in Figure 1a. In this system, fluid is removed from the bath, pumped through a filter, and returned to the bath. The design shown is for a bath with an overflow weir, although the model described in this paper is appropriate for other flow configurations as well.

The schematic shown in Figure 1a is simplistic in that it assumes a simple mixing pattern in the bath. There are two models of simple mixing patterns[7]. The first is "plug flow." This model assumes that the fluid moves uniformly from the distributor at the bottom of the tank, through the bulk of the tank, and over the weir. It provides the most effective cleansing action of the bath. It takes a single bath volume of clean fluid to remove all the particles from the bath. Plug flow would be possible if the fluid were well distributed at the bottom of the tank and there was no turbulence in the bath. The agitation in the tank would be poor and the particle removal from the wafers due to viscous forces would be minimal.

The second simple mixing model is called Constant Flow Stirred Tank Reactor (CFSTR). In this model, the tank is considered very well mixed and is at uniform concentration throughout. Because the clean fluid entering through the distributor is immediately mixed

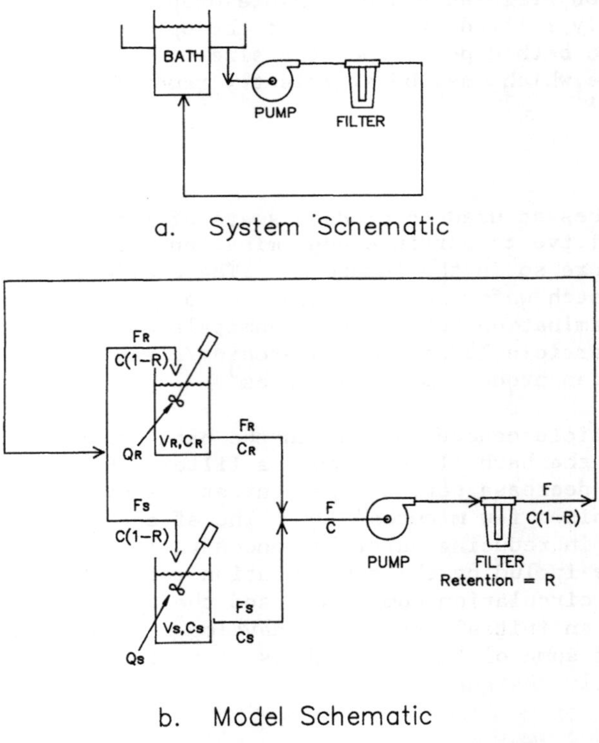


Figure 1. Recirculating Etch Bath Flow Schematic.

with the bulk fluid, the particle reduction in the tank is more gradual. The concentration of particles in the bath would follow the relationship:

$$\frac{C}{Co} = e^{-FRt/V} \quad . \quad . \quad . \quad . \quad . \quad . \quad . \quad . \quad . \quad . \quad . \quad (1)$$

where:

C = Concentration, particles/liter
Co = Initial Concentration, particles/liter
F = Flow rate, liters/minute
R = Filter fractional particle retention
V = Volume of the etch bath, liters
t = Time, min

Neither of these models is expected to be valid because recirculating etch baths are, in general, poorly mixed. Figure 2a compares the predicted performance by each model with experimental results. The experimental results were obtained from an overflow etch bath described by Willis (Figure 18 of reference 5). Willis measured the concentration of particles >0.5μm in diameter in the bath with time while the bath was being recirculated. The bath was operated using a 0.1μm filter with a flow rate such that one bath turnover occurred every 75 seconds. It is assumed that the particle retention of the filter with 0.5μm particles is 99%.

Neither simple model adequately predicts the particle performance of the bath. The plug flow model predicts much faster particle removal than measured. While the CFSTR model describes the initial particle removal adequately, it does not predict the dramatic decrease in the rate of particle removal which occurs after approximately five minutes.

A more realistic model of a recirculation etch bath particle decrease with time is one where the bath is considered poorly mixed. When poor mixing occurs, a bath can be modeled as two well mixed tanks in parallel[8]. One of the tanks has a rapid fluid turnover rate, while the other turns over slowly. The tank with the rapid turnover represents the bulk of the fluid in the center of the bath, while the tank which has a slow turnover rate represents the fluid near the walls of the bath.

If two parallel well mixed tanks are used as a model for the etch bath, the flow schematic for the bath model is shown in Figure 1b. The symbols used in the figure are as follows:

C = Concentration, particles/liter
$C_R$ = Concentration in rapid turnover region, particles/liter
$C_S$ = Concentration in slow turnover region, particles/liter
F = Flow rate, liters/minute
$F_R$ = Flow rate through rapid turnover region, lpm
$F_S$ = Flow rate through slow turnover region, lpm
$Q_R$ = Particle source in rapid turnover region, particles/minute

$Q_S$ = Particle source in slow turnover region, particles/minute
R = Filter fractional particle retention
V = Volume of the etch bath
$V_R$ = Volume in bath rapid turnover region, liters
$V_S$ = Volume in bath slow turnover region, liters

Three mass balances are required to describe this system, one for each "tank," and one for the mixing of the streams exiting the two tanks. The mass balance as a function of time, t, for the tank representing the rapid turnover region of the bath is as follows:

in - out = accumulation

$$[C(1-R)F_R + Q_R] - C_R F_R = \frac{d(C_R V_R)}{dt} = V_R \frac{dC_R}{dt} \quad . \quad . \quad . \quad (2)$$

The mass balance for the tank representing the slow turnover region is:

$$[C(1-R)F_S + Q_S] - C_S F_S = \frac{d(C_S V_S)}{dt} = V_S \frac{dC_S}{dt} \quad . \quad . \quad . \quad (3)$$

The mass balance for the mixing of the streams is:

$$C_R F_R + C_S F_S = CF \quad . \quad . \quad . \quad . \quad . \quad . \quad . \quad (4)$$

The simultaneous solution of these equations to obtain an analytical solution is not straightforward. However, the equations can be solved fairly easily using numerical approximations once the constants in the equations are known.

The form of the analytical solution of the equations is[9]:

$$C = C_\infty + (C_0 - C_\infty)e^{-t/T1} + \tau(e^{-t/T1} - e^{-t/T2}) \quad . \quad . \quad . \quad (5)$$

where:

$C_\infty$ = Final Concentration, particles/liter
T1 = "Rapid Turnover" Time Scale
T2 = "Slow Turnover" Time Scale
$\tau$ = Degree of Poor Mixing

$\tau$ measures the degree of mixing in the tank and roughly scales with Vs/V. When $\tau$ is zero, the single CFSTR model is recovered. T1 and T2 are roughly in ratio to the turnover times for the two tanks.

## EFFECTS OF SYSTEM OPERATING VARIABLES

The model described above can be used to predict the effects of various operating parameters on particle concentrations in the bath. The key parameters are the flow characteristics of the bath, the bath recirculation rate, and the particle retention characteristics of the filter.

In the rest of this paper it will be assumed that the bath has the flow characteristics of an overflow etch bath described by Willis[5]. If it is assumed that the filter used retained >99% of 0.5μm particles, the optimum fit with experimental data occurs when it is assumed that 99.4% of the flow goes through 80% of the volume of the bath (the rapid turnover portion). A fit of the model using these parameters is shown in Figure 2b.

Once the flow characteristics of the bath are known, the effects of recirculation rate and filter particle retention can be predicted. The effect of differences in filter particle retention calculated assuming a bath turnover every 75 seconds is shown in Figure 3. As indicated, the higher the filter retention, the faster the particle concentration in the bath decreases. The effect of turnover rate on the bath cleanup rate calculated assuming 100% particle retention is shown in Figure 4. The higher the turnover rate, the faster the particle concentration decreases.

Figures 3 and 4 indicate that high flow rates and high particle retention are desired to achieve rapid bath cleanup. However, filters which retain particles efficiently generally have higher pressure drops than filters with lower particle retention efficiencies. The higher pressure drop usually results in a lower recirculation flow rate. Hence, one often needs to select between high particle retention and high flow rate.

## DETERMINATION OF SYSTEM OPERATING VARIABLES

### Recirculation Rate

The recirculation rate in an etch bath is a function of the pump used in the system. If a diaphragm or other type of positive displacement pump is used, the flow rate through the system is relatively insensitive to the pressure drop characteristics of the system. However, centrifugal pumps are often used in REB applications and the flow rate through systems using pumps of this type is strongly dependent upon the pressure drop characteristics of the system.

The pump curve for a centrifugal pump relates the pump outlet pressure to the flow delivered by the pump. The curve is a function of the pump design and the horsepower of the pump motor. Examples of pump curves for a family of magnetically-coupled centrifugal pumps such as might be used in a recirculating etch bath are shown[10] in Figure 5. As shown, the pumps deliver high flow with a low outlet pressure or low flow with a high outlet pressure. The flow rate delivered at a given pressure increases with increasing motor horsepower.

The pump curves shown can be fit to a quadratic equation:

$$F = A_0 + A_1P + A_2(P)^2 \quad . . . . . . . . . . \quad (6)$$

where $A_0$, $A_1$, and $A_2$ are constants.
F is flowrate
P is pump discharge pressure

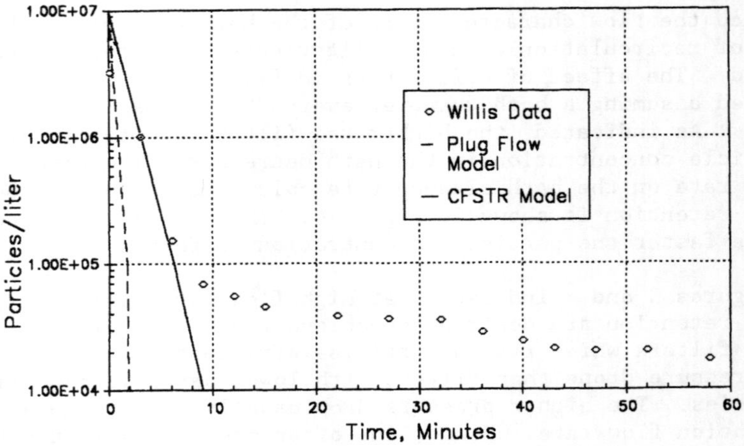


Figure 2a. Example of Simple Mixing Models Fit with Experimental Results.

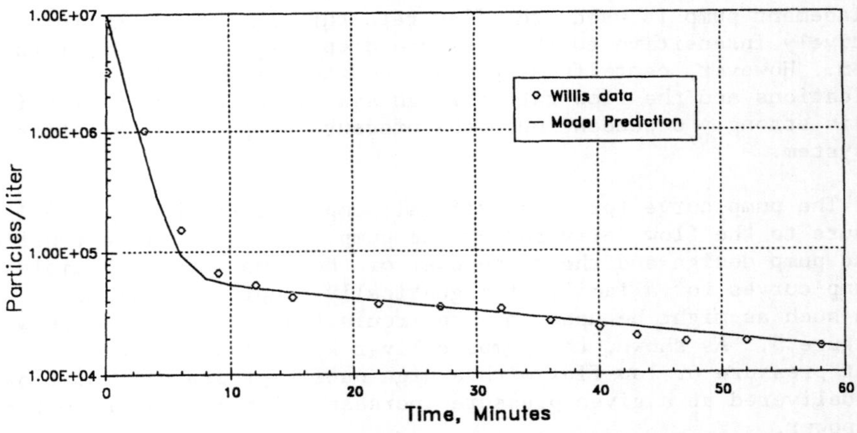


Figure 2b. Example of Parallel Well-Mixed Tank Model Fit to Experimental Results.

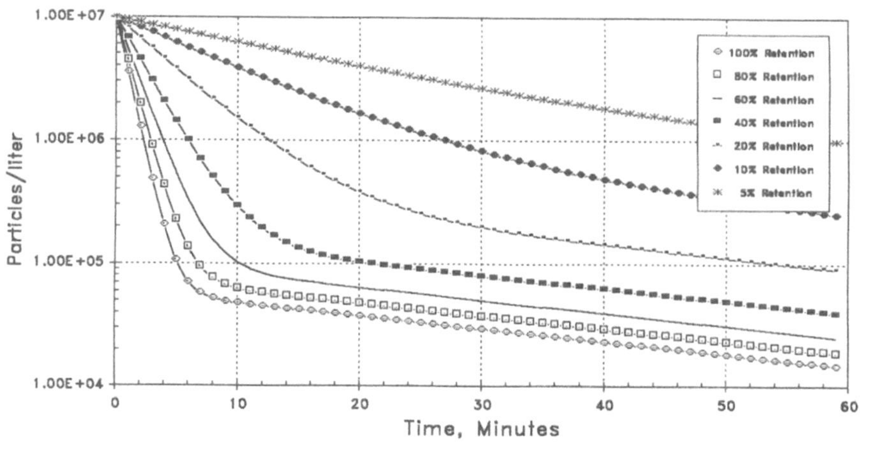


Figure 3. Model Predictions of the Effect of Filter Retention.

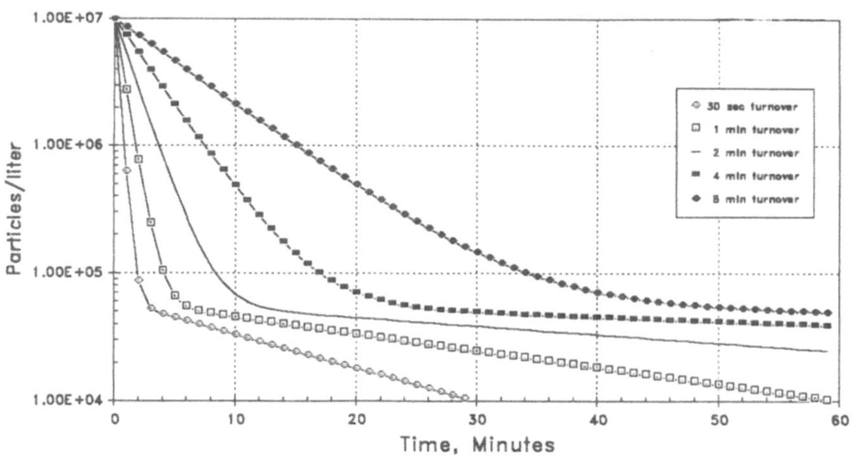


Figure 4. Model Predictions of the Effect of Bath Turnover Rate.

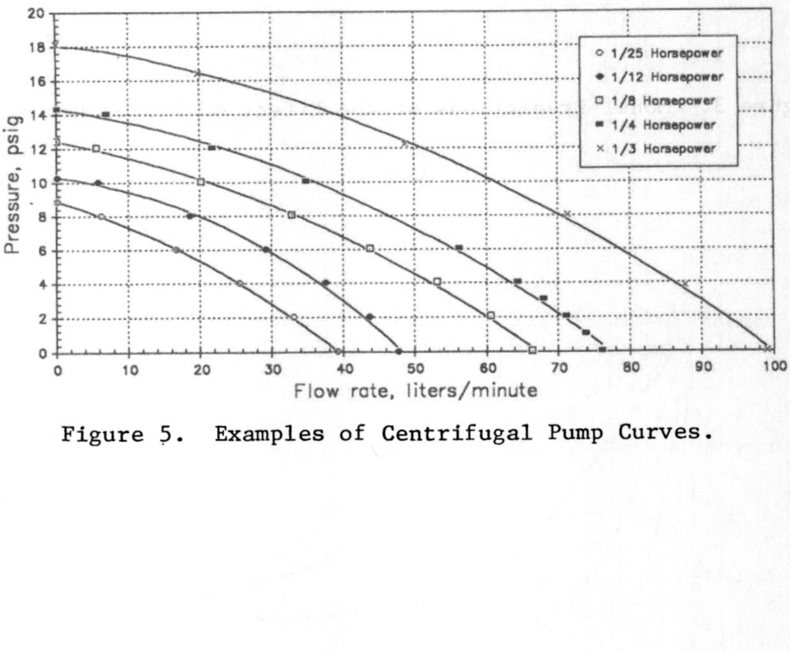


Figure 5. Examples of Centrifugal Pump Curves.

The curves shown in Figure 5 represent least squares best fits of the pump data using equation (6). The constants used in these equations are:

| Horsepower | $A_0$ lpm | $A_1$ lpm/psi | $A_2$ lpm/(psi)$^2$ |
|---|---|---|---|
| 1/25 | 39.1 | -2.68 | -0.179 |
| 1/12 | 47.6 | -1.44 | -0.275 |
| 1/8 | 66.4 | -2.46 | -0.217 |
| 1/4 | 76.1 | -2.11 | -0.202 |
| 1/3 | 97.6 | -1.70 | -0.198 |

There are three major sources of pressure drop in an REB system: the filter, frictional loss through the piping and fittings, and head loss. The pressure drop through the filter ($\Delta P_F$) is directly proportional to the recirculation rate. The pressure drop through the piping and fittings ($\Delta P_P$) is proportional to the square of the recirculation rate. The pressure drop due to head loss ($\Delta P_H$) is related to the height of the return line relative to the level of fluid in the bath and is independent of the recirculation rate. This source of pressure drop is often negligible.

The overall pressure drop in the system ($\Delta P$) is the sum of the three sources:

$$\Delta P = \Delta P_P + \Delta P_F + \Delta P_H$$

$$\Delta P = A_3F^2 + A_4F + A_5 \quad . \; . \; . \; . \; . \; . \; . \; . \; . \; . \quad (7)$$

where $A_3$, $A_4$, and $A_5$ are constants.
F is flow rate

Equation (7) can be substituted into equation (6) and the resulting equation rearranged to yield:

$$0 = A_2A_3{}^2F^4 + 2A_2A_3A_4F^3 + (A_1A_3 + 2A_2A_3A_5 + A_2A_4{}^2)F^2$$
$$+ (A_1A_4 + 2A_2A_4A_5 - 1)F + A_0 + A_1A_5 + A_2A_5{}^2 \quad . \; . \quad (8)$$

Once the constants $A_0$, $A_1$, $A_2$, $A_3$, $A_4$, and $A_5$ are known, this equation can readily be solved using numerical approximation methods.

## Filter Particle Retention

The particle retention characteristics of several types of filters have been measured using a procedure developed by Grant and Myhaver[11], which is similar to standard procedures used to characterize filter cartridge particle retention[12]. In this procedure the filters were challenged with a polydispersed suspension of polystyrene latex (PSL) particles and particle passage was measured as a function of particle size. Particle concentrations were measured using a Particle Measuring Systems (Boulder, Colorado) HSLIS sensor coupled with a μLPS counter. This instrument measures particles larger than 0.10μm in diameter.

Three types of 10" pleated filter cartridges were tested. The filters differed only in the pore size of the hydrophilic Durapore®

membrane contained in the cartridges. The pore sizes tested were 0.1μm, 0.2μm and 0.45μm.

The initial penetration of particles through the three filter types is shown in Figure 6. The data are presented as fractional penetration as a function of particle diameter. Fractional penetration (P) is defined as:

$$P = \frac{C_{OUT}}{C_{IN}}$$

where: $C_{IN}$ = particle concentration upstream of the filter
$C_{OUT}$ = particle concentration downstream of the filter

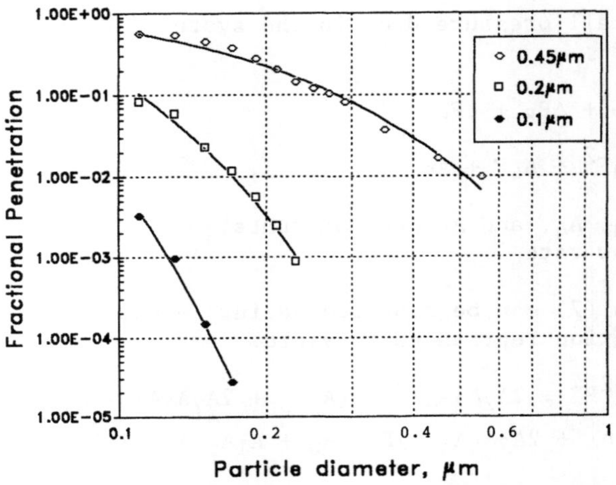


Figure 6. Initial Particle Passage of Durapore Filters.

Fractional penetration is related to retention (R) as follows:

$R = 1-P$

Hence penetration and retention have the following numerical relationship:

| <u>Penetration</u> | <u>Retention</u> |
|---|---|
| 1.00 | 0.00 |
| 0.01 | 0.99 |
| 0.0001 | 0.9999 |

Figure 6 indicates that penetration decreases rapidly with increasing particle size. For example, the fractional penetration of 0.10μm and 0.30μm diameter particles through 0.2μm Durapore cartridges is 0.15 and 0.0001, respectively (equivalent to retentions of 85% and 99.99%). Also, passage decreases with decreasing filter pore size rating. Note that the filters retain approximately 99% of particles at the rated pore size and that particles smaller than the rated pore size are also retained.

## PREDICTED BATH PERFORMANCE IN SPECIFIC CASES

The information provided above indicates that particle concentrations in an REB are a function of many variables. In this section, several specific examples will be examined. All of the examples assume the bath flow characteristics described by Willis[5]. It is also assumed that the system contains the equivalent of 10 feet of 3/4" piping and that the head loss in the system is 0.25 psi. Under these conditions $A_3$=0.0013 psi/(lpm)$^2$, and $A_5$=0.25 psi.

Three filters will be included in the examples. The pressure loss constants ($A_4$) for these filters are:

| Filter Type | Pressure loss constant |
|---|---|
| 0.45μm Durapore | 0.13 psi/lpm |
| 0.20μm Durapore | 0.24 psi/lpm |
| 0.10μm Durapore | 0.53 psi/lpm |

The first example of the use of this model concerns the removal of 0.5μm particles from a 20 liter REB with a 1/25 horsepower pump. Particle concentrations with time for three different filter types are shown in Figure 7. Concentrations are seen to decrease most quickly with the 0.45μm Durapore followed by the 0.2μm Durapore and the 0.1μm Durapore. The system with the 0.45μm Durapore cartridge removes 99% of the particles in 3.3 minutes while the systems with the 0.2μm and 0.1μm cartridges require 4.2 and 7 minutes, respectively.

The case of removal of 0.1μm particles from the bath described in the previous example is shown in Figure 8. In this case, the most rapid cleanup is achieved using the 0.2μm Durapore cartridge followed by the 0.1μm and 0.45μm cartridges. The cartridges remove 99% of the particles in 5,7 and 20 minutes, respectively. Note that although the 0.45μm filter requires a longer time to cleanup the bath than either the 0.2μm or the 0.1μm filters, it still efficiently removes particles which are significantly smaller than the rated pore size of the filter.

The effect of increasing the horsepower of the recirculation pump on the removal of 0.1μm particles in a system with a 0.2μm Durapore filter is shown in Figure 9. Increasing the horsepower from 1/25 to 1/3 decreases the time to remove 99% of the particles from 5 minutes to 2.5 minutes.

A comparison of Figures 7 and 8 indicates that for the bath configuration examined the optimum pore size of the filter used decreases with the need to remove smaller particles. The 0.45μm

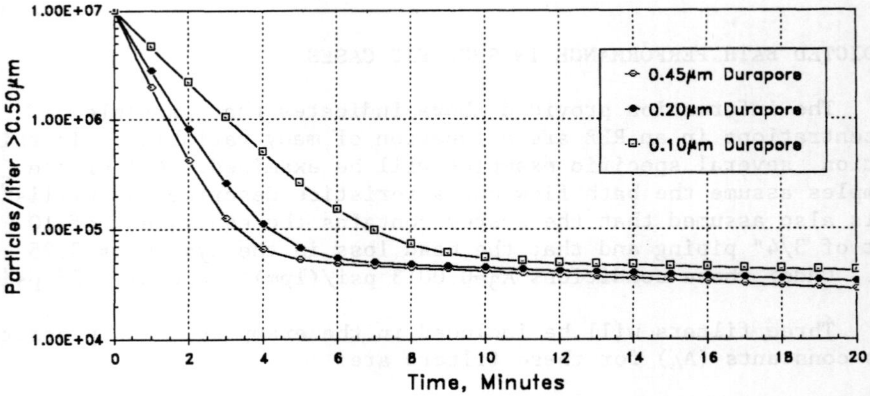


Figure 7. Model Predictions of Initial 0.5 $\mu$m Particle Removal from a 20 Liter REB with a 1/25 Horsepower Pump.

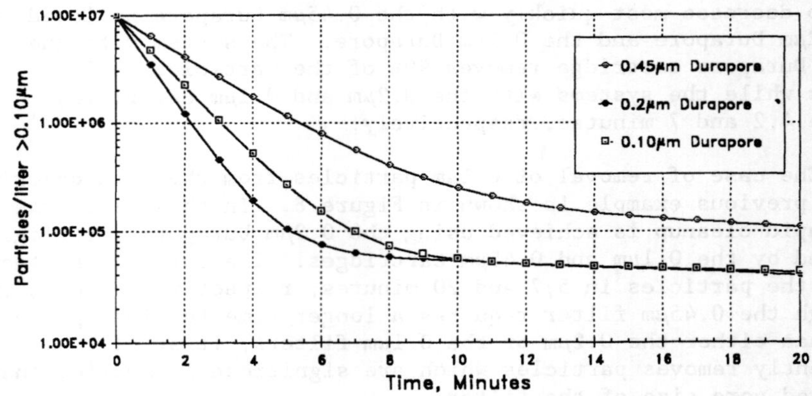


Figure 8. Model Predictions of Initial 0.1 $\mu$m Particle Removal from a 20 Liter REB with a 1/25 Horsepower Pump.

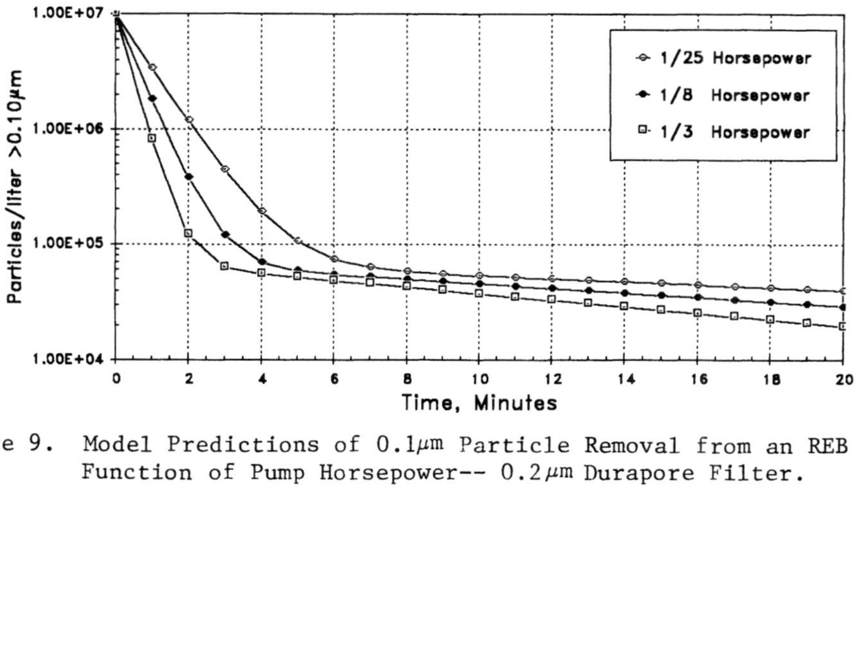


Figure 9. Model Predictions of 0.1μm Particle Removal from an REB as a Function of Pump Horsepower-- 0.2μm Durapore Filter.

filter was shown to be optimum for removal of 0.5μm particles and the 0.2μm filter was optimum for removal of 0.1μm particles. The 0.1μm filter would be optimum for even smaller particles.

Figure 7 indicates that the use of a filter with a pore size smaller than 0.45μm results in a less rapid cleanup of 0.5μm particles. Hence, if a 0.2μm filter is used to remove small particles, the larger particles will be removed more slowly than if a 0.45μm filter were used. However, there are thought to be many more 0.2μm particles in an REB than larger particles. Therefore, if the process for which the REB is being used is sensitive to small particles, a small pore size filter should be used. Alternately, if the process is less sensitive to small particles, the larger pore size filter should be used.

## FUTURE MODEL ENHANCEMENTS

The filter particle retention data presented above are for a new filter. Retention of particles smaller than the nominal pore size of the filter has been shown to decrease with filter loading[13,14]. In addition, as filters are loaded the filter pressure loss constant increases, resulting in a decreased recirculation rate. Hence, the model described above should be modified to include changes in filter retention efficiency and pressure loss constant with time as the filter becomes loaded. This will aid in selection of the optimum filter to be used and in determination of when the filter should be changed.

The filter particle retention efficiency data presented in this paper are based on retention of PSL spheres in water. Etch baths contain various chemicals and surfactants, which may affect the particle retention efficiency of the filter[15,16]. Hence, filter performance should be measured in the solutions used in the REB and the results included in the model.

## CONCLUSIONS

A model has been developed to help in the optimization of recirculating etch bath design. The model assumes that a centrifugal pump is used and that the flow characteristics of the bath can be described using two stirred tanks in parallel. The model accounts for pressure losses in the system and incorporates filter retention efficiency as a function of particle size. The parallel stirred tank assumption was shown to fit experimental data.

The model shows that high flow rate and high filter particle retention efficiency are desirable for rapid bath cleanup. However, particle retention efficiency and filter pressure drop are inversely related. Filters were shown to initially remove approximately 99% of particles the size of the filter nominal pore size rating. Smaller particles were removed less efficiently, but even 0.45μm filters were shown to initially remove >40% of 0.1μm particles. Hence, selection of the optimum filter for a specific bath depends upon the size of the smallest particle which must be efficiently removed. Examples shown in the text indicated that the smaller the particle to be removed, the smaller the optimum filter pore size.

Durapore is a registered trademark of Millipore Corporation, Bedford, Massachusetts.

## REFERENCES


1. D.W. Cooper, Particulate contamination and microelectronics manufacturing: An introduction, Aerosol Sci Technol 5(3), pp. 287-299 (1986).

2. W.G. Fisher, Particle monitoring in clean room air with the TSI 3020 Condensation Nucleus Counter, TSI J Particle Instrumentation, 2(1), pp. 3-19 (1987).

3. T.A. Milner and T.M. Brown, A model for predicting the effect of a processing bath on wafer particle concentrations, Proceedings of the Microcontamination Conference, November 1986, pp. 146-154.

4. K. Dillenbeck, The immersion wet process system, SRC Technical Report No. T88105, Research Triangle Institute, November 1988, pp. 12-1 -- 12-8.

5. C. Willis, Liquid particle counting of buffered oxide etches in bottles and in semiconductor process baths, 6th Annual Microelectronics Technical Symposium, sponsored by Millipore Corporation, San Jose, California, 1988.

6. L. Anderson, Filtered acid recirculation module development, Technical Bulletin 101, Santa Clara Plastics, 1989.

7. O. Levenspiel, "Chemical Reaction Engineering," Chapter 5, John Wiley and Sons, New York, 1962.

8. O. Levenspiel, ibid, Chapter 10.

9. C. Lawrence, University of Illinois, personal communication, 1990.

10. Gelber Pumps 1986-1987, Gelber Pumps, Inc., Lincolnwood, Illinois.

11. Millipore Test Method 0001143TM, "Membrane characterization by polystyrene (PSL) latex bead challenge."

12. I.Y. Hong and E.C. Fitch, An innovative technique in filter rating, Paper number ISSN 0148-7191, Society of Automotive Engineers, 1985.

13. H. Soo and C.J. Radke, A filtration model for the flow of dilute, stable emulsions in porous media, Chem Eng Sci, 41(2), 263-281, (1986).

14. D.C. Grant, Sieving capture of particles by microporous membrane filtration media, M.S. Thesis, University of Minnesota, Minneapolis, 1988.

15. D.C. Grant and W.R. Schmidt, Particle performance of a central chemical delivery system, 7th Annual Microelectronics Technical Symposium, sponsored by Millipore Corporation, May 1989.

16. J.A. Simonetti, H.G. Schroeder and T.H. Meltzer, A review of latex sphere retention work: Its application to membrane pore size rating, Ultrapure Water, 3(4), 46-51, (1986).

# PACKAGING HIGH PURITY CHEMICALS TO ENSURE LOW PARTICULATE CONTAMINATION AT POINT-OF-USE

Thomas B. Hackett and Stephen S. Hupp

Ashland Chemical Company
5200 Blazer Parkway
Dublin, Ohio 43017

Particulate control in high purity electronic chemicals requires that, in every step from chemical manufacture to use, a concerted effort be made to minimize contamination. In the packaging step, particulate contamination can arise from process piping and exposure to the atmosphere. Our data show that the first containers filled in each new lot of chemical contain contaminants from the process piping. Recirculation through filters can be used to reduce these particles. A clean room environment reduces introduction of airborne particulates into the products, but requires rigorous adherence to proper procedures.

The container used to transport a chemical from vendor to user can contribute greatly to particle count. Proper cleaning of the containers is necessary to remove particles before filling. In addition, containers are seldom truly inert with respect to their contents, and chemicals routinely extract from them particulates and cationic impurities. Particle shedding from containers cannot be stopped but can be controlled through container material selection. After all steps have been taken to minimize particle contamination, chemical purity and particulate contamination are verified by the quality control testing.

## INTRODUCTION

As the chip geometries decrease, the tolerance for both size of particles and number of particles decreases. Particulate and chemical

*Particles in Gases and Liquids 2,* Edited by K.L. Mittal
Plenum Press, New York, 1990

contamination must be controlled at every stage from product manufacture, packaging, distribution to the wafer fabrication facility, to delivery to point of use. Normal product specifications for chemicals, piping, valves, and packaging materials are not adequate to meet the requirements of the semiconductor industry. The semiconductor industry demands higher purity chemicals delivered at point-of-use than any other commercial application. Rigorous testing, not only of incoming raw materials and of process equipment coming in contact with the chemicals, is required to minimize particulate and metallic contamination. Packaging of semiconductor chemicals presents a unique set of problems requiring diligence to keep contamination from the product. This paper presents various aspects of the packaging process and how contamination in these areas can best be minimized.

STEPS TO OBTAIN LOW-PARTICULATE CHEMICALS

<u>The Necessity of Thoroughly Cleaning Chemical Containers</u>

One gallon containers, either polyethylene or glass, are not manufactured in clean room environments and, when received at the packaging plant, contain particles from many sources. Plastic containers pick up particles from ambient air used in blow molding and in subsequent handling, such as for trimming or leak testing. The containers are shipped to packaging plants in cardboard overpacks and some cardboard dust inevitably finds its way into the containers, providing another source of contamination, both chemical and particulate. To determine the potential for chemical contamination, a sample of typical cardboard packaging was analyzed for metal content. The results, shown in Table I, indicate that sodium, calcium, and aluminum are the greatest threat of contamination in high purity chemicals. With pressure from the semiconductor industry to continually lower impurity levels, it is imperative that all traces of cardboard be removed from containers.

The following procedures have been adopted to minimize not only cardboard contamination but also any other airborne particulate contamination from packaging containers. The bottles are inverted and blown with filtered air to remove gross contamination (Figure 1). This is followed by cleaning with filtered high purity deionized water. The water is pressurized to reach all interior areas of the bottle with sufficient force to wash out any remaining particles. The bottles are then dried by blowing filtered air into the containers. The cleaning environment is controlled to ensure that the clean bottles are not contaminated before filling and capping.

Table I. Typical Metallic Levels Found in Cardboard Packaging Material.

| Element | ppm |
|---|---|
| Aluminum (Al) | 610 |
| Barium (Ba) | 6.2 |
| Boron (B) | 21.3 |
| Cadmium (Cd) | 0.3 |
| Calcium (Ca) | 1000. + |
| Chromium (Cr) | --- |
| Cobalt (Co) | --- |
| Copper (Cu) | 0.8 |
| Gallium (Ga) | --- |
| Germanium (Ge) | --- |
| Gold (Au) | --- |
| Iron (Fe) | --- |
| Lithium (Li) | --- |
| Magnesium (Mg) | 110 |
| Manganese (Mn) | 34 |
| Nickel (Ni) | --- |
| Potassium (K) | 110 |
| Silicon (Si) | 3.7 |
| Silver (Ag) | --- |
| Sodium (Na) | 1436 |
| Strontium (Sr) | 4.6 |
| Tin (Sn) | --- |
| Zinc (Zn) | 3.6 |

Figure 1. Automatic Bottle Washing Station to Remove Contaminants from Inside the Bottles.

## Container Static Control

Static electricity can be defined as an electrical charge at rest. Containers can pick up charges during the blow molding process and any subsequent handling. Polyethylene has very high electrical resistivity and can retain electrical charges for long periods of time. This electrostatic charge is detrimental in that it increases the collection of ordinary dust, cardboard dust, and polyethylene "fines" in the surrounding air. Static charges were measured on one gallon high-density polyethylene (HDPE) and glass bottles as received from our suppliers. A Simco Type SS-2X Electrostatic Locator was used to measure the voltage level and polarity of the electrostatic charge. Typical static voltages on HDPE bottles are between 3,000 and 5,000 volts. The polarity is negative. Virtually no static charges were measured on the glass bottles.

There are several approaches to minimizing or eliminating this electrostatic charge on the polyethylene containers. One type of device is a static discharge brush made of carbon or fine metallic fibers that just touch the surface to be neutralized. The brush eliminates the charge by conducting it to ground. Nuclear static eliminators use a radioisotope to generate alpha particles which ionize oxygen molecules in air which are swept across and through the container[3]. The most widely used method to control static electricity is to provide ionized air via electrical discharge[4]. This method is used in our packaging facilities. Typically 5,000 to 10,000 volts are discharged from an emitter pin to ground. The air is ionized in this process and blown across the containers. No static charges were detected on HDPE bottles after being bathed in ionized air.

## Clean Room Facility

In order to maintain low particulates and high purity, we have found that it is a major benefit to package electronic chemicals in a clean room environment. Classification of a clean room is determined by the number of particles per cubic foot of air greater than or equal to $0.5\mu m$. In general, room cleanliness is a function of air changes per hour through a HEPA filter. We recommend that container filling and capping be performed in a Class 100 room. A Class 5000 environment is adequate for bottle washing and post filling operations (Figure 2).

As with any clean room facility, a prime source of particulate contamination is the personnel working in the clean room. Personal protective garments offering high filtration capacity to body particles are required to limit clean room contamination. Various fabrics and types of garments are available to provide free movement and comfort. The barrier and shedding properties of various clean room garments have been tested under wafer fabrication conditions[1].

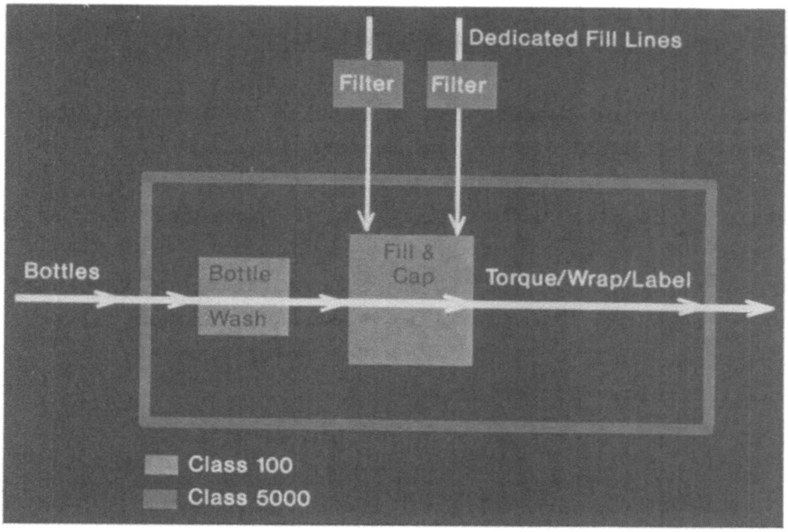


Figure 2. Clean Room Facilities for Chemical Packaging.

Air showers offer an entry way between the clean room filling and capping operation and a normal materials handling and warehousing environment. High velocity filtered air reduces surface dust and lint from garments as workers enter the packaging area and prevent contaminated air from entering as workers leave the clean room.

Each working clean room should have a contamination control engineer responsible for monitoring airborne particle levels and identifying particulate-generating sources. In a high speed packaging operation, sources of particles include: incoming containers, chain driven conveyer belt, pneumatic actuated fill station, and the automatic capping step. Each potential source of particles must be addressed to minimize particle generation.

Worker attitude and training concerning contamination control are important factors in achieving a low particulate environment. The willingness to follow established procedures to minimize particle contamination is perhaps the most important step in packaging high purity chemicals.

Recirculation Through Product Containers and Tanks

Even when a high purity chemical is filtered to give low particulates and is put into a thoroughly cleaned container, particulate levels can increase dramatically from particles shedding from the container. One method to reduce particulate level in a product container is to circulate the contents from the container through a filter and back to the container.

This recirculation process can be modeled with a simple mass balance on the number of particles in the process fluid.

Flow Rate of Particles "in" - Flow Rate of Particles "out" = Rate of Accumulation

$Q \times C = V \; dC/dt$ C = Concentration of Particles

$dC/C = Q/V \; dt$ Q = Flow Rate Through Filter

V = Volume of Drum

$C_O$= Initial Particle Concentration

By solving for concentration of particles we obtain:

$C = C_O \exp (Q/V \times t)$

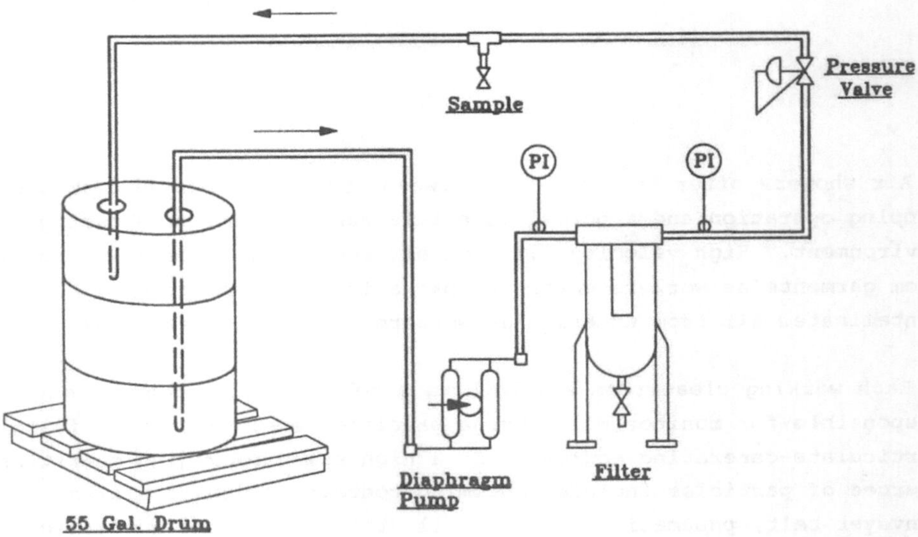


Figure 3. Product Circulation Through a Filter to Lower Particle Count to Specification Levels.

The degree of particle reduction achieved by recirculation can be calculated by the above equation. The particle decrease with time, shown in Figure 4, assumes perfect mixing in the container and filter efficiency of 100%. Neither of these assumptions is rigorously correct and more accurate models are available to characterize this process[2].

Constant recirculation of product in storage tanks through filters ensures that low particulate material reaches the filling station.

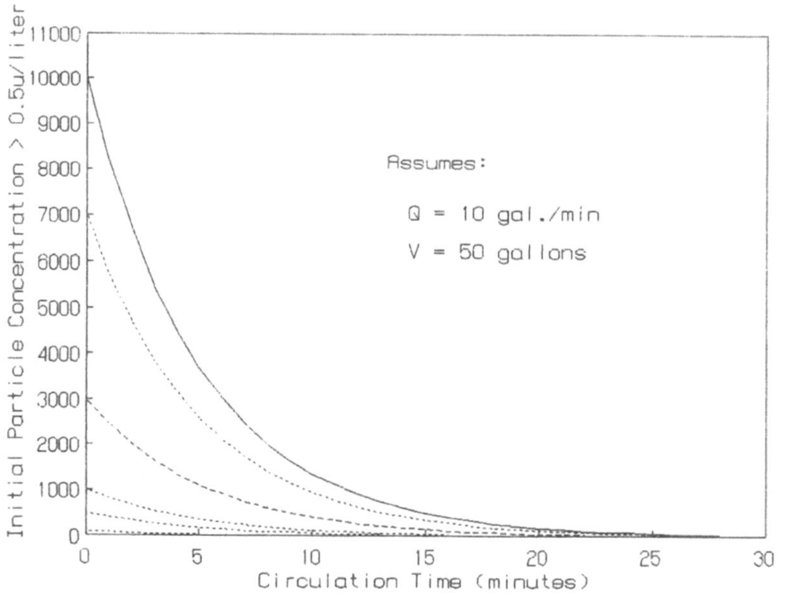


**Figure 4. Model Predicted Particle Reduction by Recirculation.**

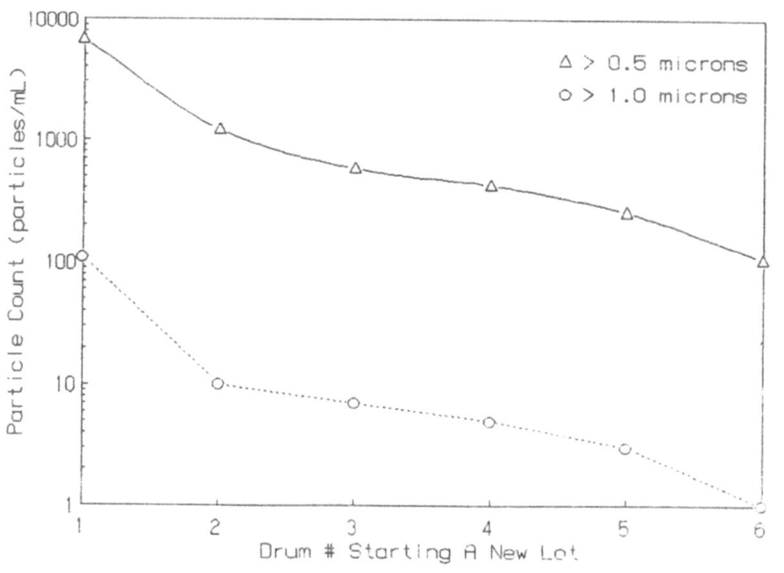


**Figure 5. Data from Sulfuric Acid Drumming Operation**

Filtration and Container Filling

Obviously, the critical steps in packaging operations are filtration and filling. Membrane filter quality has improved to the extent that the filters themselves add few particles[5] and metallic impurities to the product. Prefilters with larger pore sizes (1-2$\mu$m) precede the final polishing filters (0.1-0.2$\mu$m). The products are filtered immediately prior to bottling.

Dedicated fill lines prevent product cross-contamination; however, contamination in process piping, valves and fittings will cause initial containers in each new lot to have high particle levels. Figure 5 shows particle reduction in filling a series of 55 gallon drums with sulfuric acid. As product from a new lot is put through the system, contamination is washed from the fill lines.

Materials of construction must be properly selected for all process equipment, from storage tanks to filling manifolds, to keep contamination to a minimum.

Quality Control

Quality control begins with incoming raw materials and packaging. The packaging process cannot remove dissolved impurities from the chemicals. Chemical purity must be assured before any incoming chemical is put into the packaging system. Chemical analyses on a broad spectrum of impurities are required to meet the specifications of the semiconductor industry. As specifications for product purity become increasingly stringent, sophisticated instrumentation is required to measure low levels of impurities. As a result of the high cost associated with these chemical analyses, many wafer manufacturers rely on their chemical suppliers for accurate chemical data.

Statistical process control is used to follow key variables to monitor the packaging operation. This allows quick response to system upsets and further ensures that all products meet or exceed product specifications.

## CONCLUSIONS

The need for low particulate electronic chemicals is clear. Manufacturers, packagers and users of these materials must all be aware of contamination sources and work to minimize product contamination. In packaging high purity chemicals, not only the chemical but also the container must be handled properly so as to minimize particulate contamination. Clean room procedures are required for critical filling and capping operations. Due to the ubiquitous nature of contaminants, extensive quality control and assurance are required to ensure low particulate contamination at point-of-use.

ACKNOWLEDGEMENT

The authors would like to thank Gil Drab and Les Nichols for providing analytical data and for numerous conversations concerning the packaging of electronic chemicals.

REFERENCES

1. R. C. White and J. R. Weaver, Particle contributions of three types of cleanroom jumpsuits, in "Particles in Gases and Liquids 1: Detection, Characterization and Control", K. L. Mittal, editor, pp. 273-281, Plenun Press, New York, 1989.

2. J. G. Zahka, D.C. Grant, and C. Myhaver, Recirculating etch bath design considerations, paper presented at the 7th Annual Microelectronics Technical Symposium, San Jose, CA, May 1989.

3. R. McCraty, Electrostatics in cleanrooms, in "Handbook of Contamination Control in Microelectronics", D. Tolliver, editor, pp. 153-184, Noyes Publications, New Jersey, 1988.

4. T. Ohmi, H. Inaba, and T. Takenami, Static forces and magnetic fields - Part 1: Research on adhesion of particles to charged wafers critical in contamination control, Microcontamination, October 1989.

5. D. C. Grant, S. L. Peacock, and M. A. Accomazzo, Shedding characteristics of filters in liquids, Microcontamination, July 1987.

ABOUT THE CONTRIBUTORS

*SCOTT J. ALBERHASKY* is presently a technical engineer with Motorola in Mesa, AZ. He has an M.S. in Chemical Engineering from Clarkson University. He was a graduate "Co-op" student at IBM Corporation in Essex Junction, VT, when the reported research was performed.

*MARGARET R. ANEWALT* is the Senior Technical Leader of Microscopy and Supervisor of the Surface Science Laboratory at Air Products and Chemicals, Inc., Allentown, PA. She received M.S. in Chemistry from Lehigh University in 1986. She has published papers on the characterization of catalysts and materials used in high purity gas containment. Her interests include contaminant identification, polymer surface characterization and ceramics.

*J. SAMUEL BATCHELDER* manages applied physical measurements in IBM's manufacturing research department in Yorktown, New York. Over the past five years his group has been involved in developing and implementing defect density measurement techniques for semiconductors, disk drives, and circuit boards. These include contactless measurements for crystalline silicon defects, product wafer scanners for particulate contamination detection and characterization, and liquid inspection for contamination. He has Ph.D. in applied physics from Cal Tech.

*PETER BORDEN* is Vice President of Technology and a founder of High Yield Technology. His work involves the development and use of in-situ monitors for VLSI process control. Earlier, he worked for Varian Associates, where he held positions in R&D management, process equipment development, photovoltaics research, and III-V and silicon device development. He has Ph.D. (1978) degree in Applied Physics from Stanford University. He is the author of over 50 publications and has 12 patents.

*ERIC S. BURNETT* is currently Staff Consultant to Araclean Services, Inc., responsible for the development and evaluation of new materials, instrumentation and cleaning processes for the microelectronics, pharmaceutical and bioengineering industries. Prior to his involvement with cleanroom and protective clothing development and processing, he spent many years as an aerospace systems engineer and project manager with TRW, GE, RCA and British Aerospace Corporation. He received his Ph.D. in Environmental Engineering from California Coast University. He is a Senior Member of the Institute of Environmental Sciences (IES), an Associate Fellow of the Institute of Aeronautics and Astronautics, and is an active member on several IES RP committees. He has authored numerous papers on contamination and pollution control, remote sensing of the environment and energy conservation.

*AHMED A. BUSNAINA* is an Associate Professor of Mechanical Engineering and Director of the Microcontamination Laboratory at Clarkson University, Potsdam, New York. Dr. Busnaina is the author of over 96 publications and presentations on computational fluid mechanics, microcontamination and environmental flows. He is also the author of three software packages for fluid flow simulation on personal computers and main frames.

*YU CHEN CHANG* is a graduate student in the Chemical Engineering Department at the University of Maryland. She is conducting research in aerosol generation and sampling for the last 3 years after receiving her B.S. in Chemical Engineering at the Tamkong University in Taiwan in 1986.

*LISA D. DETTER-HOSKIN* is currently employed as a microanalytical Research Scientist with McCrone Associates-Atlanta. She received a Ph.D. in Inorganic Chemistry from Purdue University, West Lafayette, IN, in 1987. She has published and presented more than a dozen articles concerning issues in desorption ionization mass spectrometry, scanning electron microscopy, and automated scanning electron microscopy.

*EDWARD EZELL* is a Section Leader in the Chemical and Structural Analysis Group at the Group Technical Center of The BOC Group, Inc., in Murray Hill, NJ. He joined BOC in 1979 after receiving his Ph.D. in Physics from Stevens Institute of Technology, He has authored and co-authored a number of papers on materials characterization and gas analysis.

*CLIFFORD F. FRITH* is President and Chief Executive Officer of Anatel Corporation, Boulder, CO., which he founded in 1983. Prior to Anatel, he held the position Executive Vice President of Vaponics. Additionally, he held senior management positions during his 14 years with the Millipore Corp. He has pioneered testing methods and has served on international committees for the development of contamination control stadards. He coauthored The Air Force Technical Order for cleanrooms and clean workstations and served on the Federal Standard 209 Committee for several years. He has published many technical articles, lectured internationally and presently serves as an associate editor to several technical journals specializing in microcontamination. He holds a BS from the Virginia Military Institute with graduate work in Engineering Management and Sanitary Engineering at Northeastern University.

*JAMES W. GENTRY* is Professor of Chemical Engineering at the University of Maryland. He received his doctoral degree in 1969 at the University of Texas at Austin. He is the author of over 130 papers in all aspects of aerosol science including sampling and measurement of submicrometer particles and the electrical properties of aerosols.

*WILLIAM GERRISTEAD* is a Staff Scientist at The Group Technical Center of The BOC Group, Inc, in Murray Hill, NJ, which he joined in 1979. During his tenure at BOC he has been responsible for electron microscopy for materials characterization as well as microcontaminationa. He holds a bachelor's degree in Biological Sciences from Rutgers College. He is the author of several papers in microcontamination and materials characterization.

*SUSAN H. GOLDSMITH* is Co-founder and Director of Engineering of IBR, Inc. in Ann Arbor, MI. She is active in the design and validation of high purity testing and test equipment for filters, valves, and cleanroom products. She has numerous publications in the microcontamination field for the semiconductor, aerospace, and fine chemicals industries. She received a B.S.Ch.E. from The University of Michigan at Ann Arbor.

*ROBERT GOOZNER* is currently a Technical Vice President at Chemstreams, Inc., in Matthews, NC. Prior to his current position, he was employed as a Staff Scientist with Pall Corporation specializing in contamination control problems in the semiconductor industry. He has a Ph.D. in Chemistry from Temple University and was a postdoctoral fellow at the TU-Clausthal (FRG) and U-Freiburg (FRG).

*BARRY GOTLINSKY* is presently an Associate Director of Pall Corporation's Scientific and Laboratory Services Department, responsible for technical support in the semiconductor market. He is involved in numerous studies of contamination control, and has published many papers on the subject. He is involved with SEMI, Sematech, and other trade organizations. He received his Ph.D. degree in Organic Chemistry from the City College of New York.

*DONALD C. GRANT* is a senior staff engineer at FSI International where he manages the Process Research and Development activities for the company. Previously, he was a senior consulting engineer at Millipore Corporation. He received an M.S. in Mechanical Engineering from the University of Minnesota. He has authored more than 40 technical papers.

*RANDY GREB* is a Q/R Engineer with the Intel Materials Technology Laboratory in Aloha, OR. He is responsible for particle analysis and defect reduction technique development and implementation. He earned a B.S. in geology from Oregon State University, Corvallis, OR. He is currently working on an M.S. degree in electrical engineering from the University of Portland, Portland, OR.

*JOHN GREGG* is Applications Manager at High Yield Technology. His work involves the development of applications for in-situ monitors used in VLSI process control. Earlier, he worked for Intel Corporation, where he was a process engineer specializing in thin film and etch processes. He has a BA (1981) in Chemistry from Ripon College and a BA (1984) in Chemical Engineering from the Univeristy of Wisconsin. He is the author of four publications.

*JOHN GREGORY* is Reader in Water Chemistry Department of Civil and Municipal Engineering, University College, London, where he is engaged in teaching and research in the area of water pollution and water treatment processes. He has spent sabbatical periods at the Universities of Pittsburgh, and Delaware. His main research interests are in physico-chemical aspects of water and wastewater treatment processes, including filtration, flocculation and ion exchange, and has many publications to his credit. Recent work has been concerned with the development of novel techniques for monitoring particles and aggregates in flowing suspensions. A new monitor for low levels of particulate contaminants has also been developed. He is an Advisor to the American Water Works Association Research Committee on Coagulation and serves on the Editorial Board of Colloids and Surfaces.

*GEORGE P. GRUNDELMAN* is Co-founder and President of IBR, Inc., in Ann Arbor, MI. He is active in the design and fabrication of high purity particle counting sampling systems, cleanliness test stands, and particulate flushing stands. He studied engineering at the City College in New York, and was Director of Engineering for two device manufacturers before joining IBR.

*THOMAS B. HACKETT* is currently a Senior Research Engineer at Ashland Chemical Company in Dublin, OH, where he is evaluating containers for electronic chemicals and is using advanced separation techniques to upgrade chemical purity. At Ashland Chemical Co., he has worked several years developing thermoplastic alloys and blends including polyester elastomers and paintable TPO's. Previously he was working at Union Carbide Corporation in South Charleston, WV. He received his Ph.D. in Chemical Engineering at the Ohio State University.

*BRUCE HULING* is a Staff Engineer assigned to the Motorola Mos-5 equipment engineering group in Mesa, Arizona, where he is currently project manager for a diffusion rebuild program, and is a member of Motorola scientific and technical society. He has a patent on a purge block and flow switch used on gas delivery systems, and is the author of several technical papers. Prior to joining Motorola, he worked at Intel. In 1980 he received a B.S. degree in Electrical Engineering Technology from Arizona State University.

*STEVEN S. HUPP* is with Ashland Chemical Company in Dublin, OH. Prior to joining Ashland Chemical Co., he worked as a Research Chemist at Gulf Science and Technology, Pittsburgh. He received his Ph.D. in Physical Chemistr from the University of Cincinnati. He has numerous patents and publications in the areas of heterogeneous catalysis and polyester/fiberglass composite automotive body panels.

*DUANE A. JACKSON* is a Ph.D. candidate in the department of Electronics and Instrumentation at the University of Arkansas at Little Rock, Little Rock, Arkansas. He is a recipient of a minority student training grant from the National Aeronautics and Space Administration.

*GERHARD KASPER* is director of R&D at the Chicago Research Center of American Air Liquide. He received his Ph.D. from the University of Vienna, Austria, where he currently has an Adjunct Professor position.

*BRIAN H. KAYE* is Professor of Physics at Laurentian University, Canada. He received his Ph.D. degree in Physics from London University, England. He has made many contributions to the subject of Fineparticle Characterization and recently has been very active in developing applications of Fractal Geometry in the area of Aerosol Science and other branches of Powder Science and Technology. He has published over 100 scientific publications on Powder Science and Technology and is the author of Direct Characterization of Fineparticle Systems (J. Wiley, 1981) and A Randomwalk Through Fractal Dimensions (VCH Publishers, 1989).

*ULRICH H. KOCH* is a Senior Engineer with Nupro Co., Willoughby, OH, one of the Swagelok group of companies. Previously he was manager of new product development for Whitey Co., also a Swagelok group company. During his 30-year career with the company, he has had 27 U.S. patents issued and has published articles in numerous technical journals and magazines. He received a BS (1955) in Naval Architecture and Marine Engineering from the Webb Institute of Naval Architecture, Glen Cove, NY.

*MICHAEL C. KOVACH* is the Contamination Control Manager for Nupro Co., Willoughby, OH. He is responsible for researching, establishing, and implementing processes and quality criteria for the control of particulate and chemical contamination. He graduated from Cleveland, State University with a B.S. in Industrial Engineering (1984).

*ALVIN LIEBERMAN* is presently a Technical Specialist at Particle Measuring Systems, Inc., where he had served in the same capacity from 1983 to 1985. He was Chief Scientist at Hiac/Royco Instrument Division of Pacific Scientific from 1980 to 1983 and from 1985-1987. From 1968 to 1980 he was with Royco Instruments as R&D Vice President. Prior to that he was at the IIT Research Institute (1951-1968) in Chicago. He received an M.S. in Chemical Engineering from Illinois Institute of Technology in

1949. He has been active in several technical societies, particularly in work involved with particle technology. He is past priesident of Fine Particle Society, director of American Association for Aerosol Research, and a fellow of the Institute of Environmental Sciences. He has authored many publications dealing with aerosols, hydrosols and powders, and has worked on several standards and specifications for particle characterization and control, including Federal Standard 209 (A through E). He is regional editor of the journal, Powder Technology, is on the control board of Particulate Science and Technology, and is on the editorial review board of J. Environmental Sciences.

*CHARLES E. LYMAN* is Associate Professor Materials Science and Engineering at Lehigh University, where he has been since 1984. Prior to that he was with The Central Research and Development Department of E.I. du Pont de Nemours & Co. (1980-1984). During 1976-1980 he was an Assistant Professor and Director of the Electron Optics Laboratory at Rensselaer Polytechnic Institute. He received his Ph.D. in 1974 from MIT. He is a director of the Electron Microscopy Society of America and served as national program chairman for the 1984 annual meeting. He serves on the Editorial Board of Review for Metallurgical Transactions. He has published more than 25 technical papers on the characterization and development of Analytical Electron Microscopy to catalysts and other materials.

*CHARLES F. MATTINA* formed in 1988 his own company -- GDE Analytical in Lenox, MA -- which specializes in analytical chemistry and consulting to high-technology industries. He is presently engaged as a consultant to The Texwipe Company as Director of Contamination Control. He received his Ph.D. degree in 1969 from Yale University. After teaching chemistry for three years at Albertus Magnus College in New Haven, CT, he joined the Kimberly-Clark Corporation in 1971, where he was Technical Director for the Specialty Products Division. He is the author of several scientific publications and the holder of five patents. He is the Technical Director of the Contamination Control Division of the Institute of Environmental Sciences.

*KEITH MAY* is currently employed as an Engineer for Eastman Kodak Co., in Rochester, NY. He received his M.S. in Electrical Engineering from Clarkson University in 1987.

*MALAY K. MAZUMDER* is Head of the Department of Electronics and Instrumentation and the Interim Director of the Graduate Institute of Technology at the University of Arkansas at Little Rock. Prior to his appointment at the University of Arkansas, he served with the Atomic Energy Commission as an Instrument Engineer, from 1961 to 1967. He is the inventor of the E-SPART Analyzer--a laser-based instrument for simultaneous measurement of size and charge distribution of powders. Among his other achievements is the development of a dual-beam laser Doppler velocimeter. He received his Ph.D. in instrumental sciences from the University of Arkansas, and has 50 publications and presentations in the area of laser applications and particle technology.

*RICHARD J. McCLUSKEY* is an Associate Professor of Chemical Engineering at Clarkson University, where he is a member of the Center for Particulate Control. He received his Ph.D. in Chemical Engineering from The University of Minnesota.

*MICHAEL McKEAG* is a Senior Engineer with the Intel Materials Technology Laboratory in Aloha, OR. He is responsible for development and implementation of analytical techniques for IC defect analysis. He earned a B.S. in Physics from the University of Washington, Seattle, WA.

*PAUL CHARLES MELANSON* has held the position of lead chemist and instrument design.engineer within the Engineering Department of Anatel Corporation since 1985. After offering consulting services to numerous companies within the water purification industry, he began his Anatel career by designing ultrapure water monitors and analyzers. Prior to consulting, he worked as a process control engineer for Pure Cycle Corporation from 1976 to 1983. He graduated from The Colorado College in 1976 as a Barnes Scholar with a BA in Chemistry.

*KASHMIRI LAL MITTAL* * is presently employed at the IBM US Technical Education in Thornwood, N.Y. He received his M.Sc. (First Class First) in 1966 from Indian Institute of Technology, New Delhi, and Ph.D. in Colloid Chemistry in 1970 from the University of Southern California. He has organized and chaired a number of very successful international symposia and in addition to this volume, he has edited 33 more books as follows: Adsorption at Interfaces, and Colloidal Dispersions and Micellar Behavior (1975); Micellization, Solubilization, and Microemulsions, Volumes 1 & 2 (1977); Adhesion Measurement of Thin Films, Thick Films and Bulk Coatings (1978); Surface Contamination: Genesis, Detection, and Control, Volumes 1 & 2(1979); Solution Chemistry of Surfactants, Volumes 1 & 2 (1979); Solution Behavior of Surfactants: Theoretical and Applied Aspects, Volumes 1 & 2 (1982); Adhesion Aspects of Polymeric Coatings, (1983); Physicochemical Aspects of Polymer Surfaces, Volumes 1 & 2 (1983); Surfactants in Solution, Volumes 1, 2 & 3 (1984), Adhesive Joints: Formation, Characteristics, and Testing (1984), Polyimides: Synthesis, Characterization and Applications, Volumes 1 & 2 (1984); Surfactants in Solution, Volumes 4, 5 & 6 (1986); Surface and Colloid Science in Computer Technology (1987); Particles on Surfaces 1: Detection, Adhesion and Removal, (1988); Particles in Gases and Liquids 1: Detection, Characterization and Control (1989); Particles on Surfaces 2: Detection, Adhesion and Removal (1989); Surfactants in Solution, Volumes 7-10 (1989); Metallized Plastics 1: Fundamental and Applied Aspects (1989); and Polymers in Information Storage Technology (1989). Also he is Editor of the Series, Treatise on Clean Surface Technology, the premier volume appeared in 1987. In addition to these books he has published about 60 papers in the areas of surface and colloid chemistry, adhesion, polymers, etc. He has given many invited talks on the multifarious facets of surface science, particularly adhesion, on the invitation of various societies and organizations in many countries all over the world, and is always a sought-after speaker. He is a Fellow of the American Institute of Chemists and Indian Chemical Society, is listed in American Men and Women of Science, Who's Who in the East, Men of Achievement and many other reference works. He is or has been a member of the Editorial Boards of a number of scientific and technical journals, and is the Editor of the Journal of Adhesion Science and Technology, which made its debut in 1987. In 1990 he received the Charles B. Dudley Award of the American Society for Testing and Materials.

*MARC W. MITTELMAN* is an industrial microbiologist specializing in the analysis and control of biological contamination in industrial and environmental water systems. He is currently researching bacterial adhesion processes at the Institute for Applied Microbiology/University of Tennessee, Knoxville, where he is completing doctoral studies. From 1982-1986, he was president of Applied Microbiological Services, Inc. In 1987, he formed Water Micro Associates, a consulting and technical

---

* As the editor of this volume.

publishing firm, in San Diego, CA. He holds M.S. degree in microbiology from CSU-Long Beach, CA. He is the editor and co-author of the text Biological Fouling of Industrial Water Systems: A Problem Solving Approach (Water Micro Associates, 1987) and has authored or coathored over 30 publications in the fields of contamination control and environmental microbiology.

*CYNTHIA MYHAVER* is an Applications Engineer in Research, Development and Engineering at Millipore Corporation, Bedford, MA. She is currently attending Keene State College, working towards a BS in Industrial Chemistry. Her research interests include particle counting methodology, hard particle retention of liquid filters, chemical filtration, and vacuum outgassing of gas filters and purifiers.

*STEVEN J. PALEY* is the Director of Research and Development for The Texwipe Company in Upper Saddle River, New Jersey. He holds an M.S. degree in mechanical engineering/product design from Stanford University. His previous employment includes the IBM Corporation where he worked on the System 38 computer development, and AT&T Bell Laboratories where he was involved in the development of specialized flat screen computer terminals for medical information systems. He holds two United States patents and has several others pending.

*TERESA PFEIFER* is currently working on her M.S. degree in Electrical Engineering at Clarkson University, Potsdam, NY.

*DENIS J. POLEY* is presently a Staff Engineer working on development of 16 Mbit DRAMs for the General Technology Division of the IBM Corporation, Essex Junction, VT. He has over twenty years experience in the analysis of high purity gases.

*B. RAGENT* is currently a lecturer at Stanford University and a member of San Jose State University Foundation. Dr. Ragent served as senior staff scientist, Space Science Division of NASA Ames Research Center. Some of his research projects on aerosols include his work at NASA as Co-PI/PI on nephelometer experiments related to Pioneer-Venus Probe, Galileo Jupiter Mission Probe. He also served as Acting Chief, Atmospheric Experiments Branch, in charge of aircraft flight experiments to measure atmospheric aerosol.

*M.B. (ARUN) RANADE* is an Adjunct Professor in the Chemical Engineering Department at the University of Maryland at College Park, MD and President of Particle Technology, Inc. He received his doctoral degree in Chemical Engineering at the Illinois Institute of Technology in 1974. He has over 20 years experience in Research and Development in Aerosols and Fine Particles.

*JOHN T. ROBERTSON* is a contamination control specialist for Vectech Inc., an environmental control and consulting firm in Farmington Hills, MI. He has 5 years experience in the contamination control field, and has worked with both semiconductor and pharmaceutical manufacturers. His previous responsibilities have included: cleanroom materials evaluation and selection; process equipment particle reduction and monitoring; and cleanroom garment evaluation, selection, and field testing. He is a former employee of both Motorola, Inc., and Delco Electronics Corp. The work described in this paper was completed while he was an employee of Delco Electronics Corp, Kokomo, IN.

*CAMILLA M. SAVIZ* is a graduate student in the Mechanical Engineering Department at Clarkson University. She received her MSME from Clarkson University in 1988.

*PABLO SCHABES-RETCHKIMAN* is Associate Professor in the Instituto de Fisica, Universidad Nacional Autonoma de Mexico, where he has worked since 1980. He received D. Phil. degree from University of Sussex, England in 1980 and was a Visiting Scientist at Lehigh University (on sabbatical). In 1984 he was admitted as a member of the Mexican Academy of Sciences (Scientific Research) and received an award of the Mexican government "Investigador Nacional." He has been involved in research of physical adsorption of rare gas monolayers, small metallic particles and catalytic materials by the use of electron microscopy and diffraction, and has published more than 15 technical papers in these areas.

*ROBERT SCHILLING* is Associate Professor in Electrical Engineering Department at Clarkson University, which he joined in 1978. He received his Ph.D. degree from the University of California at Berkeley in 1973. He worked as a Lecturer in Electrical and Computer Science Department at the University of California, Santa Barbara from 1974 to 1978. He has published two textbooks including one in the field of robotics.

*ROBERT SHERMAN* is a Lead Scientist at the Group Technical Center of The BOC Group, Inc., in Murray Hill, NJ. He obtained his Ph.D. in Metallurgical Engineering from the University of Illinois in 1982. He has authored and co-authored over twenty papers related to materials science, surface analysis, and electron microscopy.

*M.P. SINHA* has been at the Caltech Jet Propulsion Laboratory, Pasadena, CA since 1978 when he joined as a Member of the Technical Staff. At JPL he has been interested in the chemical analysis of aerosol particles, and in the development of a GC-MS system for the field-measurement of air pollutants. He has developed a new technique known as Particle Analysis by Mass Spectrometry (PAMS) for the analysis of individual particles on a continous, real time basis. He received his Ph.D. degree in Chemistry in 1974 from Columbia University, New York, and did postdoctoral work at Yale University. He was awarded a NASA Exceptional Service Medal in 1988. He holds a U.S. patent on the particle analysis by PAMS technique.

*MARK SMITH* is Supervisor of Testing Services at IBR, Inc., in Ann Arbor, MI. He is responsible for cleanliness and performance testing of critical filters, valves, cleanroom products, and flow controllers. He studied engineering at Eastern Michigan University.

*T.C. SMITH* is a Member of Technical Staff in Motorola's Semiconductor Sector and is section manager for ion implantation in the Bipolar Technology Center in Mesa, Arizona. He has been working with ions since receiving his M.S. degree in Physics from the University of Notre Dame. Since 1967, he has specialized in various aspects of ion implantation technology and has published numerous articles in this field. He has directed his efforts recently toward reducing particle contamination in implanters. He has served on the faculty of the School on Ion Implantation which accompanies the International Conference on Ion Implantation Technology.

*HOLGER T. SOMMER* is Vice President of Engineering and Development for Particle Measuring Technology of Hiac/Royco., a division of Pacific Scientific. Prior to this position he held positions at the University of Maryland and Carnegie Mellon University. His engineering and research interests include, among other topics, the development of electro-optical measurement instrumentation for velocity measurements in flames, and particle and droplet sizing in multiphase sprays. He holds a Ph.D. in Mechanical Engineering from the Technical University in Aachen, FRG.

*ARNOLD STEINMAN* is director of engineering for Ion Systems, Berkeley, CA, where he oversees the design and production engineering of all static control products. He was chief engineer at Staefa Controls, San Leandro, CA, from 1981 to 1983, and he has also worked as a consulting engineer with Electronics Workshop, Berkeley, from 1972 to 1981, and as an electronics engineer at the Lawrence Radiation Laboratory, Livermore, CA, from 1966 to 1971. He serves on the EOS/ESD Associations's ionization standards committee and as a panel member for its annual ionization workshop.

*MARC A. TAUBENBLATT* has been working at the IBM T.J. Watson Research Center in Yorktown Hts., N.Y. in the area of optical inspection and measurement since 1985. He received his Ph.D. in Electrical Engineering from Stanford University in 1985. His specific interests include light scattering instruments and theory, materials analysis and semiconductor materials.

*KEVIN B. TENNAL* is a research associate in the Department of Electronics and Instrumentation, University of Arkansas at Little Rock. He received his M.S. degree in Physics from the University of Arkansas.

*SAIED TOUSI* is presently a Senior Staff Scientist with the Scientific and Laboratory Services Department of Pall Corporation. He is involved in aerosol research related to filter applications in the semiconductor industry. He is active in various professional organizations supporting the semiconductor industry and has published many papers on contamination control. He received his Ph.D. degree in Chemical Engineering from the University of Tulsa.

*JAY TURNER* is a graduate student in the Chemical Reaction Engineering Laboratory at Washington University, St. Louis, MO. He received an M.S. in Chemical Engineering in 1987 from the University of California at Los Angeles, where he conducted research in the Air Quality/Aerosol Technology Laboratory. Models to describe particle transport in cleanroom environments were developed during his tenure at the University of Duisburg, FRG, as a fellow of the Deutscher Akademischer Austauschdienst (DAAD).

*TIM B. VANDER WOOD* is currently Director of Services, McCrone Associates-Atlanta and he had joined McCrone Associates, Inc. in 1983 as an Analytical Chemist. He received his Ph.D. from the University of Chicago, and did post-doctoral research at Arizona State University. He is on the Editorial Board of Microcontamination.

*HWA-CHI WANG* is a Research Scientist at the Chicago Research Center of American Air Liquide, where he is involved in aerosol research. He received his Ph.D. in 1984 from the University of Illinois and studied particle-surface impact phenomena at Air & Industrial Hygiene Laboratory in Berkeley, CA before he accepted the present position in 1987.

*HORNG-YUAN WEN* is a Senior Scientist at the Chicago Research Center of American Air Liquide, where he is responsible for particle and surface contamination research. He received his Ph.D. in 1983 from SUNY, Buffalo.

*DAVID C. WHITE* is currently director of the Institute for Applied Microbiology, a University of Tennessee/Oak Ridge National Laboratory Distinguished Scientist, and professor of microbiology and ecology at the University of Tennessee, Knoxville. He is the founding editor of the Journal of Microbiological Methods and has served as an editor for several journals including Applied and Environmental Microbiology and the Journal of Contaminant Hydrology. In addition to authoring over 270 papers in the area of molecular characterization of microbial communities, he has served on a number of scientific advisory panels including the NRC Space Science Board. White received an M.D. degree from Tufts University and subsequently a Ph.D. in biochemistry/microbiology from Rockefeller University

*J. DOUGLAS WILSON* is on the faculty of the Department of Electronics and Instrumentation at the University of Arkansas at Little Rock. He received his M.S. in Instrumental Sciences from the University of Arkansas. His research interests include electronic system design, aerosol technology, laser doppler velocimetry, and biomedical instrumentation.

*CHAOPING WU* is a graduate student of the Department of Electronics and Instrumentation at the University of Arkansas at Little Rock.

*MICHAEL YOST* is the director of research for Ion Systems, Berkeley, CA, where he develops and tests new products and advanced applications for air ionization. He received Ph.D. from the University of California, Berkeley, and is on the faculty of the School of Public Health. He has written extensively on air ionization and holds several patents in the field.

*JOSEPH ZAHKA* is the Manager of the Applications Section of Millipore's Process Group Research, Development, and Engineering function, and has worked for Millipore for ten years. Previous to Millipore, he worked for Koch Membrane Systems and Durham Associates developing membrane separation processes. He has an SM in Chemical Engineering from MIT.

# INDEX

MIX
Papier aus verantwortungsvollen Quellen
Paper from responsible sources
FSC® C105338

If you have any concerns about our products,
you can contact us on
**ProductSafety@springernature.com**

In case Publisher is established outside the EU,
the EU authorized representative is:
**Springer Nature Customer Service Center GmbH**
**Europaplatz 3, 69115 Heidelberg, Germany**

Printed by Libri Plureos GmbH
in Hamburg, Germany